高效Go语言（影印版）

[波兰] 巴特洛梅耶·普洛特卡（Bartłomiej Płotka）著

Beijing · Boston · Farnham · Sebastopol · Tokyo

O'Reilly Media, Inc.授权东南大学出版社出版

南京 东南大学出版社

图书在版编目(CIP)数据

高效 Go 语言 = Efficient Go：影印版：英文 /（波）巴特洛梅耶·普洛特卡 (Bartłomiej Płotka)著 . —南京：东南大学出版社，2023.3

ISBN 978 - 7 - 5766 - 0659 - 1

Ⅰ. ①高… Ⅱ. ①巴… Ⅲ. ①程序语言-程序设计-英文 Ⅳ. ①TP312

中国国家版本馆 CIP 数据核字(2023)第 002302 号

图字：10 - 2022 - 485 号

© 2022 by O'Reilly Media, Inc.

Reprint of the English Edition, jointly published by O'Reilly Media, Inc. and Southeast University Press, 2023. Authorized reprint of the original English edition, 2022 O'Reilly Media, Inc., the owner of all rights to publish and sell the same.

All rights reserved including the rights of reproduction in whole or in part in any form.

英文原版由 O'Reilly Media, Inc.出版 2022。

英文影印版由东南大学出版社出版 2023。此影印版的出版和销售得到出版权和销售权的所有者 —— O'Reilly Media, Inc.的许可。

版权所有，未得书面许可，本书的任何部分和全部不得以任何形式重制。

高效 Go 语言（影印版）

著　　者：[波兰]巴特洛梅耶·普洛特卡 (Bartłomiej Płotka)
责任编辑：张　烨　　封面设计：Karen Montgomery，张　健　　责任印制：周荣虎
出版发行：东南大学出版社
社　　址：南京四牌楼 2 号　　邮编：210096　　电话：025-83793330
网　　址：http://www.seupress.com
电子邮件：press@ seupress.com
经　　销：全国各地新华书店
印　　刷：常州市武进第三印刷有限公司
开　　本：787mm×1000mm　1/16
印　　张：31
字　　数：607 千
版　　次：2023 年 3 月第 1 版
印　　次：2023 年 3 月第 1 次印刷
书　　号：ISBN 978 - 7 - 5766 - 0659 - 1
定　　价：128.00 元

本社图书若有印装质量问题，请直接与营销部联系。电话(传真)：025 - 83791830

Table of Contents

Preface

Welcome to the pragmatic software development world, where engineers are not afraid of ambitious performance goals. Where the change in requirements or unexpected efficiency issues is handled without stress, where code is optimized tactically and effectively, based on data, yet the codebase is kept simple and easy to read, maintain, and extend. Wait, is this even possible?

Yes, and I will show you how! The good news is that if you bought this book, you are already halfway there—it means you acknowledge the problem and are open to learning more! The bad news is that, while I tried to distill the knowledge to only what's necessary, there are still 11 chapters to go through. I think *Efficient Go* is unique in this regard as it is not a quick tutorial. Instead, it is a complete guide to writing efficient yet pragmatic software that goes through all aspects I wish I had known when I started my career.

In this book, you will undoubtedly learn a lot about my favorite programming language, Go, and how to optimize it. But don't let the title of this book fool you. While I use Go as the example language to show the optimization mindset and observability patterns, 8 out of 11 chapters of this book are language agnostic. You can use the same techniques to improve software written in any other language like Java, C#, Scala, Python, C++, Rust, or Haskell.

Finally, if you expected a full list of low-level optimization tricks, this is not the right book. Firstly, optimizations do not generalize well. The fact that someone unrolled the loop or used a pointer in their struct field and achieved better efficiency does not mean it will be helpful if we do the same! We will go through some optimization tricks, but I emphasize complete knowledge about efficiency in pragmatic software development instead.

Secondly, "low-level" dangerous tricks are often not needed. In most cases, an awareness of simple points where your program wastes time and resources is enough to fulfill your efficiency and scalability goals cheaply and effectively. Furthermore, you will learn that in most cases, there is no need to rewrite your program to C++, Rust, or Assembly to have an efficient solution!

Before we start, let's go through the main goals behind this book and why I found it necessary to focus my time on the subject of efficiency. You will also learn how to get the most out of this book and effectively use it in your software development tasks.

Why I Wrote This Book

I spent around 1,200 hours writing *Efficient Go*, so the choice to deliver such a book was not spur-of-the-moment. In the era of social media, YouTube, and TikTok, book writing and reading might feel outdated, but in my experience, modern media tend to oversimplify topics. You have to condense those to an absolute minimum not to lose viewers and monetization. It leads to the wrong incentives (*https://oreil.ly/A8dCv*), which generally collide with what I wanted to achieve with this book.

My mission here is straightforward: I want the software I use or depend on to be better! I want software project contributors and maintainers to understand their code's efficiency and how to assess it. I want them to reliably review my or others' pull requests with efficiency improvements. I want people around me to know how to handle performance issues professionally instead of building a stressful atmosphere. I want users and stakeholders to be cautious with the benchmarks and cheap marketing we see in the industry. Finally, I want leaders, directors, and product managers to approach software efficiency topics maturely with the awareness of how to form pragmatic efficiency requirements that help engineers to deliver excellent products.

I also consider this book a small contribution toward more sustainable software. Every wasted CPU time and memory wastes a significant amount of your business's money. However, it also wastes energy and hardware, which has a serious environmental effect. So saving money and the planet at the same time while enabling better value for your business is not a bad outcome of the skills you will learn here.

I figured out that writing a book is the best way to achieve this goal. It's easier than continuously explaining the same nuances, tooling, and techniques in my daily work, open source, and conferences!

How I Gathered This Knowledge

I built my experience toward efficiency topics and high-quality software development through a lot of practice, mistakes, experiments, implicit mentors (*https://oreil.ly/7IFBd*), and research.

I was 29 years old when I started writing this book. That might not feel like much experience, but I started a full-time, professional software development career when I was 19. I did full-time computer science studies in parallel to work at Intel around software-defined infrastructure (SDI). I initially coded in Python around the OpenStack project (*https://www.openstack.org*), then in C++ including contributions to the popular-back-then Mesos (*https://mesos.apache.org*) project under the supervision of amazing engineers from Mesosphere (*https://oreil.ly/yUHzn*) and Twitter. Finally, I moved to develop Go around Kubernetes (*https://kubernetes.io*) and fell in love with this language.

I spent a nontrivial amount of time at Intel on node oversubscription feature (*https://oreil.ly/uPnb7*) with noisy neighbor mitigations. Generally, oversubscription allows running more programs on a single machine than would be otherwise possible. This can work since statistically, all programs rarely use all of their reserved resources simultaneously. Looking at this now from a later perspective, it is usually easier and more effective to save money by starting with software optimization than by using complex algorithms like this.

In 2016, I moved to London to work for a gaming start-up. I worked with past employees of Google, Amazon, Microsoft, and Facebook to develop and operate a global gaming platform. We were developing microservices, mostly in Go running on dozens of Kubernetes clusters worldwide. This is where I learned a lot about distributed systems, site reliability engineering, and monitoring. Perhaps this was when I got addicted to amazing tooling around observability, which is key to achieving pragmatic efficiency and explained in Chapter 6.

My passion for good visibility of the running software translated to becoming an expert in using and developing a popular, open source, time-series database for monitoring purposes called Prometheus (*https://prometheus.io*). Eventually, I became an official maintainer and started multiple other Go open source projects and libraries. Finally, I had an opportunity to cocreate with Fabian Reinartz a large distributed time-series database in the open source called Thanos (*https://thanos.io*). I would not be surprised if some of my code runs in your company infrastructure!

In 2019, I moved to Red Hat, where I work full-time on observability systems in open source. This is when I also dived more into continuous profiling solutions, which you will learn in this book too.

I am also active in the Cloud Native Computing Foundation (CNCF) (*https://cncf.io*) as the ambassador and observability Technical Advisory Group (TAG) (*https://oreil.ly/f9UYG*) tech lead. In addition, I co-organize conferences and meetups. Finally, with the Prometheus and Thanos projects, with the team, we mentor multiple engineers every year via the CNCF mentoring initiatives (*https://oreil.ly/rU0bg*).[1]

1 If you are new to software development or open source, talk to us, start contributing, and apply for two months paid mentorship. Let me know if you would like to have fun while mentoring others! We need good mentors too—it's important to teach another generation of open source maintainers.

I wrote or reviewed thousands of code lines for various software that had to run on production, be reliable, and scale. I have taught and mentored over two dozen engineers so far. However, perhaps the most insightful was the open source work. You interact with diverse people, from different companies and places worldwide, with different backgrounds, goals, and needs.

Overall, I believe we achieved amazing things with the fantastic people I had a chance to work with. I was lucky to work in environments where high-quality code was more important than decreasing code review delays or reducing time spent addressing style issues. We thrived on good system design, code maintainability, and readability. We tried to bring those values to open source, and I think we did a good job there. However, there is one important thing I would improve if I had a chance to write, for instance, the Thanos project again: I would try to focus more on the pragmatic efficiency of my code and the algorithms we chose. I would focus on having clearer efficiency requirements from the start and invest more in benchmarking and profiling.

And don't get me wrong, the Thanos system nowadays is faster and uses much fewer resources than some competitors, but it took a lot of time, and there is still a massive amount of hardware resources we could use less. We still have many bottlenecks that await community attention. However, if I applied the knowledge, techniques, and suggestions that you will learn in this book, I believe we could have cut the development cost in half, if not more, to have Thanos in the state we have today (I hope my ex-boss who paid for this work won't read that!).

My journey showed me how much a book like this was needed. With more people programming overall, often without a computer science background, there are plenty of mistakes and misconceptions, especially regarding software efficiency. Not much literature was available to give us practical answers to our efficiency or scaling questions, especially for Go. Hopefully, this book fills that literature gap.

Who This Book Is For

Efficient Go focuses on giving the tools and knowledge necessary to answer when and how to apply efficiency optimization, depending strongly on circumstances and your organization's goals. As a result, the primary audience for this book is software developers designing, creating, or changing programs written in Go and any other modern language. It should be a software engineer's job to be an expert on ensuring the software they create works within both functional and efficiency requirements. Ideally, you have some basic programming skills when starting this book.

I believe this book is also useful to those who primarily operate software somebody else writes, e.g., DevOps engineers, SRE, sysadmins, and platform teams. There are many optimization design levels (as discussed in "Optimization Design Levels" on page 98). Sometimes it makes sense to invest in software optimizations, and sometimes we might need to address it on other levels! Moreover, to achieve reliable efficiency, software engineers have to benchmark and experiment a lot with production-like environments (as explained in Chapter 6), which usually means close collaboration with platform teams. Finally, the observability practices explained in Chapter 6 are state-of-the-art tools recommended for modern platform engineering. I am a strong proponent of avoiding differentiating between application performance monitoring (APM) and observability for SRE. If you hear that differentiation, it's mostly coming from vendors who want you to pay more or feel like they have more features. As I will explain, we can reuse the same tools, instrumentations, and signals across all software observations.[2] Generally, we are on the same team—we want to build better products!

Finally, I would like to recommend this book to managers, product managers, and leaders who want to stay technical and understand how to ensure you are not wasting millions of dollars on easy-to-fix efficiency issues within your team!

How This Book Is Organized

This book is organized into 11 chapters. In Chapter 1, we discuss efficiency and why it matters. Then, in Chapter 2, I briefly introduce Go with efficiency in mind. Then, in Chapter 3, we will talk about optimizations and how to think about them and approach those. Efficiency improvements can take enormous amounts of your time, but systematic approaches help you save a lot of time and effort.

In Chapters 4 and 5, I will explain all you need to know about latency, CPU, and memory resources, as well as how OS and Go abstract them.

Then we will move on to what it means to perform data-driven decisions around software efficiency. We will start with Chapter 6. Then we will discuss the reliability of experiments and complexity analysis in Chapter 7. Finally, I will explain benchmarking and profiling techniques in Chapters 8 and 9.

Last but not least, I will show you various examples of different optimization situations in Chapter 10. Finally, in Chapter 11, we will take a few learnings and summarize various efficiency patterns and tricks we see in the Go community.

2 I've already gotten feedback from some experienced people that they did not know you could use metrics to work on efficiency and performance improvements! It's possible, and you will learn how here.

Conventions Used in This Book

The following typographical conventions are used in this book:

Italic
: Indicates new terms, URLs, email addresses, filenames, and file extensions.

`Constant width`
: Used for program listings, as well as within paragraphs to refer to program elements such as variable or function names, databases, data types, environment variables, statements, and keywords.

`Constant width bold`
: Shows commands or other text that should be typed literally by the user.

`Constant width italic`
: Shows text that should be replaced with user-supplied values or by values determined by context.

This element signifies a tip or suggestion.

This element signifies a general note.

This element indicates a warning or caution.

Using Code Examples

This book contains code examples that should help you understand the tools, techniques, and good practices. All of them are in the Go programming language and work with Go version 1.18 and above.

You can find all the examples from this book in the executable and tested open source GitHub repository `efficientgo/examples` (*https://github.com/efficientgo/examples*). You are welcome to fork it, use it, and play with the examples I share in this book. Everybody learns differently. For some people, it is helpful to import some examples into their favorite IDE and play with it by modifying it, running, testing, or debugging. Find the way that works for you and feel free to ask questions or propose improvements through GitHub issues or pull requests (*https://github.com/efficientgo/examples/issues*)!

Note that the code examples in this book are simplified for a clear view and smaller size. Particularly, the following rules apply:

- If the `Go` package is not specified, assume `package main`.
- If the filename or extension of the example is not specified, assume the file has a *.go* extension. If it's a functional test or microbenchmark, the file name has to end with *_test.go*.
- `import` statements are not always provided. In such cases, assume standard library or previously introduced packages are imported.
- Sometimes, I don't provide imports in the `import` statement but in a comment (`// import <URL>`). This is when I want to explain a single nontrivial import out of many needed in this code example.
- A comment with three dots (`// ...`) specifies that some unrelated content was removed. This highlights that some logic is there for a function to make sense.
- A comment with the `handle error` statement (`// handle error`) indicates that error handling was removed for readability. Always handle errors in your code!

This book is here to help you get your job done. In general, if this book offers an example code, you may use it in your programs and documentation. You do not need to contact us for permission unless you're reproducing a significant portion of the code. For example, writing a program that uses several chunks of code from this book does not require permission. Selling or distributing examples from O'Reilly books does require permission. Answering a question by citing this book and quoting example code does not require permission. However, incorporating a significant amount of example code from this book into your product's documentation does require permission.

We appreciate but generally do not require attribution. An attribution usually includes the title, author, publisher, and ISBN. For example, "*Efficient Go* by Bartłomiej Płotka (O'Reilly). Copyright 2023 Alloc Limited, 978-1-098-10571-6."

If you feel your use of code examples falls outside fair use or the permission given above, feel free to contact us at *permissions@oreilly.com*.

Acknowledgments

As they say, "the greatness is in the agency of others" (*https://oreil.ly/owETM*). This book is no different. Numerous people helped me directly or indirectly in my *Efficient Go* book-writing journey and my career.

First of all, I would love to thank my wife, Kasia—without her support, this wouldn't be possible.

Thanks to my main tech reviewers, Michael Bang and Saswata Mukherjee, for relentlessly checking all the content in detail. Thanks to others who looked at some parts of the early content and gave amazing feedback: Matej Gera, Felix Geisendörfer, Giedrius Statkevičius, Björn Rabenstein, Lili Cosic, Johan Brandhorst-Satzkorn, Michael Hausenblas, Juraj Michalak, Kemal Akkoyun, Rick Rackow, Goutham Veeramachaneni, and more!

Furthermore, thanks to the many talented people from the open source community who share enormous knowledge in their public content! They might not realize it, but they help with such work, including my writing of this book. You will see quotes from some of them in this book: Chandler Carruth, Brendan Gregg, Damian Gryski, Frederic Branczyk, Felix Geisendörfer, Dave Cheney, Bartosz Adamczewski, Dominik Honnef, William (Bill) Kennedy, Bryan Boreham, Halvar Flake, Cindy Sridharan, Tom Wilkie, Martin Kleppmann, Rob Pike, Russ Cox, Scott Mayers, and more.

Finally, thanks to the O'Reilly team, especially Melissa Potter, Zan McQuade, and Clare Jensen, for amazing help and understanding of delays, moving deadlines, and sneaking more content into this book than planned! :)

Feedback Is Welcome!

If you are interested in following my work or the groups I work with or want to learn even more in this area, follow me on Twitter (*https://twitter.com/bwplotka*) or check my blog (*https://www.bwplotka.dev*).

Do not hesitate to reach out to me if you have feedback on my work or the content I have produced. I am always open to learning more!

O'Reilly Online Learning

For more than 40 years, *O'Reilly Media* has provided technology and business training, knowledge, and insight to help companies succeed.

Our unique network of experts and innovators share their knowledge and expertise through books, articles, and our online learning platform. O'Reilly's online learning platform gives you on-demand access to live training courses, in-depth learning paths, interactive coding environments, and a vast collection of text and video from O'Reilly and 200+ other publishers. For more information, visit *https://oreilly.com*.

How to Contact Us

Please address comments and questions concerning this book to the publisher:

O'Reilly Media, Inc.
1005 Gravenstein Highway North
Sebastopol, CA 95472
800-998-9938 (in the United States or Canada)
707-829-0515 (international or local)
707-829-0104 (fax)

We have a web page for this book, where we list errata, examples, and any additional information. You can access this page at *https://oreil.ly/efficient-go*.

Email *bookquestions@oreilly.com* to comment or ask technical questions about this book.

For news and information about our books and courses, visit *https://oreilly.com*.

Find us on LinkedIn: *https://linkedin.com/company/oreilly-media*.

Follow us on Twitter: *https://twitter.com/oreillymedia*.

Watch us on YouTube: *https://www.youtube.com/oreillymedia*.

CHAPTER 1

Software Efficiency Matters

> The primary task of software engineers is the cost-effective development of maintainable and useful software.
>
> —Jon Louis Bentley, *Writing Efficient Programs* (Prentice Hall, 1982)

Even after 40 years, Jon's definition of development is fairly accurate. The ultimate goal for any engineer is to create a useful product that can sustain user needs for the product lifetime. Unfortunately, nowadays not every developer realizes the significance of the software cost. The truth can be brutal; stating that the development process can be expensive might be an underestimation. For instance, it took 5 years and 250 engineers for Rockstar to develop the popular Grand Theft Auto 5 video game, which was estimated to cost $137.5 million (*https://oreil.ly/0CRW2*). On the other hand, to create a usable, commercialized operating system, Apple had to spend way over $500 million before the first release of macOS (*https://oreil.ly/hQhiv*) in 2001.

Because of the high cost of producing software, it's crucial to focus our efforts on things that matter the most. Ideally, we don't want to waste engineering time and energy on unnecessary actions, for example, spending weeks on code refactoring that doesn't objectively reduce code complexity, or deep micro-optimizations of a function that rarely runs. Therefore, the industry continually invents new patterns to pursue an efficient development process. Agile Kanban methods that allow us to adapt to ever-changing requirements, specialized programming languages for mobile platforms like Kotlin, or frameworks for building websites like React are only some examples. Engineers innovate in these fields because every inefficiency increases the cost.

What makes it even more difficult is that when developing software now, we should also be aware of the future costs. Some sources even estimate that running and maintenance costs can be higher than the initial development costs (*https://oreil.ly/59Zqe*). Code changes to stay competitive, bug fixing, incidents, installations, and finally, compute cost (including electricity consumed) are only a few examples of the total software cost of ownership (TCO) (*https://oreil.ly/ZzUCx*) we have to take into account. Agile methodologies help reveal this cost early by releasing software often and getting feedback sooner.

However, is that TCO higher if we descope efficiency and speed optimizations from our software development process? In many cases, waiting a few more seconds for our application execution should not be a problem. On top of that, the hardware is getting cheaper and faster every month. In 2022, buying a smartphone with a dozen GBs of RAM was not difficult. Finger-sized, 2 TB SSD disks capable of 7 GBps (*https://oreil.ly/eVcPQ*) read and write throughput are available. Even home PC workstations are hitting never-before-seen performance scores. With 8 CPUs or more that can perform billions of cycles per second each, and with 2 TB of RAM (*https://oreil.ly/eUzNh*), we can compute things fast. Plus, we can always add optimizations later, right?

> Machines have become increasingly cheap compared to people; any discussion of computer efficiency that fails to take this into account is short-sighted. "Efficiency" involves the reduction of overall cost—not just machine time over the life of the program, but also time spent by the programmer and by the users of the program.
>
> —Brian W. Kernighan and P. J. Plauger, *The Elements of Programming Style* (McGraw-Hill, 1978)

After all, improving the runtime or space complexity of the software is a complicated topic. Especially when you are new, it's common to lose time optimizing without significant program speedups. And even if we start caring about the latency introduced by our code, things like Java Virtual Machine or Go compiler will apply their optimizations anyway. Spending more time on something tricky, like efficiency on modern hardware that can also sacrifice our code's reliability and maintainability, may sound like a bad idea. These are only a few reasons why engineers typically put performance optimizations at the lowest position of the development priority list.

Unfortunately, as with every extreme simplification, there is some risk in such performance de-prioritization. Don't be worried, though! In this book, I will not try to convince you that you should now measure the number of nanoseconds each code line introduces or every bit it allocates in memory before adding it to your software. You should not. I am far from trying to motivate you to put performance at the top of your development priority list.

However, there is a difference between consciously postponing optimizations and making silly mistakes, causing inefficiencies and slowdowns. As the common saying goes, "Perfect is the enemy of good" (*https://oreil.ly/OogZF*), but we have to find that balanced good first. So I want to propose a subtle but essential change to how we, as software engineers, should think about application performance. It will allow you to bring small but effective habits to your programming and development management cycle. Based on data and as early as possible in the development cycle, you will learn how to tell when you can safely ignore or postpone program inefficiencies. Finally, when you can't afford to skip performance optimizations, where and how to apply them effectively, and when to stop.

In "Behind Performance" on page 3, we will unpack the word *performance* and learn how it is related to *efficiency* in this book's title. Then in "Common Efficiency Misconceptions" on page 7, we will challenge five serious misconceptions around efficiency and performance, often descoping such work from developer minds. You will learn that thinking about efficiency is not reserved only for "high-performance" software.

Some of the chapters, like this one, Chapter 3, and parts of other chapters, are fully language agnostic, so they should be practical for non-Go developers too!

Finally, in "The Key to Pragmatic Code Performance" on page 32, I will teach you why focusing on efficiency will allow us to think about performance optimizations effectively without sacrificing time and other software qualities. This chapter might feel theoretical, but trust me, the insights will train your essential programming judgment on how and if to adopt particular efficiency optimizations, algorithms, and code improvements presented in other parts of this book. Perhaps it will also help you motivate your product manager or stakeholder to see that more efficient awareness of your project can be beneficial.

Let's start by unpacking the definition of efficiency.

Behind Performance

Before discussing why software efficiency or optimizations matter, we must first demystify the overused word *performance*. In engineering, this word is used in many contexts and can mean different things, so let's unpack it to avoid confusion.

When people say, "This application is performing poorly," they usually mean that this particular program is executing slowly.[1] However, if the same people say, "Bartek is not performing well at work," they probably don't mean that Bartek is walking too slowly from the computer to the meeting room. In my experience, a significant number of people in software development consider the word *performance* a synonym of *speed*. For others, it means the overall quality of execution, which is the original

1 I even did a small experiment on Twitter (*https://oreil.ly/997J5*), proving this point.

definition of this word.[2] This phenomenon is sometimes called a "semantic diffusion" (*https://oreil.ly/Qx9Ft*), which occurs when a word starts to be used by larger groups with a different meaning than it originally had.

> The word performance in computer performance means the same thing that performance means in other contexts, that is, it means "How well is the computer doing the work it is supposed to do?"
>
> — Arnold O. Allen, *Introduction to Computer Performance Analysis with Mathematica* (Morgan Kaufmann, 1994)

I think Arnold's definition describes the word *performance* as accurately as possible, so it might be the first actionable item you can take from this book. Be specific.

Clarify When Someone Uses the Word "Performance"

When reading the documentation, code, bug trackers, or attending conference talks, be careful when you hear that word, *performance*. Ask follow-up questions and ensure what the author means.

In practice, performance, as the quality of overall execution, might contain much more than we typically think. It might feel picky, but if we want to improve software development's cost-effectiveness, we must communicate clearly, efficiently, and effectively!

I suggest avoiding the *performance* word unless we can specify its meaning. Imagine you are reporting a bug in a bug tracker like GitHub Issues. Especially there, don't just mention "bad performance," but specify exactly the unexpected behavior of the application you described. Similarly, when describing improvements for a software release in the changelog,[3] don't just mention that a change "improved performance." Describe what, exactly, was enhanced. Maybe part of the system is now less prone to user input errors, uses less RAM (if yes, how much less, in what circumstances?), or executes something faster (how many seconds faster, for what kinds of workloads?). Being explicit will save time for you and your users.

I will be explicit in my book about this word. So whenever you see the word *performance* describing the software, remind yourself about this visualization in Figure 1-1.

---

2 The UK Cambridge Dictionary defines (*https://oreil.ly/AXq4Q*) the noun *performance* as "How well a person, machine, etc. does a piece of work or an activity."

3 I would even recommend, with your changelog, sticking to common standard formats like you can see here (*https://oreil.ly/rADTI*). This material also contains valuable tips on clean release notes.

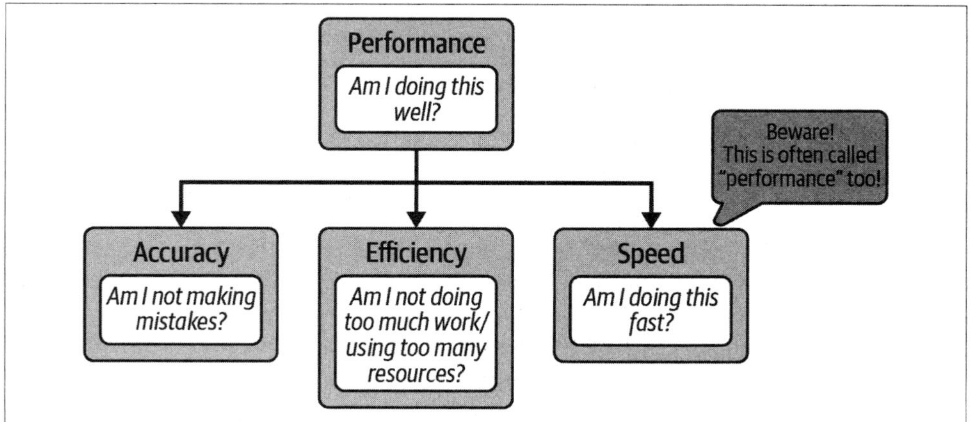

Figure 1-1. Performance definition

In principle, software performance means "how well software runs" and consists of three core execution elements you can improve (or sacrifice):

Accuracy
: The number of errors you make while doing the work to accomplish the task. This can be measured for software by the number of wrong results your application produces. For example, how many requests finished with non-200 HTTP status codes in a web system.

Speed
: How fast you do the work needed to accomplish the task—the timeliness of execution. This can be observed by operation latency or throughput. For example, we can estimate that typical compression of 1 GB of data in memory typically takes around 10 s (latency), allowing approximately 100 MBps throughput (*https://oreil.ly/eOJdK*).

Efficiency
: The ratio of the useful energy delivered by a dynamic system to the energy supplied to it. More simply, this is the indicator of how many extra resources, energy, or work were used to accomplish the task. In other words, how much effort we wasted. For instance, if our operation of fetching 64 bytes of valuable data from disk allocates 420 bytes on RAM, our memory efficiency would equal 15.23%.

  This does not mean our operation is 15.23% efficient in absolute measure. We did not calculate energy, CPU time, heat, and other efficiencies. For practical purposes, we tend to specify what efficiency we have in mind. In our example, that was memory space.

To sum up, performance is a combination of at least those three elements:

performance = (accuracy * efficiency * speed)

Improving any of those enhances the performance of the running application or system. It can help with reliability, availability, resiliency, overall latency, and more. Similarly, ignoring any of those can make our software less useful.[4] The question is, at what point should we say "stop" and claim it is good enough? Those three elements might also feel disjointed, but in fact, they are connected. For instance, notice that we can still achieve better reliability and availability without changing accuracy (not reducing the number of bugs). For example, with efficiency, reducing memory consumption decreases the chances of running out of memory and crashing the application or host operating system. This book focuses on knowledge, techniques, and methods, allowing you to increase the efficiency and speed of your running code without degrading accuracy.

It's No Mistake That the Title of My Book Is "Efficient Go"

My goal is to teach you pragmatic skills, allowing you to produce high-quality, accurate, efficient, and fast code with minimum effort. For this purpose, when I mention the overall efficiency of the code (without saying a particular resource), I mean both speed and efficiency, as shown in Figure 1-1. Trust me, this will help us to get through the subject effectively. You will learn more about why in "The Key to Pragmatic Code Performance" on page 32.

Misleading use of the *performance* word might be the tip of the misconceptions iceberg in the efficiency subject. We will now walk through many more serious stereotypes and tendencies that are causing the development of our software to worsen. Best case, it results in more expensive to run or less valuable programs. Worse case, it causes severe social and financial organizational problems.

---

4 Can we say "less performant" in this sentence? We can't, because the word *performant* does not exist in English vocabulary. Perhaps it indicates that our software can't be "performant"—there is always room to improve things. In a practical sense, there are limits to how fast our software can be. H. J. Bremermann in 1962 suggested (*https://oreil.ly/1sl3f*) there is a computational physical limit that depends on the mass of the system. We can estimate that 1 kg of the ultimate laptop can process ~10^{50} bits per second, while the computer with the mass of the planet Earth can process at a maximum of 10^{75} bits per second. While those numbers feel enormous, even such a large computer would take ages to force all chess movements estimated to 10^{120} complexity (*https://oreil.ly/6qS1T*). Those numbers have practical use in cryptography to assess the difficulty of cracking certain encryption algorithms.

Common Efficiency Misconceptions

The number of times when I was asked, in code reviews or sprint plannings, to ignore the efficiency of the software "for now" is staggering. And you have probably heard that too! I also rejected someone else's change set for the same reasons numerous times. Perhaps our changes were dismissed at that time for good reasons, especially if they were micro-optimizations that added unnecessary complexity.

On the other hand, there were also cases where the reasons for rejection were based on common, factual misconceptions. Let's try to unpack some of the most damaging misunderstandings. Be cautious when you hear some of these generalized statements. Demystifying them might help you save enormous development costs long-term.

Optimized Code Is Not Readable

Undoubtedly, one of the most critical qualities of software code is its readability.

> It is more important to make the purpose of the code unmistakable than to display virtuosity.... The problem with obscure code is that debugging and modification become much more difficult, and these are already the hardest aspects of computer programming. Besides, there is the added danger that a too clever program may not say what you thought it said.
>
> —Brian W. Kernighan and P. J. Plauger, *The Elements of Programming Style* (McGraw-Hill, 1978)

When we think about ultrafast code, the first thing that sometimes comes to mind is those clever, low-level implementations with a bunch of byte shifts, magic byte paddings, and unrolled loops. Or worse, pure assembly code linked to your application.

Yes, low-level optimizations like that can make our code significantly less readable, but as you will learn in this book, such extreme changes are rare in practice. Code optimizations might produce extra complexity, increase cognitive load, and make our code harder to maintain. The problem is that engineers tend to associate optimization with complexity to the extreme and avoid efficiency optimization like fire. In their minds, it translates to an immediate negative readability impact. The point of this section is to show you that there are ways to make efficiency-optimized code clear. Efficiency and readability can coexist.

Similarly, the same risk exists if we add any other functionality or change the code for different reasons. For example, refusing to write more efficient code because of a fear of decreasing readability is like refusing to add vital functionality to avoid complexity. So, again, this is a fair question, and we can consider descoping the feature, but we should evaluate the consequences first. The same should be applied to efficiency changes.

For example, when you want to add extra validation to the input, you can naively paste a complex 50-line code waterfall of `if` statements directly into the handling function. This might make the next reader of your code cry (or yourself when you revisit this code months later). Alternatively, you can encapsulate everything to a single `func validate(input string) error` function, adding only slight complexity. Furthermore, to avoid modifying the handling block of code, you can design the code to validate it on the caller side or in the middleware. We can also rethink our system design and move validation complexity to another system or component, thus not implementing this feature. There are many ways to compose a particular feature without sacrificing our goals.

How are performance improvements in our code different from extra features? I would argue they are not. You can design efficiency optimizations with readability in mind as you do with features. Both can be entirely transparent to the readers if hidden under abstractions.[5]

Yet we tend to mark optimizations as the primary source of readability problems. The foremost damaging consequence of this and other misconceptions in this chapter is that it's often used as an excuse to ignore performance improvements completely. This often leads to something called *premature pessimization*, the act of making the program less efficient, the opposite of optimization.

> Easy on yourself, easy on the code: All other things being equal, notably code complexity and readability, certain efficient design patterns and coding idioms should just flow naturally from your fingertips and are no harder to write than the pessimized alternatives. This is not premature optimization; it is avoiding gratuitous [unnecessary] pessimization.
>
> —H. Sutter and A. Alexandrescu, *C++ Coding Standards: 101 Rules, Guidelines, and Best Practices* (Addison-Wesley, 2004)

Readability is essential. I would even argue that unreadable code is rarely efficient over the long haul. When software evolves, it's easy to break previously made, too-clever optimization because we misinterpret or misunderstand it. Similar to bugs and mistakes, it's easier to cause performance issues in tricky code. In Chapter 10, you will see examples of deliberate efficiency changes, with a focus on maintainability and readability.

5 It's worth mentioning that hiding features or optimization can sometimes lead to lower readability. Sometimes explicitness is much better and avoids surprises.

Readability Is Important!

It's easier to optimize readable code than make heavily optimized code readable. This is true for both humans and compilers that might attempt to optimize your code!

Optimization often results in less readable code because we don't design good efficiency into our software from the beginning. If you refuse to think about efficiency now, it might be too late to optimize the code later without impacting readability. It's much easier to find a way to introduce a simpler and more efficient way of doing things in the fresh modules where we just started to design APIs and abstractions. As you will learn in Chapter 3, we can do performance optimizations on many different levels, not only via nitpicking and code tuning. Perhaps we can choose a more efficient algorithm, faster data structure, or a different system trade-off. These will likely result in much cleaner, maintainable code and better performance than improving efficiency after releasing the software. Under many constraints, like backward compatibility, integrations, or strict interfaces, our only way to improve performance would be to introduce additional, often significant, complexity to the code or system.

Code after optimization can be more readable

Surprisingly, code after optimization can be more readable! Let's look at a few Go code examples. Example 1-1 is a naive use of a getter pattern that I have personally seen hundreds of times when reviewing student or junior developer Go code.

Example 1-1. Simple calculation for the ratio of reported errors

```
type ReportGetter interface {
   Get() []Report
}

func FailureRatio(reports ReportGetter) float64 { ❶
   if len(reports.Get()) == 0 { ❷
      return 0
   }

   var sum float64
   for _, report := range reports.Get() { ❷
      if report.Error() != nil {
         sum++
      }
   }
   return sum / float64(len(reports.Get())) ❷
}
```

❶ This is a simplified example, but there is quite a popular pattern of passing a function or interface to get the elements needed for operation instead of passing them directly. It is useful when elements are dynamically added, cached, or fetched from remote databases.

❷ Notice we execute `Get` to retrieve reports three times.

I think you would agree that code from Example 1-1 would work for most cases. It is simple and quite readable. Yet, I would most likely not accept such code because of potential efficiency and accuracy issues. I would suggest simple modification as in Example 1-2 instead.

Example 1-2. Simple, more efficient calculation for the ratio of reported errors

```
func FailureRatio(reports ReportGetter) float64 {
   got := reports.Get() ❶
   if len(got) == 0 {
      return 0
   }

   var sum float64
   for _, report := range got {
      if report.Error() != nil {
         sum++
      }
   }
   return sum / float64(len(got))
}
```

❶ In comparison with Example 1-1, instead of calling `Get` in three places, I do it once and reuse the result via the `got` variable.

Some developers could argue that the `FailureRatio` function is potentially used very rarely; it's not on a critical path, and the current `ReportGetter` implementation is very cheap and fast. They could argue that without measuring or benchmarking we can't decide what's more efficient (which is mostly true!). They could call my suggestion a "premature optimization."

However, I deem it a very popular case of premature pessimization. It is a silly case of rejecting more efficient code that doesn't speed up things a lot right now but doesn't harm either. On the contrary, I would argue that Example 1-2 is superior in many aspects:

Without measurements, the Example 1-2 code is more efficient.
: Interfaces allow us to replace the implementation. They represent a certain contract between users and implementations. From the point of view of the `FailureRatio` function, we cannot assume anything beyond that contract. Most likely, we cannot assume that the `ReportGetter.Get` code will always be fast and

cheap.[6] Tomorrow, someone might swap the `Get` code with the expensive I/O operation against a filesystem, implementation with mutexes, or call to the remote database.[7]

We, of course, can iterate and optimize it later with a proper efficiency flow that we will discuss in "Efficiency-Aware Development Flow" on page 102, but if it's a reasonable change that actually improves other things too, there is no harm in doing it now.

Example 1-2 code is safer.
: It is potentially not visible in plain sight, but the code from Example 1-1 has a considerable risk of introducing race conditions. We may hit a problem if the `ReportGetter` implementation is synchronized with other threads that dynamically change the `Get()` result over time. It's better to avoid races and ensure consistency within a function body. Race errors are the hardest to debug and detect, so it's better to be safe than sorry.

Example 1-2 code is more readable.
: We might be adding one more line and an extra variable, but at the end, the code in Example 1-2 is explicitly telling us that we want to use the same result across three usages. By replacing three instances of the `Get()` call with a simple variable, we also minimize the potential side effects, making our `FailureRatio` purely functional (except the first line). By all means, Example 1-2 is thus more readable than Example 1-1.

Such a statement might be accurate, but evil is in the "premature" part. Not every performance optimization is premature. Furthermore, such a rule is not a license for rejecting or forgetting about more efficient solutions with comparable complexity.

Another example of optimized code yielding clarity is visualized by the code in Examples 1-3 and 1-4.

---

6 As the part of the interface "contract," there might be a comment stating that implementations should cache the result. Hence, the caller should be safe to call it many times. Still, I would argue that it's better to avoid relying on something not assured by a type system to prevent surprises.

7 All three examples of `Get` implementations could be considered costly to invoke. Input-output (I/O) operations against the filesystem are significantly slower than reading or writing something from memory. Something that involves mutexes means you potentially have to wait on other threads before accessing it. Call to database usually involves all of them, plus potentially communication over the network.

Example 1-3. Simple loop without optimization

```
func createSlice(n int) (slice []string) { ❶
    for i := 0; i < n; i++ {
        slice = append(slice, "I", "am", "going", "to", "take", "some", "space") ❷
    }
    return slice
}
```

❶ Returning named parameter called `slice` will create a variable holding an empty `string` slice at the start of the function call.

❷ We append seven `string` items to the slice and repeat that `n` times.

Example 1-3 shows how we usually fill slices in Go, and you might say nothing is wrong here. It just works. However, I would argue that this is not how we should append in the loop if we know exactly how many elements we will append to the slice up front. Instead, in my opinion, we should always write it as in Example 1-4.

Example 1-4. Simple loop with pre-allocation optimization. Is this less readable?

```
func createSlice(n int) []string {
    slice := make([]string, 0, n*7) ❶
    for i := 0; i < n; i++ {
        slice = append(slice, "I", "am", "going", "to", "take", "some", "space") ❷
    }
    return slice
}
```

❶ We are creating a variable holding the string `slice`. We are also allocating space (capacity) for `n * 7` strings for this slice.

❷ We append seven `string` items to the slice and repeat that `n` times.

We will talk about efficiency optimizations like those in Examples 1-2 and 1-4 in "Pre-Allocate If You Can" on page 441, with the more profound Go runtime knowledge from Chapter 4. In principle, both allow our program to do less work. In Example 1-4, thanks to initial pre-allocation, the internal `append` implementation does not need to extend slice size in memory progressively. We do it once at the start. Now, I would like you to focus on the following question: is this code more or less readable?

Readability can often be subjective, but I would argue the more efficient code from Example 1-4 is more understandable. It adds one more line, so we could say the code is a bit more complex, but at the same time, it is explicit and clear in the message. Not only does it help Go runtime perform less work, but it also hints to the reader about the purpose of this loop and how many iterations we expect exactly.

If you have never seen raw usage of the built-in `make` function in Go, you probably would say that this code is less readable. That is fair. However, once you realize the benefit and start using this pattern consistently across the code, it becomes a good habit. Even more, thanks to that, any slice creation without such pre-allocation tells you something too. For instance, it could say that the number of iterations is unpredictable, so you know to be more careful. You know one thing before you even looked at the loop's content! To make such a habit consistent across the Prometheus and Thanos codebase, we even added a related entry to the Thanos Go coding style guide (*https://oreil.ly/Nq6tY*).

Readability Is Not Written in Stone; It Is Dynamic

The ability to understand certain software code can change over time, even if the code never changes. Conventions come and go as the language community tries new things. With strict consistency, you can help the reader understand even more complex pieces of your program by introducing a new, clear convention.

Readability now versus past

Generally, developers often apply Knuth's "premature optimization is the root of all evil" quote[8] to reduce readability problems with optimizations. However, this quote was made a long time ago. While we can learn a lot about general programming from the past, there are many things we have improved enormously from 1974. For example, back then it was popular to add information about the type of the variable to its name, as showcased in Example 1-5.[9]

Example 1-5. Example of Systems Hungarian notation applied to Go code

```
type structSystem struct {
   sliceU32Numbers []uint32
   bCharacter      byte
   f64Ratio        float64
}
```

8 This famous quote is used to stop someone from spending time on optimization effort. Generally overused, it comes from Donald Knuth's "Structured Programming with `goto` statements" (*https://oreil.ly/m3P50*) (1974).

9 This type of style is usually referred to as Hungarian notation, which is used extensively in Microsoft. There are two types of this notation too: App and Systems. Literature indicates that Apps Hungarian can still give many benefits (*https://oreil.ly/rYLX4*).

Hungarian notation was useful because compilers and Integrated Development Environments (IDEs) were not very mature at that point. But nowadays, on our IDEs or even repository websites like GitHub, we can hover over the variable to immediately know its type. We can go to the variable definition in milliseconds, read the commentary, and find all invocations and mutations. With smart code suggestions, advanced highlighting, and dominance of object-oriented programming developed in the mid-1990s, we have tools in our hands that allow us to add features and efficiency optimizations (complexity) without significantly impacting the practical readability.[10] Furthermore, the accessibility and capabilities of the observability and debugging tools have grown enormously, which we will explore in Chapter 6. It still does not permit clever code but allows us to more quickly understand bigger codebases.

To sum up, performance optimization is like another feature in our software, and we should treat it accordingly. It can add complexity, but there are ways to minimize the cognitive load required to understand our code.[11]

How to Make Efficient Code More Readable

- Remove or avoid unnecessary optimization.
- Encapsulate complex code behind clear abstraction (e.g., interface).
- Keep the "hot" code (the critical part that requires better efficiency) separate from the "cold" code (rarely executed).

As we learned in this chapter, there are even cases when a more efficient program is often a side effect of the simple, explicit, and understandable code.

You Aren't Going to Need It

You Aren't Going to Need It (YAGNI) is a powerful and popular rule that I use often while writing or reviewing any software.

> One of the most widely publicized principles of XP [Extreme Programming] is the You Aren't Going to Need It (YAGNI) principle. The YAGNI principle highlights the value of delaying an investment decision in the face of uncertainty about the return on the

10 It is worth highlighting that these days, it is recommended to write code in a way that is easily compatible with IDE functionalities; e.g., your code structure should be a "connected" graph (*https://oreil.ly/mFzH9*). This means that you connect functions in a way that IDE can assist. Any dynamic dispatching, code injection, and lazy loading disables those functionalities and should be avoided unless strictly necessary.

11 Cognitive load is the amount of "brain processing and memory" a person must use to understand a piece of code or function (*https://oreil.ly/5CJ9X*).

> investment. In the context of XP, this implies delaying the implementation of fuzzy features until uncertainty about their value is resolved.
>
> —Hakan Erdogmu and John Favaro, "Keep Your Options Open: Extreme Programming and the Economics of Flexibility"

In principle, it means avoiding doing the extra work that is not strictly needed for the current requirements. It relies on the fact that requirements constantly change, and we have to embrace iterating rapidly on our software.

Let's imagine a potential situation where Katie, a senior software engineer, is assigned the task of creating a simple web server. Nothing fancy, just an HTTP server that exposes some REST endpoint. Katie is an experienced developer who has created probably a hundred similar endpoints in the past. She goes ahead, programs functionality, and tests the server in no time. With some time left, she decides to add extra functionality: a simple bearer token authorization layer (*https://oreil.ly/EuKD0*). Katie knows that such change is outside the current requirements, but she has written hundreds of REST endpoints, and each had a similar authorization. Experience tells her it's highly likely such requirements will come soon, too, so she will be prepared. Do you think such a change would make sense and should be accepted?

While Katie has shown good intention and solid experience, we should refrain from merging such change to preserve the quality of the web server code and overall development cost-effectiveness. In other words, we should apply the YAGNI rule. Why? In most cases, we cannot predict a feature. Sticking to requirements allows us to save time and complexity. There is a risk that the project will never need an authorization layer, for example, if the server is running behind a dedicated authorization proxy. In such a case, the extra code Katie wrote can bring a high cost even if not used. It is additional code to read, which adds to the cognitive load. Furthermore, it will be harder to change or refactor such code when needed.

Now, let's step into a grayer area. We explained to Katie why we needed to reject the authorization code. She agreed, and instead, she decided to add some critical monitoring to the server by instrumenting it with a few vital metrics. Does this change violate the YAGNI rule too?

If monitoring is part of the requirements, it does not violate the YAGNI rule and should be accepted. If it's not, without knowing the full context, it's hard to say. Critical monitoring should be explicitly mentioned in the requirements. Still, even if it is not, web server observability is the first thing that will be needed when we run such code anywhere. Otherwise, how will we know that it is even running? In this case, Katie is technically doing something important that is immediately useful. In the end, we should apply common sense and judgment, and add or explicitly remove monitoring from the software requirements before merging this change.

Later, in her free time, Katie decided to add a simple cache to the necessary computation that enhances the performance of the separate endpoint reads. She even wrote and performed a quick benchmark to verify the endpoint's latency and resource consumption improvements. Does that violate the YAGNI rule?

The sad truth about software development is that performance efficiency and response time are often missing from stakeholders' requirements. The target performance goal for an application is to "just work" and be "fast enough," without details on what that means. We will discuss how to define practical software efficiency requirements in "Resource-Aware Efficiency Requirements" on page 86. For this example, let's assume the worst. There was nothing in the requirements list about performance. Should we then apply the YAGNI rule and reject Katie's change?

Again, it is hard to tell without full context. Implementing a robust and usable cache is not trivial, so how complex is the new code? Is the data we are working on easily "cachable"?[12] Do we know how often such an endpoint will be used (is it a critical path)? How far should it scale? On the other hand, computing the same result for a heavily used endpoint is highly inefficient, so cache is a good pattern.

I would suggest Katie take a similar approach as she did with monitoring change: consider discussing it with the team to clarify the performance guarantees that the web service should offer. That will tell us if the cache is required now or is violating the YAGNI rule.

As a last change, Katie went ahead and applied a reasonable efficiency optimization, like the slice pre-allocation improvement you learned in Example 1-4. Should we accept such a change?

I would be strict here and say yes. My suggestion is to always pre-allocate, as in Example 1-4 when you know the number of elements up front. Isn't that violating the core statement behind the YAGNI rule? Even if something is generally applicable, you shouldn't do it before you are sure you *are* going to need it?

I would argue that small efficiency habits that do not reduce code readability (some even improve it) should generally be an essential part of the developer's job, even if not explicitly mentioned in the requirements. We will cover them as "Reasonable Optimizations" on page 74. Similarly, no project requirements state basic best practices like code versioning, having small interfaces, or avoiding big dependencies.

---

12 *Cachability* is often defined (*https://oreil.ly/WNaRz*) as the ability to be cached. It is possible to cache (save) any information to retrieve it later, faster. However, the data might be valid only for a short time or only for a tiny amount of requests. If the data depends on external factors (e.g., user or input) and changes frequently, it's not well cachable.

The main takeaway here is that using the YAGNI rule helps, but it is not permission for developers to completely ignore performance efficiency. Thousands of small things usually make up excessive resource usage and latency of an application, not just a single thing we can fix later. Ideally, well-defined requirements help clarify your software's efficiency needs, but they will never cover all the details and best practices we should try to apply.

Hardware Is Getting Faster and Cheaper

> When I started programming we not only had slow processors, we also had very limited memory—sometimes measured in kilobytes. So we had to think about memory and optimize memory consumption wisely.
>
> —Valentin Simonov, "Optimize for Readability First" (*https://oreil.ly/I2NPk*)

Undoubtedly, hardware is more powerful and less expensive than ever before. We see technological advancement on almost every front every year or month. From single-core Pentium CPUs with a 200-MHz clock rate in 1995, to smaller, energy-efficient CPUs capable of 3- to 4-GHz speeds. RAM sizes increased from dozens of MB in 2000 to 64 GB in personal computers 20 years later, with faster access patterns. In the past, small capacity hard disks moved to SSD, then 7 GBps fast NVME SSD disks with a few TB of space. Network interfaces have achieved 100 gigabits throughput. In terms of remote storage, I remember floppy disks with 1.44 MB of space, then read-only CD-ROMs with a capacity of up to 553 MB; next we had Blu-Ray, read-write capability DVDs, and now it's easy to get SD cards with TB sizes.

Now let's add to the preceding facts the popular opinion that the amortized hourly value of typical hardware is cheaper than the developer hour. With all of this, one would say that it does not matter if a single function in code takes 1 MB more or does excessive disk reads. Why should we delay features, and educate or invest in performance-aware engineers if we can buy bigger servers and pay less overall?

As you can probably imagine, it's not that simple. Let's unpack this quite harmful argument descoping efficiency from the software development to-do list.

First of all, stating that spending more money on hardware is cheaper than investing expensive developer time into efficiency topics is very shortsighted. It is like claiming that we should buy a new car and sell an old one every time something breaks, because repairing is nontrivial and costs a lot. Sometimes that might work, but in most cases it's not very efficient or sustainable.

Let's assume a software developer's annual salary oscillates around $100,000. With other employment costs (*https://oreil.ly/AxI0Y*), let's say the company has to pay $120,000 yearly, so $10,000 monthly. For $10,000 in 2021, you could buy a server with 1 TB of DDR4 memory, two high-end CPUs, 1-gigabit network card, and 10 TB of hard disk space. Let's ignore for now the energy consumption cost. Such a deal means that our software can overallocate terabytes of memory every month, and we would still be better off than hiring an engineer to optimize this, right? Unfortunately, it doesn't work like this.

It turns out that terabytes of allocation are more common than you think, and you don't need to wait a whole month! Figure 1-2 shows a screenshot of the heap memory profile of a single replica (of six total) of a single Thanos (*https://thanos.io*) service (of dozens) running in a single cluster for five days. We will discuss how to read and use profiles in Chapter 9, but Figure 1-2 shows the total memory allocated by some `Series` function since the last restart of the process five days before.

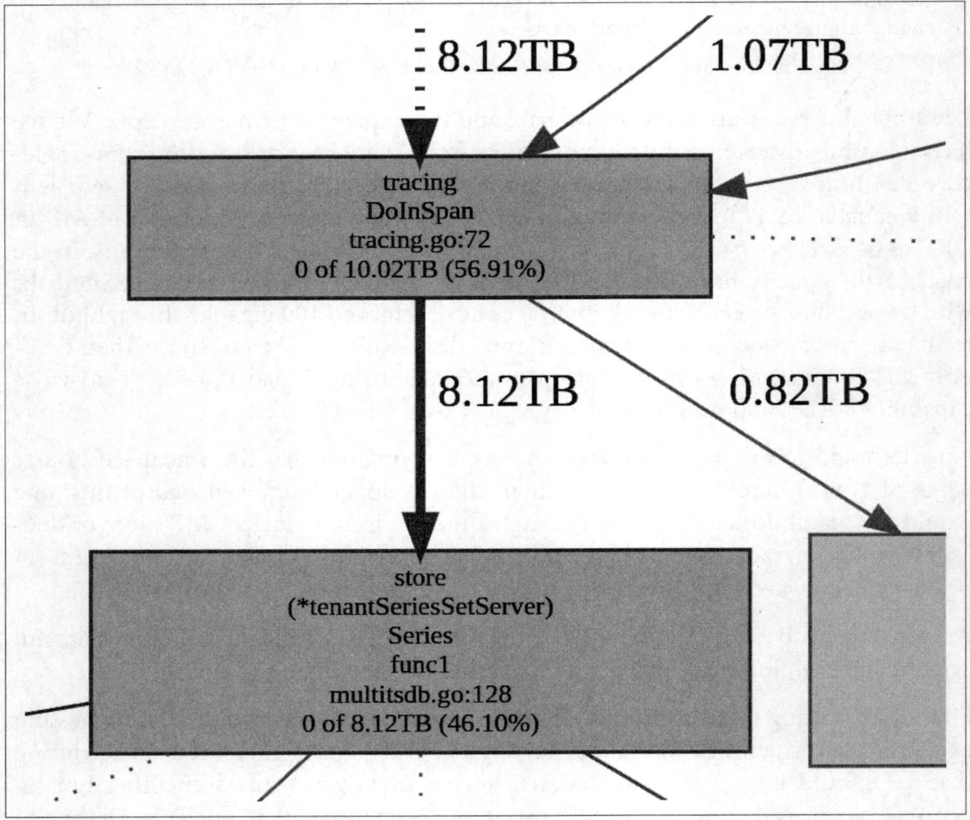

Figure 1-2. Snippet of memory profile showing all memory allocations within five days made by high-traffic service

Most of that memory was already released, but notice that this software from the Thanos project used 17.61 TB in total for only five days of running.[13] If you write desktop applications or tools instead, you will hit a similar scale issue sooner or later. Taking the previous example, if one function is overallocating 1 MB, that is enough to run it 100 times for critical operation in our application with only 100 desktop users to get to 10 TB wasted in total. Not in a month, but on a single run done by 100 users. As a result, slight inefficiency can quickly create overabundant hardware resources.

There is more. To afford an overallocation of 10 TB, it is not enough to buy a server with that much memory and pay for energy consumption. The amortized cost, among other things, has to include writing, buying, or at least maintaining firmware, drivers, operating systems, and software to monitor, update, and operate the server. Since for extra hardware we need additional software, by definition, this requires spending money on engineers, so we are back where we were. We might have saved engineering costs by avoiding focusing on performance optimizations. In return, we would spend more on other engineers required to maintain overused resources, or pay a cloud provider that already calculated such extra cost, plus a profit, into the cloud usage bill.

On the other hand, today 10 TB of memory costs a lot, but tomorrow it might be a marginal cost due to technological advancements. What if we ignore performance problems and wait until server costs decrease or more users replace their laptops or phones with faster ones? Waiting is easier than debugging tricky performance issues!

Unfortunately, we cannot skip software development efficiency and expect hardware advancements to mitigate needs and performance mistakes. Hardware is getting faster and more powerful, yes. But, unfortunately, not fast enough. Let's go through three main reasons behind this nonintuitive effect.

Software expands to fill the available memory

This effect is known as Parkinson's Law.[14] It states that no matter how many resources we have, the demands tend to match the supply. For example, Parkinson's Law is heavily visible in universities. No matter how much time the professor gives

13 That is a simplification, of course. The process might have used more memory. Profiles do not show memory used by memory maps, stacks, and many other caches required for modern applications to work. We will learn more about this in Chapter 4.

14 Cyril Northcote Parkinson was a British historian who articulated the management phenomenon that is now known as Parkinson's Law. Stated as "Work expands to fill the time available for its completion," it was initially referred to as the government office efficiency that highly correlates to the official's number in the decision-making body.

for assignments or exam preparations, students will always use all of it and probably do most of it last-minute.[15] We can see similar behavior in software development too.

Software gets slower more rapidly than hardware becomes faster

Niklaus Wirth mentions a "fat software" term that explains why there will always be more demand for more hardware.

> Increased hardware power has undoubtedly been the primary incentive for vendors to tackle more complex problems.... But it is not the inherent complexity that should concern us; it is the self-inflicted complexity. There are many problems that were solved long ago, but for the same problems, we are now offered solutions wrapped in much bulkier software.
>
> —Niklaus Wirth, "A Plea for Lean Software" (*https://oreil.ly/bctyb*)

Software is getting slower faster than hardware is getting more powerful because products have to invest in a better user experience to get profitable. These include prettier operating systems, glowing icons, complex animations, high-definition videos on websites, or fancy emojis that mimic your facial expression, thanks to facial recognition techniques. It's a never-ending battle for clients, which brings more complexity, and thus increased computational demands.

On top of that, rapid democratization of software occurs thanks to better access to computers, servers, mobile phones, IoT devices, and any other kind of electronics. As Marc Andreessen said, "Software is eating the world" (*https://oreil.ly/QUND4*). The COVID-19 pandemic that started in late 2019 accelerated digitalization even more as remote, internet-based services became the critical backbone of modern society. We might have more computation power available every day, but more functionalities and user interactions consume all of it and demand even more. In the end, I would argue that our overused 1 MB in the aforementioned single function might become a critical bottleneck on such a scale pretty quickly.

If that still feels very hypothetical, just look at the software around you. We use social media, where Facebook alone generates 4 PB (*https://oreil.ly/oowCN*)[16] of data per day. We search online, causing Google to process 20 PB of data per day. However, one would say those are rare, planet-scale systems with billions of users. Typical developers don't have such problems, right? When I looked at most of the software

15 At least that's what my studying looked like. This phenomenon is also known as the "student syndrome" (*https://oreil.ly/4Vpqb*).

16 *PB* means petabyte. One petabyte is 1,000 TB. If we assume an average two-hour-long 4K movie takes 100 GB, this means with 1 PB, we could store 10,000 movies, translating to roughly two to three years of constant watching.

co-created or used, they hit some performance issues related to significant data usage sooner or later. For example:

- A Prometheus UI page, written in React, was performing a search on millions of metric names or tried to fetch hundreds of megabytes of compressed samples, causing browser latencies and explosive memory usage.
- With low usage, a single Kubernetes cluster at our infrastructure generated 0.5 TB of logs daily (most of them never used).
- The excellent grammar checking tool I used to write this book was making too many network calls when the text had more than 20,000 words, slowing my browser considerably.
- Our simple script for formatting our documentation in Markdown and link checking took minutes to process all elements.
- Our Go static analysis job and linting exceeded 4 GB of memory and crashed our CI jobs.
- My IDE used to take 20 minutes to index all code from our mono-repo, despite doing it on a top-shelf laptop.
- I still haven't edited my 4K ultrawide videos from GoPro because the software is too laggy.

I could go on forever with examples, but the point is that we live in a really "big data" world. As a result, we have to optimize memory and other resources wisely.

It will be much worse in the future. Our software and hardware have to handle the data growing at extreme rates, faster than any hardware development. We are just on the edge of introducing 5G networks capable of transfers up to 20 gigabits per second (*https://oreil.ly/CWvFG*). We introduce mini-computers in almost every item we buy, like TVs, bikes, washing machines, freezers, desk lamps, or even deodorants (*https://oreil.ly/DvZil*)! We call this movement the "Internet of Things" (IoT). Data from these devices is estimated to grow from 18.3 ZB in 2019 to 73.1 ZB by 2025 (*https://oreil.ly/J1o6D*).[17] The industry can produce 8K TVs, rendering resolutions of 7,680 × 4,320, so approximately 33 million pixels. If you have written computer games, you probably understand this problem well—it will take a lot of efficient effort to render so many pixels in highly realistic games with immersive, highly destructive environments at 60+ frames per second. Modern cryptocurrencies and blockchain algorithms also pose challenges in computational energy efficiencies; e.g., Bitcoin energy consumption during the value peak was using roughly 130 Terawatt-hours of energy (0.6% of global electricity consumption) (*https://oreil.ly/NfnJ9*).

17 1 zettabyte is 1 million PB, one billion of TB. I won't even try to visualize this amount of data. :)

Technological limits

The last reason, but not least, behind not fast enough hardware progression is that hardware advancement has stalled on some fronts like CPU speed (clock rate) or memory access speeds. We will cover some challenges of that situation in Chapter 4, but I believe every developer should be aware of the fundamental technological limits we are hitting right now.

It would be odd to read a modern book about efficiency that doesn't mention Moore's Law, right? You've probably heard of it somewhere before. It was first stated in 1965 by former CEO and cofounder of Intel, Gordon Moore.

> The complexity for minimum component costs [the number of transistors, with minimal manufacturing cost per chip] has increased at a rate of roughly a factor of two per year. ... Over the longer term, the rate of increase is a bit more uncertain, although there is no reason to believe it will not remain nearly constant for at least 10 years. That means by 1975, the number of components per integrated circuit for minimum cost will be 65,000.
>
> —Gordon E. Moore, "Cramming More Components onto Integrated Circuits" (*https://oreil.ly/WhuWd*), *Electronics* 38 (1965)

Moore's observation had a big impact on the semiconductor industry. But decreasing the transistors' size would not have been that beneficial if not for Robert H. Dennard and his team. In 1974, their experiment revealed that power use stays proportional to the transistor dimension (constant power density).[18] This means that smaller transistors were more power efficient. In the end, both laws promised exponential performance per watt growth of transistors. It motivated investors to continuously research and develop ways to decrease the size of MOSFET[19] transistors. We can also fit more of them on even smaller, more dense microchips, which reduced manufacturing costs. The industry continuously decreased the amount of space needed to fit the same amount of computing power, enhancing any chip, from CPU through RAM and flash memory, to GPS receivers and high-definition camera sensors.

In practice, Moore's prediction lasted not 10 years as he thought, but nearly 60 so far, and it still holds. We continue to invent tinier, microscopic transistors, currently oscillating around ~70 nm. Probably we can make them even smaller. Unfortunately,

18 Robert H. Dennard et al., "Design of Ion-Implanted MOSFET's with Very Small Physical Dimension" (*https://oreil.ly/OAGPC*), *IEEE Journal of Solid-State Circuits* 9, no. 5 (October 1974): 256–268.

19 MOSFET (*https://oreil.ly/mhc5k*) stands for "metal–oxide–semiconductor field-effect transistor," which is, simply speaking, an insulated gate allowing to switch electronic signals. This particular technology is behind most memory chips and microprocessors produced between 1960 and now. It has proven to be highly scalable and capable of miniaturization. It is the most frequently manufactured device in history, with 13 sextillion pieces produced between 1960 and 2018.

as we can see on Figure 1-3, we reached the physical limit of Dennard's scaling around 2006.[20]

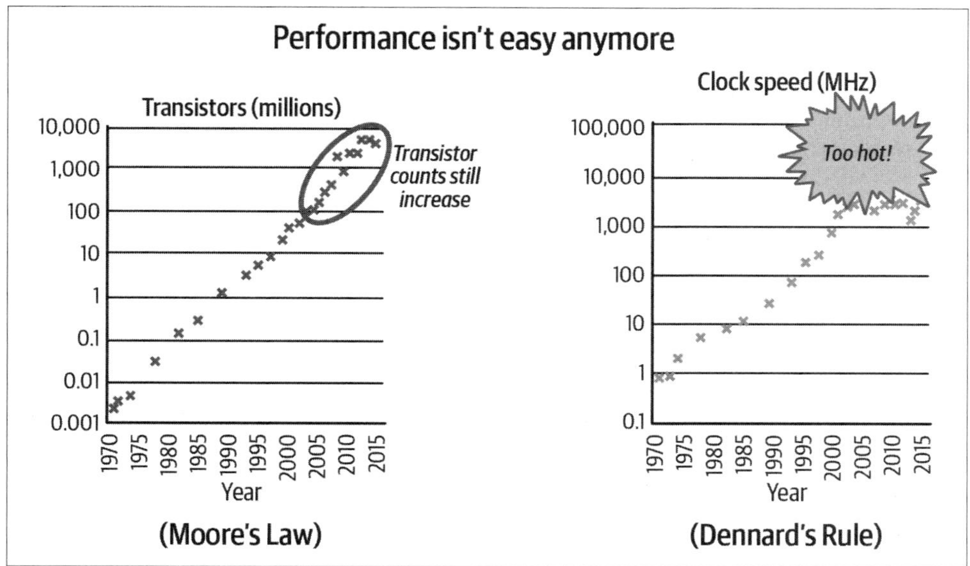

Figure 1-3. Image inspired by "Performance Matters" by Emery Berger (https://oreil.ly/Tyfog): Moore's Law versus Dennard's Rule

While technically, power usage of the higher density of tiny transistors remains constant, such dense chips heat up quickly. Beyond 3–4 GHz of clock speed, it takes significantly more power and other costs to cool the transistors to keep them running. As a result, unless you plan to run software on the bottom of the ocean,[21] you are not getting CPUs with faster instruction execution anytime soon. We only can have more cores.

Faster execution is more energy efficient

So, what we have learned so far? Hardware speed is getting capped, the software is getting bulkier, and we have to handle continuous growth in data and users. Unfortunately, that's not the end. There is a vital resource we tend to forget about while developing the software: power. Every computation of our process takes electricity,

---

20 Funnily enough, marketing reasons led companies to hide the inability to reduce the size of transistors effectively by switching the CPU generation naming convention from transistor gate length to the size of the process. 14 nm generation CPUs still have 70 nm transistors, similar to 10, 7, and 5 nm processes.

21 I am not joking. Microsoft has proven (*https://oreil.ly/nJzkN*) that running servers 40 meters underwater is a great idea that improves energy efficiency.

which is heavily constrained on many platforms like mobile phones, smartwatches, IoT devices, or laptops. Nonintuitively there is a strong correlation between energy efficiency and software speed and efficiency. I love the Chandler Carruth presentation, which explained this surprising relation well:

> If you ever read about "power-efficient instructions" or "optimizing for power usage," you should become very suspicious. ... This is mostly total junk science. Here is the number one leading theory about how to save battery life: Finish running the program. Seriously, race to sleep. The faster your software runs, the less power it consumes. ... Every single general-usage microprocessor you can get today, the way it conserves power is by turning itself off. As rapidly and as frequently as possible.
>
> —Chandler Carruth, "Efficiency with Algorithms, Performance with Data Structures" (*https://oreil.ly/9OftP*), CppCon 2014

To sum up, avoid the common trap of thinking about hardware as a continuously faster and cheaper resource that will save us from optimizing our code. It's a trap. Such a broken loop makes engineers gradually lower their coding standards in performance, and demand more and faster hardware. Cheaper and more accessible hardware then creates even more mental room to skip efficiency and so on. There are amazing innovations like Apple's M1 silicons,[22] RISC-V standard,[23] and more practical Quantum computing appliances, which promise a lot. Unfortunately, as of 2022, hardware is growing slower than software efficiency needs.

Efficiency Improves Accessibility and Inclusiveness

Software developers are often "spoiled" and detached from typical human reality in terms of the machines we use. It's often the case that engineers create and test software on premium, high-end laptop or mobile devices. We need to realize that many people and organizations are utilizing older hardware or worse internet connections.[24] People might have to run your applications on slower computers. It might be worth considering efficiency in our development process to improve overall software accessibility and inclusiveness.

22 The M1 chip (*https://oreil.ly/emBke*) is a great example of an interesting trade-off: choosing speed and both energy and performance efficiency over the flexibility of hardware scaling.

23 RISC-V is an open standard for the instruction set architecture, allowing easier manufacturing of compatible "reduced instruction set computer" chips. Such a set is much simpler and allows more optimized and specialized hardware than general-usage CPUs.

24 To ensure developers understand and empathize with users who have a slower connection, Facebook introduced "2G Tuesdays" (*https://oreil.ly/fZSoQ*) that turn on the simulated 2G network mode on the Facebook application.

We Can Scale Horizontally Instead

As we learned in the previous sections, we expect our software to handle more data sooner or later. But it's unlikely your project will have billions of users from day one. We can avoid enormous software complexity and development cost by pragmatically choosing a much lower target number of users, operations, or data sizes to aim for at the beginning of our development cycle. For example, we usually simplify the initial programming cycle by assuming a low number of notes in the mobile note-taking app, fewer requests per second in the proxy being built, or smaller files in the data converter tool the team is working on. It's OK to simplify things. It's also important to roughly predict performance requirements in the early design phase.

Similarly, finding the expected load and usage in the mid to long term of software deployment is essential. The software design that guarantees similar performance levels, even with increased traffic, is *scalable*. Generally, scalability is very difficult and expensive to achieve in practice.

> Even if a system is working reliably today, that doesn't mean it will necessarily work reliably in the future. One common reason for degradation is increased load: perhaps the system has grown from 10,000 concurrent users to 100,000 concurrent users, or from 1 million to 10 million. Perhaps it is processing much larger volumes of data than it did before. Scalability is the term we use to describe a system's ability to cope with increased load.
>
> —Martin Kleppmann, *Designing Data-Intensive Applications* (O'Reilly, 2017)

Inevitably, while talking about efficiency, we might touch on some scalability topics in this book. However, for this chapter's purpose, we can distinguish the scalability of our software into two types, presented in Figure 1-4.

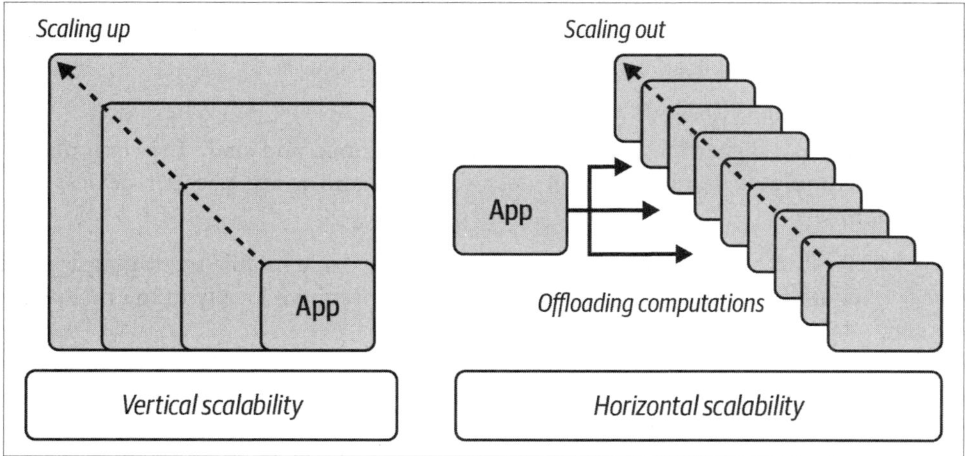

Figure 1-4. Vertical versus horizontal scalability

Vertical scalability

The first and sometimes simplest way of scaling our application is by running the software on hardware with more resources—"vertical" scalability. For example, we could introduce parallelism for software to use not one but three CPU cores. If the load increases, we provide more CPU cores. Similarly, if our process is memory intensive, we might bump up running requirements and ask for bigger RAM space. The same with any other resource, like disk, network, or power. Obviously, that does not come without consequences. In the best case, you have that room in the target machine. Potentially, you can make that room by rescheduling other processes to different machines (e.g., when running in the cloud) or closing them temporarily (useful when running on a laptop or smartphone). Worst case, you may need to buy a bigger computer, or a more capable smartphone or laptop. The latter option is usually very limited, especially if you provide software for customers to run on their noncloud premises. In the end, the usability of resource-hungry applications or websites that scale only vertically is much lower.

The situation is slightly better if you or your customers run your software in the cloud. You can "just" buy a bigger server. As of 2022, you can scale up your software on the AWS platform to 128 CPU cores, almost 4 TB of RAM, and 14 GBps of bandwidth.[25] In extreme cases, you can also buy an IBM mainframe with 190 cores and 40 TB of memory (*https://oreil.ly/P0auH*), which requires different programming paradigms.

Unfortunately, vertical scalability has its limits on many fronts. Even in the cloud or datacenters, we simply cannot infinitely scale up the hardware. First of all, giant machines are rare and expensive. Secondly, as we will learn in Chapter 4, bigger machines run into complex issues caused by many hidden single points of failures. Pieces like memory bus, network interfaces, NUMA nodes, and the operating system itself can be overloaded and too slow.[26]

Horizontal scalability

Instead of a bigger machine, we might try to offload and share the computation across multiple remote, smaller, less complex, and much cheaper devices. For example:

- To search for messages with the word "home" in a mobile messaging app, we could fetch millions of past messages (or store them locally in the first place)

25 That option is not as expensive as we might think. Instance type x1e.32xlarge costs $26.60 per hour (*https://oreil.ly/9fw5G*), so "only" $19,418 per month.

26 Even hardware management has to be different for machines with extremely large hardware. That's why Linux kernels have the special hugemem (*https://oreil.ly/tlWh3*) type of kernels that can manage up to four times more memory and ~eight times more logical cores for x86 systems.

and run regex matching on each. Instead, we can design an API and remotely call a backend system that splits the search into 100 jobs matching 1/100 of the dataset.

- Instead of building "monolith" software, we could distribute different functionalities to separate components and move to a "microservice" design.
- Instead of running a game that requires expensive CPUs and GPUs on a personal computer or gaming console, we could run it in a cloud and stream the input and output in high resolution (*https://oreil.ly/FKmTE*).

Horizontal scalability is easier to use as it has fewer limitations, and usually allows great dynamics. For instance, if the software is used only in a certain company, you might have almost no users at night, and large traffic during the day. With horizontal scalability it's easy to implement autoscaling that scales out and back in seconds based on demand.

On the other hand, horizontal scalability is much harder to implement on the software side. Distributed systems, network impacts, and hard problems that cannot be sharded are some of the many complications in the development of such systems. That's why it's often better to stick to vertical scalability in some cases.

With horizontal and vertical scalability in mind, let's look at a specific scenario from the past. Many modern databases rely on compaction to efficiently store and look up data. We can reuse many indices during this process, deduplicate the same data, and gather fragmented pieces into the sequential data stream for faster reads. At the beginning of the Thanos project, we decided to reuse a very naive compaction algorithm for simplicity. We calculated that, in theory, we don't need to make the compaction process parallel within a single block of data. Given a steady stream of 100 GB (or more) of eventually compacted data from a single source, we could rely on a single CPU, a minimal amount of memory, and some disk space. The implementation was initially very naive and unoptimized, following the YAGNI rule and avoiding premature optimization patterns. We wanted to avoid the complexity and effort of optimizing the project's reliability and functionality features. As a result, users who deployed our project quickly hit compaction problems: too slow to cope with incoming data or to consume hundreds of GB of memory per operation. The cost was the first problem, but not the most urgent. The bigger issue was that many Thanos users did not have bigger machines in their datacenters to scale the memory vertically.

At first glance, the compaction problem looked like a scalability problem. The compaction process depended on resources that we could not just add up infinitely. As users wanted a solution fast, together with the community, we started brainstorming potential horizontal scalability techniques. We talked about introducing a compactor scheduler service that would assign compaction jobs to different machines, or intelligent peer networks using a gossip protocol. Without going into details, both solutions would add enormous complexity, probably doubling or tripling the complication of developing and running the whole system. Luckily, it took a few days of brave and experienced developer time to redesign the code for efficiency and performance. It allowed the newer version of Thanos to make compactions twice as fast, and stream data directly from the disk, allowing minimal peak memory consumption. A few years later, the Thanos project still doesn't have any complex horizontal scalability for compaction, besides simple sharding, even with thousands of successful users running it with billions of metrics.

It might feel funny now, but in some ways, this story is quite scary. We were so close to bringing enormous, distributed system-level complexity, based on social and customer pressure. It would be fun to develop, but it could also risk collapsing the project's adoption. We might add it someday, but first we will make sure there is no other efficiency optimization to compaction. A similar situation has been repeated in my career in both open and closed sources for smaller and bigger projects.

Premature Scalability Is Worse than Premature Efficiency Optimizations!

Make sure you consider improving the efficiency on the algorithm and code level before introducing complex scalable patterns.

As presented by the "lucky" Thanos compaction situation, if we don't focus on the efficiency of our software, we can quickly be forced to introduce premature horizontal scalability. It is a massive trap because, with some optimization effort, we might completely avoid jumping into scalability method complications. In other words, avoiding complexity can bring even bigger complexity. This appears to me as an unnoticed but critical problem in the industry. It is also one of the main reasons why I wrote this book.

The complications come from the fact that complexity has to live somewhere (*https://oreil.ly/0PcmN*). We don't want to complicate code, so we have to complicate the system, which, if built from inefficient components, wastes resources and an enormous amount of developer or operator time. Horizontal scalability is especially complex. By design, it involves network operations. As we might know from the CAP Theorem,[27] we inevitably hit either availability or consistency issues as soon as we start distributing our process. Trust me, mitigating these elemental constraints, dealing with race conditions, and understanding the world of network latencies and unpredictability is a hundred times more difficult than adding small efficiency optimization, e.g., hidden behind the `io.Reader` interface.

27 CAP (*https://oreil.ly/EYqPI*) is a core system design principle. Its acronym comes from Consistency, Availability, and Partition tolerance. It defines a simple rule that only two of the three can be achieved.

It might seem to you that this section touches only on infrastructure systems. That's not true. It applies to all software. For example, if you write a frontend software or dynamic website, you might be tempted to move small client computations to the backend. We should probably only do that if the computation depends on the load and grows out of user space hardware capabilities. Moving it to the server prematurely might cost you the complexity caused by extra network calls, more error cases to handle, and server saturations causing Denial of Service (DoS).[28]

Another example comes from my experience. My master's thesis was about a "Particle Engine Using Computing Cluster." In principle, the goal was to add a particle engine to a 3D game in a Unity engine (*https://unity.com*). The trick was that the particle engine was not supposed to operate on client machines, instead offloading "expensive" computation to a nearby supercomputer in my university called "Tryton."[29] Guess what? Despite the ultrafast InfiniBand network,[30] all particles I tried to simulate (realistic rain and crowd) were much slower and less reliable when offloaded to our supercomputer. It was not only less complex but also much faster to compute all on client machines.

Summing up, when someone says, "Don't optimize, we can just scale horizontally," be very suspicious. Generally, it is simpler and cheaper to start from efficiency improvements before we escalate to a scalability level. On the other hand, a judgment should tell you when optimizations are becoming too complex and scalability might be a better option. You will learn more about that in Chapter 3.

Time to Market Is More Important

Time is expensive. One aspect of this is that software developer time and expertise cost a lot. The more features you want your application or system to have, the more time is needed to design, implement, test, secure, and optimize the solution's performance. The second aspect is that the more time a company or individual spends to deliver the product or service, the longer their "time to market" is, which can hurt the financial results.

28 Denial of Service is a state of the system that makes the system unresponsive, usually due to malicious attack. It can also be trigged "accidentally" by an unexpectedly large load.

29 Around 2015, it was the fastest supercomputer in Poland, offering 1.41 PFlop/s and over 1,600 nodes, most of them with dedicated GPUs.

30 InfiniBand is a high-performance network communication standard, especially popular before fiber optic was invented.

> Once time was money. Now it is more valuable than money. A McKinsey study reports that, on average, companies lose 33% of after-tax profit when they ship products six months late, as compared with losses of 3.5% when they overspend 50% on product development.
>
> —Charles H. House and Raymond L. Price, "The Return Map: Tracking Product Teams" (*https://oreil.ly/SmLFQ*)

It's hard to measure such impact, but your product might no longer be pioneering when you are "late" to market. You might miss valuable opportunities or respond too late to a competitor's new product. That's why companies mitigate this risk by adopting Agile methodologies or proof of concept (POC) and minimal viable product (MVP) patterns.

Agile and smaller iterations help, but in the end, to achieve faster development cycles, companies try other things too: scale their teams (hire more people, redesign teams), simplify the product, do more automation, or do partnerships. Sometimes they try to reduce the product quality. As Facebook's proud initial motto was "Move fast and break things,"[31] it's very common for companies to descope software quality in areas like code maintainability, reliability, and efficiency to "beat" the market.

This is what our last misconception is all about. Descoping your software's efficiency to get to the market faster is not always the best idea. It's good to know the consequences of such a decision. Know the risk first.

> Optimization is a difficult and expensive process. Many engineers argue that this process delays entry into the marketplace and reduces profit. This may be true, but it ignores the cost associated with poor-performing products (particularly when there is competition in the marketplace).
>
> —Randall Hyde, "The Fallacy of Premature Optimization" (*https://oreil.ly/mMjHb*)

Bugs, security issues, and poor performance happen, but they might damage the company. Without looking too far, let's look at a game released in late 2020 by the biggest Polish game publisher, CD Projekt. *Cyberpunk 2077* (*https://oreil.ly/ohJft*) was known to be a very ambitious, open world, massive, and high-quality production. Well marketed, from a publisher with a good reputation, despite the delays, excited players around the world bought eight million preorders. Unfortunately, when released in December 2020, the otherwise excellent game had massive performance issues. It had bugs, crashes, and a low frame rate on all consoles and most PC setups. On some older consoles like PS4 or Xbox One, the game was claimed to be unplayable. There were, of course, updates with plenty of fixes and drastic improvements over the following months and years.

31 Funny enough, Mark Zuckerberg at an F8 conference in 2014 announced a change of the famous motto to "Move fast with stable infra" (*https://oreil.ly/Yt2VI*).

Unfortunately, it was too late. The damage was done. The issues, which for me felt somewhat minor, were enough to shake CD Projekt's financial perspectives. Five days after launch, the company lost one-third of its stock value, costing the founders more than $1 billion (*https://oreil.ly/x5Qd8*). Millions of players asked for game refunds. Investors sued (*https://oreil.ly/CRKg4*) CD Projekt over game issues, and famous lead developers left the company (*https://oreil.ly/XwcX9*). Perhaps the publisher will survive and recover. Still, one can only imagine the implications of a broken reputation impacting future productions.

More experienced and mature organizations know well the critical value of software performance, especially the client-facing ones. Amazon found that if its website loaded one second slower, it would lose $1.6 billion annually (*https://oreil.ly/cHT2j*). Amazon also reported that 100 ms of latency costs 1% of profit (*https://oreil.ly/Bod7k*). Google realized that slowing down their web search from 400 ms to 900 ms caused a 20% drop in traffic (*https://oreil.ly/hHmYJ*). For some businesses, it's even worse. It was estimated that if a broker's electronic trading platform is 5 milliseconds slower than the competition, it could lose 1% of its cash flow, if not more. If 10 milliseconds slower, this number grows to a 10% drop in revenue (*https://oreil.ly/fK7mE*).

Realistically speaking, it's true that millisecond-level slowness might not matter in most software cases. For example, let's say we want to implement a file converter from PDF to DOCX. Does it matter if the whole experience lasts 4 seconds or 100 milliseconds? In many cases, it does not. However, when someone puts that as a market value and a competitor's product has a latency of 200 milliseconds, code efficiency and speed suddenly become a matter of winning or losing customers. And if it's physically possible to have such fast file conversion, competitors will try to achieve it sooner or later. This is also why so many projects, even open source, are very loud about their performance results. While sometimes it feels like a cheap marketing trick, this works because if you have two similar solutions with similar feature sets and other characteristics, you will pick the fastest one. It's not all about the speed, though—resource consumption matters as well.

Efficiency Is Often More Important in Market than Features!

During my experience as a consultant for infrastructure systems, I saw many cases where customers migrated away from solutions requiring a larger amount of RAM or disk storage, even if that meant some loss in functionalities.[32]

32 One example I see often in the cloud-native world is moving logging stack from Elasticsearch to simpler solutions like Loki. Despite the lack of configurable indexing, the Loki project can offer better logging read performance with a smaller amount of resources.

To me, the verdict is simple. If you want to win the market, skipping efficiency in your software might not be the best idea. Don't wait with optimization until the last moment. On the other hand, time to market is critical, so balancing a good enough amount of efficiency work into your software development process is crucial. One way of doing this is to set the nonfunctional goals early (discussed in "Resource-Aware Efficiency Requirements" on page 86). In this book, we will focus a lot on finding that healthy balance and reducing the effort (thus the time) required to improve the efficiency of your software. Let's now look at what is the pragmatic way to think about the performance of our software.

The Key to Pragmatic Code Performance

In "Behind Performance" on page 3, we learned that performance splits into accuracy, speed, and efficiency. I mentioned that in this book when I use the word *efficiency*, it naturally means efficient resource consumption, but also our code's speed (latency). A practical suggestion is hidden in that decision regarding how we should think about our code performing in production.

The secret here is to stop focusing strictly on the speed and latency of our code. Generally, for nonspecialized software, speed matters only marginally; the waste and unnecessary consumption of resources are what introduce slowdowns. And achieving high speed with bad efficiency will always introduce more problems than benefits. As a result, we should generally focus on efficiency. Sadly, it is often overlooked.

Let's say you want to travel from city A to city B across the river. You can grab a fast car and drive over a nearby bridge to get to city B quickly. But if you jump into the water and slowly swim across the river, you will get to city B much faster. Slower actions can still be faster when done efficiently, for example, by picking a shorter route. One could say that to improve travel performance and beat the swimmer, we could get a faster car, improve the road surface to reduce drag, or even add a rocket engine. We could potentially beat the swimmer, yes, but those drastic changes might be more expensive than simply doing less work and renting a boat instead.

Similar patterns exist in software. Let's say our algorithm does search functionality on certain words stored on disk and performs slowly. Given that we operate on persistent data, the slowest operation is usually the data access, especially if our algorithm does this extensively. It's very tempting to not think about efficiency and instead find a way to convince users to use SSD instead of HDD storage. This way, we could potentially reduce latency up to 10 times. That would improve performance by increasing the speed element of the equation. On the contrary, if we could find a way to enhance the current algorithm to read data only a few times instead of a million, we could achieve even lower latencies. That would mean we can have the same or even better effect by keeping the cost low.

I want to propose focusing our efforts on efficiency instead of mere execution speed. That is also why this book's title is *Efficient Go*, not something more general and catchy[33] like *Ultra Performance Go* or *Fastest Go Implementations*.

It's not that speed is less relevant. It is important, and as you will learn in Chapter 3, you can have more efficient code that is much slower and vice versa. Sometimes it's a trade-off you will need to make. Both speed and efficiency are essential. Both can impact each other. In practice, when the program is doing less work on the critical path, it will most likely have lower latency. In the HDD versus SDD example, changing to a faster disk might allow you to remove some caching logic, which results in better efficiency: less memory and CPU time used. The other way around works sometimes too—as we learned in "Hardware Is Getting Faster and Cheaper" on page 17, the faster your process is, the less energy it consumes, improving battery efficiency.

I would argue that we generally should focus on improving efficiency before speed as the first step when improving performance. As you will see in "Optimizing Latency" on page 383, only by changing efficiency was I able to reduce latency seven times, with just one CPU core. You might be surprised that sometimes after improving efficiency, you have achieved desired latency! Let's go through some further reasons why efficiency might be superior:

It is much harder to make efficient software slow.
: This is similar to the fact that readable code is easier to optimize. However, as I mentioned before, efficient code usually performs better simply because less work has to be done. In practice, this also translates to the fact that slow software is often inefficient.

Speed is more fragile.
: As you will learn in "Reliability of Experiments" on page 256, the latency of the software process depends on a huge amount of external factors. One can optimize the code for fast execution in a dedicated and isolated environment, but it can be much slower when left running for a longer time. At some point, CPUs might be throttled due to thermal issues with the server. Other processes (e.g., periodic backup) might surprisingly slow your main software. The network might be throttled. There are tons of hidden unknowns to consider when we program for mere execution speed. This is why efficiency is usually what we, as programmers, can control the most.

33 There is also another reason. The "Efficient Go" name is very close to one of the best documentation pieces you might find about the Go programming language: "Effective Go" (*https://oreil.ly/OHbMt*)! It might also be one of the first pieces of information I have read about Go. It's specific, actionable, and I recommend reading it if you haven't.

Speed is less portable.

If we optimize only for speed, we cannot assume it will work the same when moving our application from the developer machine to a server or between various client devices. Different hardware, environments, and operating systems can diametrically change the latency of our application. That's why it's critical to design software for efficiency. First of all, there are fewer things that can be affected. Secondly, if you make two calls to the database on your developer machine, chances are that you will do the same number of calls, no matter if you deploy it to an IoT device in the space station or an ARM-based mainframe.

Generally, efficiency is something we should do right after or together with readability. We should start thinking about it from the very beginning of the software design. A healthy efficiency awareness, when not taken to the extreme, results in robust development hygiene. It allows us to avoid silly performance mistakes that are hard to improve on in later development stages. Doing less work also often reduces the overall complexity of the code, and improves code maintainability and extensibility.

Summary

I think it's very common for developers to start their development process with compromises in mind. We often sit down with the attitude that we must compromise certain software qualities from the beginning. We are often taught to sacrifice qualities of our software, like efficiency, readability, testability, etc., to accomplish our goals.

In this chapter, I wanted to encourage you to be a bit more ambitious and greedy for software quality. Hold out and try not to sacrifice any quality until you have to—until it is demonstrated that there is no reasonable way you can achieve all of your goals. Don't start your negotiations with default compromises in mind. Some problems are hard without simplifications and compromises, but many have solutions with some effort and appropriate tools.

Hopefully, at this point, you are aware that we have to think about efficiency, ideally from the early development stages. We learned what performance consists of. In addition, we learned that many misconceptions are worth challenging when appropriate. We need to be aware of the risk of premature pessimization and premature scalability as much as we need to consider avoiding premature optimizations.

Finally, we learned that efficiency in the performance equation might give us an advantage. It is easier to improve performance by improving efficiency first. It helped my students and me many times to effectively approach the subject of performance optimizations.

In the next chapter, we will walk through a quick introduction to Go. Knowledge is key to better efficiency, but it's extra hard if we are not proficient with the basics of the programming language we use.

CHAPTER 2
Efficient Introduction to Go

> Go is efficient, scalable, and productive. Some programmers find it fun to work in; others find it unimaginative, even boring. ... Those are not contradictory positions. Go was designed to address the problems faced in software development at Google, which led to a language that is not a breakthrough research language but is nonetheless an excellent tool for engineering large software projects.
>
> —Rob Pike, "Go at Google: Language Design in the Service of Software Engineering" (*https://oreil.ly/3EItq*)

I am a huge fan of the Go programming language. The number of things developers around the world have been able to achieve with Go is impressive. For a few years in a row, Go has been on the list of top five languages people love or want to learn (*https://oreil.ly/la9bx*). It is used in many businesses, including bigger tech companies like Apple, American Express, Cloudflare, Dell, Google, Netflix, Red Hat, Twitch, and others (*https://oreil.ly/DSM73*). Of course, as with everything, nothing is perfect. I would probably change, remove, or add a few things to Go, but if you would wake me in the middle of the night and ask me to quickly write reliable backend code, I would write it in Go. CLI? In Go. Quick, reliable script? In Go as well. The first language to learn as a junior programmer? Go. Code for IoT, robots, and microprocessors? The answer is also Go.[1] Infrastructure configuration? As of 2022, I don't think there is a better tool for robust templating than Go.[2]

---

1 New frameworks on tools for writing Go on small devices are emerging, e.g., GoBot (*https://gobot.io*) and TinyGo (*https://tinygo.org*).

2 It's a controversial topic. There is quite a battle in the infrastructure industry for the superior language for configuration as code. For example, among HCL, Terraform, Go templates (Helm), Jsonnet, Starlark, and Cue. In 2018, we even open sourced a tool for writing configuration in Go, called "mimic" (*https://oreil.ly/FNjYD*). Arguably, the loudest arguments against writing configuration in Go are that it feels too much like "programming" and requires programming skills from system administrators.

Don't get me wrong, there are languages with specialized capabilities or ecosystems that are superior to Go. For example, think about graphical user interfaces (GUIs), advanced rendering parts of the game industry, or code running in browsers.[3] However, once you realize the many advantages of the Go language, it is pretty painful to jump back to others.

In Chapter 1, we spent some time establishing an efficiency awareness for our software. As a result, we learned that our goal is to write efficient code with the least development effort and cost. This chapter will explain why the Go programming language can be a solid option to achieve this balance between performance and other software qualities.

We will start with "Basics You Should Know About Go" on page 36, then continue with "Advanced Language Elements" on page 55. Both sections list the short but essential facts everyone should know about Go, something I wish I had known when I started my journey with Go in 2014. These sections will cover much more than just basic information about efficiency and can be used as an introduction to Go. However, if you are entirely new to the language, I would still recommend reading those sections, then checking other resources mentioned in the summary, perhaps writing your first program in Go, and then getting back to this book. On the other hand, if you consider yourself a more advanced user or expert, I suggest not skipping this chapter. I explain a few lesser-known facts about Go that you might find interesting or controversial (it's OK, everyone can have their own opinions!).

Last but not least, we will finish by answering the tricky question about the overall Go efficiency capabilities in "Is Go 'Fast'?" on page 67, as compared to other languages.

Basics You Should Know About Go

Go is an open source project maintained by Google within a distributed team called the "Go team." The project consists of the programming language specification, compiler, tooling, documentation, and standard libraries.

Let's go through some facts and best practices to understand Go basics and its characteristics in fast-forward mode. While some advice here might feel opinionated, this is based on my experience working with Go since 2014—a background full of incidents, past mistakes, and lessons learned the hard way. I'm sharing them here so you don't need to make those errors.

3 WebAssembly is meant to change this, though, but not soon (*https://oreil.ly/rZqtp*).

Imperative, Compiled, and Statically Typed Language

The central part of the Go project is the general-purpose language with the same name, primarily designed for systems programming. As you will notice in Example 2-1, Go is an imperative language, so we have (some) control over how things are executed. In addition, it's statically typed and compiled, which means that the compiler can perform many optimizations and checks before the program runs. These characteristics alone are an excellent start to make Go suitable for reliable and efficient programs.

Example 2-1. Simple program printing "Hello World" and exiting

```
package main

import "fmt"

func main() {
   fmt.Println("Hello World!")
}
```

Both project and language are called "Go," yet sometimes you can refer to them as "Golang."

Go Versus Golang

As a rule of thumb, we should always use the "Go" name everywhere, unless it's clashing with the English word *go* or an ancient game called "Go." "Golang" came from the domain choice (*https://golang.org*) since "go" was unavailable to its authors. So use "Golang" when searching for resources about this programming language on the web.

Go also has its mascot, called the "Go gopher" (*https://oreil.ly/SbxVX*). We see this cute gopher in various forms, situations, and combinations, such as conference talks, blog posts, or project logos. Sometimes Go developers are called "gophers" too!

Designed to Improve Serious Codebases

It all started when three experienced programmers from Google sketched the idea of the Go language around 2007:

Rob Pike
: Cocreator of UTF-8 and the Plan 9 operating system. Coauthor of many programming languages before Go, such as Limbo for writing distributed systems

and Newsqueak for writing concurrent applications in graphical user interfaces. Both were inspired by Hoare's Communicating Sequential Processes (CSP).[4]

Robert Griesemer
: Among other work, Griesemer developed the Sawzall language (*https://oreil.ly/gYKMj*) and did a doctorate with Niklaus Wirth. The same Niklaus wrote "A Plea for Lean Software" quoted in "Software gets slower more rapidly than hardware becomes faster" on page 20.

Ken Thompson
: One of the original authors of the first Unix system. Sole creator of the `grep` command-line utility. Ken cocreated UTF-8 and Plan 9 with Rob Pike. He wrote a couple of languages, too, e.g., the Bon and B programming languages.

These three aimed to create a new programming language that was meant to improve mainstream programming, led by C++, Java, and Python at that point. After a year, it became a full-time project, with Ian Taylor and Russ Cox joining in 2008 what was later referenced as the Go team (*https://oreil.ly/Nnj6N*). The Go team announced the public Go project in 2009, with version 1.0 released in March 2012.

The main frustrations[5] related to C++ mentioned in the design of Go were:

- Complexity, many ways of doing the same thing, too many features
- Ultralong compilation times, especially for bigger codebases
- Cost of updates and refactors in large projects
- Not easy to use and memory model prone to errors

These elements are why Go was born, from the frustration of existing solutions and the ambition to allow more by doing less. The guiding principles were to make a language that does not trade safety for less repetition, yet allows simpler code. It does not sacrifice execution efficiency for faster compilation or interpreting, yet ensures that build times are quick enough. Go tries to compile as fast as possible, e.g., thanks to explicit imports (*https://oreil.ly/qxuUS*). Especially with caching enabled by default, only changed code is compiled, so build times are rarely longer than a minute.

4 CSP is a formal language that allows describing interactions in concurrent systems. Introduced by C.A.R. Hoare in *Communications of the ACM* (1978), it was an inspiration for the Go language concurrency system.

5 Similar frustrations triggered another part of Google to create yet another language—Carbon (*https://oreil.ly/ijFPA*) in 2022. Carbon looks very promising, but it has different goals than Go. It is, by design, more efficiency aware and focused on familiarity with C++ concepts and interoperability. So let's see how adoption will catch up for Carbon!

You Can Treat Go Code as Script!

While technically Go is a compiled language, you can run it like you would run JavaScript, Shell, or Python. It's as simple as invoking `go run <executable package> <flags>`. It works great because the compilation is ultrafast. You can treat it like a scripting language while maintaining the advantages of compilation.

In terms of syntax, Go was meant to be simple, light on keywords, and familiar. Syntax is based on C with type derivation (automatic type detection, like `auto` in C++), and no forward declarations, no header files. Concepts are kept orthogonal, which allows easier combination and reasoning about them. Orthogonality for elements means that, for example, we can add methods to any type or data definition (adding methods is separate from creating types). Interfaces are orthogonal to types too.

Governed by Google, Yet Open Source

Since announcing Go, all development has been done in open source (*https://oreil.ly/ZeKm6*), with public mailing lists and bug trackers. Changes go to the public, authoritative source code, held under the BSD style license (*https://oreil.ly/XBDEK*). The Go team reviews all contributions. The process is the same if the change or idea is coming from Google or not. The project road maps and proposals are developed in public too.

Unfortunately, the sad truth is that there are many open source projects, but some projects are less open than others. Google is still the only company stewarding Go and has the last decisive control over it. Even if anyone can modify, use, and contribute, projects coordinated by a single vendor risk selfish and damaging decisions like relicensing or blocking certain features. While there were some controversial cases where the Go team decision surprised the community,[6] overall the project is very reasonably well governed. Countless changes came from outside of Google, and the Go 2.0 draft proposal process has been well respected and community driven. In the end, I believe consistent decision-making and stewarding from the Go team bring many benefits too. Conflicts and different views are inevitable, and having one consistent overview, even if not perfect, might be better than no decision or many ways of doing the same thing.

6 One notable example is the controversy behind dependency management work (*https://oreil.ly/3gB9m*).

So far, this project setup has proven to work well for adoption and language stability. For our software efficiency goals, such alignment couldn't be better too. We have a big company invested in ensuring each release doesn't bring any performance regressions. Some internal Google software depends on Go, e.g., Google Cloud Platform (*https://oreil.ly/vjyOc*). And many people rely on the Google Cloud Platform to be reliable. On the other hand, we have a vast Go community that gives feedback, finds bugs, and contributes ideas and optimizations. And if that's not enough, we have open source code, allowing us, mere mortal developers, to dive into the actual Go libraries, runtime (see "Go Runtime" on page 59), etc., to understand the performance characteristics of the particular code.

Simplicity, Safety, and Readability Are Paramount

Robert Griesemer mentioned in GopherCon 2015 (*https://oreil.ly/s3ZZ5*) that first of all, they knew when they first started building Go what things NOT to do. The main guiding principle was simplicity, safety, and readability. In other words, Go follows the pattern of "less is more." This is a potent idiom that spans many areas. In Go, there is only one *idiomatic* coding style,[7] and a tool called `gofmt` ensures most of it. In particular, code formatting (next to naming) is an element that is rarely settled among programmers. We spend time arguing about it and tuning it to our specific needs and beliefs. Thanks to a single style enforced by tooling, we save enormous time. As one of the Go proverbs (*https://oreil.ly/ua2G8*) goes, "Gofmt's style is no one's favorite, yet gofmt is everyone's favorite." Overall, the Go authors planned the language to be minimal so that there is essentially one way to write a particular construct. This takes away a lot of decision-making when you are writing a program. There is one way of handling errors, one way of writing objects, one way of running things concurrently, etc.

A huge number of features might be "missing" from Go, yet one could say it is more expressive than C or C++ (*https://oreil.ly/CPkvV*). Such minimalism allows for maintaining the simplicity and readability of the Go code, which improves software reliability, safety, and overall higher velocity toward application goals.

7 Of course, there are some inconsistencies here and there; that's why the community created more strict formatters (*https://oreil.ly/RKUme*), linters (*https://oreil.ly/VnQSC*), or style guides (*https://oreil.ly/ETWSq*). Yet the standard tools are good enough to feel comfortable in every Go codebase.

> **Is My Code Idiomatic?**
>
> The word *idiomatic* is heavily overused in the Go community. Usually, it means Go patterns that are "often" used. Since Go adoption has grown a lot, people have improved the initial "idiomatic" style in many creative ways. Nowadays, it's not always clear what's idiomatic and what's not.
>
> It's like the "This is the way" saying from the *Mandalorian* series. It makes us feel more confident when we say, "This code is idiomatic." So the conclusion is to use this word with care and avoid it unless you can elaborate the reasoning why some pattern is better (*https://oreil.ly/dAAKz*).

Interestingly, the "less is more" idiom can help our efficiency efforts for this book's purpose. As we learned in Chapter 1, if you do less work at runtime, it usually means faster, lean execution and less complex code. In this book, we will try to maintain this aspect while improving our code performance.

Packaging and Modules

The Go source code is organized into directories representing either packages or modules. A package is a collection of source files (with the *.go* suffix) in the same directory. The package name is specified with the `package` statement at the top of each source file, as seen in Example 2-1. All files in the same directory must the same package name[8] (the package name can be different from the directory name). Multiple packages can be part of a single Go module. A module is a directory with a *go.mod* file that states all dependent modules with their versions required to build the Go application. This file is then used by the dependency management tool Go Modules (*https://oreil.ly/z5GqG*). Each source file in a module can import packages from the same or external modules. Some packages can also be "executable." For example, if a package is called `main` and has `func main()` in some file, we can execute it. Sometimes such a package is placed in the cmd directory for easier discovery. Note that you cannot import the executable package. You can only build or run it.

Within the package, you can decide what functions, types, interfaces, and methods are exported to package users and which are accessible only in the package scope. This is important because exporting the minimal amount of API possible for readability, reusability, and reliability is better. Go does not have any `private` or `public` keywords for this. Instead, it takes a slightly new approach. As Example 2-2 shows, if the construct name starts with an uppercase letter, any code outside the package can

8 There is one exception: unit test files that have to end with *_test.go*. These files can have either the same package name or the `<package_name>_test` name allowing to mimic external users of the package.

use it. If the element name begins with a lowercase letter, it's private. It's worth noting that this pattern works for all constructs equally, e.g., functions, types, interfaces, variables, etc. (orthogonality).

Example 2-2. Construct accessibility control using naming case

```
package main

const privateConst = 1
const PublicConst = 2

var privateVar int
var PublicVar int

func privateFunc() {}
func PublicFunc()  {}

type privateStruct struct {
   privateField int
   PublicField  int ❶
}

func (privateStruct) privateMethod() {}
func (privateStruct) PublicMethod()  {} ❶

type PublicStruct struct {
   privateField int
   PublicField  int
}

func (PublicStruct) privateMethod() {}
func (PublicStruct) PublicMethod()  {}

type privateInterface interface {
   privateMethod()
   PublicMethod() ❶
}

type PublicInterface interface {
   privateMethod()
   PublicMethod()
}
```

❶ Careful readers might notice tricky cases of exported fields or methods on private type or `interface`. Can someone outside the package use them if the `struct` or `interface` is private? This is quite rarely used, but the answer is yes, you can return a private `interface` or type in a public function, e.g., `func New() privateStruct { return privateStruct{}}`. Despite the `privateStruct` being private, all its public fields and methods are accessible to package users.

Internal Packages

You can name and structure your code directories as you want to form packages, but one directory name is reserved for special meaning. If you want to ensure that only the given package can import other packages, you can create a package subdirectory named internal. Any package under the internal directory can't be imported by any package other than the ancestor (and other packages in internal).

Dependencies Transparency by Default

In my experience, it is common to import precompiled libraries, such as in C++, C#, or Java, and use exported functions and classes defined in some header files. However, importing compiled code has some benefits:

- It relieves engineers from making an effort to compile particular code, i.e., find and download correct versions of dependencies, special compilation tooling, or extra resources.
- It might be easier to sell such a prebuilt library without exposing the source code and worrying about the client copying the business value-providing code.[9]

In principle, this is meant to work well. Developers of the library maintain specific programmatic contracts (APIs), and users of such libraries do not need to worry about implementation complexities.

Unfortunately, in practice, this is rarely that perfect. Implementation can be broken or inefficient, the interfaces can mislead, and documentation can be missing. In such cases, access to the source code is invaluable, allowing us to more deeply understand implementation. We can find issues based on specific source code, not by guessing. We can even propose a fix to the library or fork the package and use it immediately. We can extract the required pieces and use them to build something else.

Go assumes this imperfection by requiring each library's parts (in Go: module's packages) to be explicitly imported using a package URI called "import path." Such import is also strictly controlled, i.e., unused imports or cyclic dependencies cause a compilation error. Let's see different ways to declare these imports in Example 2-3.

9 In practice, you can quickly obtain the C++ or Go code (even when obfuscated) from the compiled binary anyway, especially if you don't strip the binary from the debugging symbols.

Example 2-3. Portion of `import` statements from `github.com/prometheus/prometheus` module, main.go *file*

```
import (
    "context" ❶
    "net/http"
    _ "net/http/pprof" ❷

    "github.com/oklog/run" ❸
    "github.com/prometheus/common/version"
    "go.uber.org/atomic"

    "github.com/prometheus/prometheus/config" ❹
    promruntime "github.com/prometheus/prometheus/pkg/runtime"
    "github.com/prometheus/prometheus/scrape"
    "github.com/prometheus/prometheus/storage"
    "github.com/prometheus/prometheus/storage/remote"
    "github.com/prometheus/prometheus/tsdb"
    "github.com/prometheus/prometheus/util/strutil"
    "github.com/prometheus/prometheus/web"
)
```

❶ If the import declaration does not have a domain with a path structure, it means the package from the "standard"[10] library is imported. This particular import allows us to use code from the `$(go env GOROOT)/src/context/` directory with `context` reference, e.g., `context.Background()`.

❷ The package can be imported explicitly without any identifier. We don't want to reference any construct from this package, but we want to have some global variables initialized. In this case, the `pprof` package will add debugging endpoints to the global HTTP server router. While allowed, in practice we should avoid reusing global, modifiable variables.

❸ Nonstandard packages can be imported using an import path in the form of an internet domain name and an optional path to the package in a certain module. For example, the Go tooling integrates well with `https://github.com`, so if you host your Go code in a Git repository, it will find a specified package. In this case, it's the `https://github.com/oklog/run` Git repository with the `run` package in the `github.com/oklog/run` module.

10 Standard library means packages that are shipped together with the Go language tooling and runtime code. Usually, only mature and core functionalities are provided, as Go has strong compatibility guarantees. Go also maintains an experimental `golang.org/x/exp` (*https://oreil.ly/KBTwn*) module that contains useful code that must be proven to graduate to the standard library.

❹ If the package is taken from the current module (in this case, our module is `github.com/prometheus/prometheus`), packages will be resolved from your local directory. In our example, `<module root>/config`.

This model focuses on open and clearly defined dependencies. It works exceptionally well with the open source distribution model, where the community can collaborate on robust packages in the public Git repositories. Of course, a module or package can also be hidden using standard version control authentication protocols. Furthermore, the official tooling does not support distributing packages in binary form (*https://oreil.ly/EnkBT*), so the dependency source is highly encouraged to be present for compilation purposes.

The challenges of software dependency are not easy to solve. Go learned from the mistakes of C++ and others, and takes a careful approach to avoid long compilation times, and an effect commonly called "dependency hell."

> Through the design of the standard library, great effort was spent on controlling dependencies. It can be better to copy a little code than to pull in a big library for one function. (A test in the system build complains if new core dependencies arise.) Dependency hygiene trumps code reuse. One example of this in practice is that the (low-level) net package has its own integer-to-decimal conversion routine to avoid depending on the bigger and dependency-heavy formatted I/O package. Another is that the string conversion package strconv has a private implementation of the definition of "printable" characters rather than pull in the large Unicode character class tables; that strconv honors the Unicode standard is verified by the package's tests.
>
> —Rob Pike, "Go at Google: Language Design in the Service of Software Engineering" (*https://oreil.ly/wqKGT*)

Again, with efficiency in mind, potential minimalism in dependencies and transparency brings enormous value. Fewer unknowns means we can quickly detect main bottlenecks and focus on the most significant value optimizations first. We don't need to work around it if we notice potential room for optimization in our dependency. Instead, we are usually welcome to contribute the fix directly to the upstream, which helps both sides!

Consistent Tooling

From the beginning, Go had a powerful and consistent set of tools as part of its command-line interface tool, called `go`. Let's enumerate a few utilities:

- `go bug` opens a new browser tab with the correct place where you can file an official bug report (Go repository on GitHub).
- `go build -o <output path> <packages>` builds given Go packages.

- `go env` shows all Go-related environment variables currently set in your terminal session.
- `go fmt <file, packages or directories>` formats given artifacts to the desired style, cleans whitespaces, fixes wrong indentations, etc. Note that the source code does not need to be even valid and compilable Go code. You can also install an extended official formatter.
- `goimports` (*https://oreil.ly/6fDcy*) also cleans and formats your `import` statements.

For the best experience, set your programming IDE to run `goimports -w $FILE` on every file to not worry about the manual indentation anymore!

- `go get <package@version>` allows you to install the desired dependency with the expected version. Use the `@latest` suffix to get the latest version of `@none` to uninstall the dependency.
- `go help <command/topic>` prints documentation about the command or given topic. For example, `go help environment` tells you all about the possible environment variables Go uses.
- `go install <package>` is similar to `go get` and installs the binary if the given package is "executable."
- `go list` lists Go packages and modules. It allows flexible output formatting using Go templates (explained later), e.g., `go list -mod=readonly -m -f '{{ if and (not .Indirect) (not .Main)}}{{.Path}}{{end}}' all` lists all direct nonexecutable dependent modules.
- `go mod` allows managing dependent modules.
- `go test` allows running unit tests, fuzz tests, and benchmarks. We will discuss the latter in detail in Chapter 8.
- `go tool` hosts a dozen more advanced CLI tools. We will especially take a close look at `go tool pprof` in "pprof Format" on page 332 for performance optimizations.
- `go vet` runs basic static analysis checks.

In most cases, the Go CLI is all you need for effective Go programming.[11]

Single Way of Handling Errors

Errors are an inevitable part of every running software. Especially in distributed systems, they are expected by design, with advanced research and algorithms for handling different types of failures.[12] Despite the need for errors, most programming languages do not recommend or enforce a particular way of failure handling. For example, in C++ you see programmers using all means possible to return an error from a function:

- Exceptions
- Integer return codes (if the returned value is nonzero, it means error)
- Implicit status codes[13]
- Other sentinel values (if the returned value is `null`, then it's an error)
- Returning potential error by argument
- Custom error classes
- Monads[14]

Each option has its pros and cons, but just the fact that there are so many ways of handling errors can cause severe issues. It causes surprises by potentially hiding that some statements can return an error, introduces complexity and, as a result, makes our software unreliable.

Undoubtedly, the intention for so many options was good. It gives a developer choices. Maybe the software you create is noncritical, or is the first iteration, so you want to make a "happy path" crystal clear. In such cases, masking some "bad paths"

---

11 While Go is improving every day, sometimes you can add more advanced tools like `goimports` (*https://oreil.ly/pS9MI*) or `bingo` (*https://oreil.ly/mkjO2*) to improve the development experience further. In some areas, Go can't be opinionated and is limited by stability guarantees.

12 The CAP Theorem (*https://oreil.ly/HyBdB*) mentions an excellent example of treating failures seriously. It states that you can only choose two from three system characteristics: consistency, availability, and partition. As soon as you distribute your system, you must deal with network partition (communication failure). As an error-handling mechanism, you can either design your system to wait (lose availability) or operate on partial data (lose consistency).

13 `bash` has many methods for error handling (*https://oreil.ly/Tij9n*), but the default one is implicit. The programmer can optionally print or check `${?}` that holds the exit code of the last command executed before any given line. An exit code of 0 means the command is executed without any issues.

14 In principle, a monad is an object that holds some value optionally, for example, some object `Option<Type>` with methods `Get()` and `IsEmpty()`. Furthermore, an "error monad" is an `Option` object that holds an error if the value is not set (sometimes referred to as `Result<Type>`).

sounds like a good short-term idea, right? Unfortunately, as with many shortcuts, it poses numerous dangers. Software complexity and demand for functionalities cause the code to never go out of the "first iteration," and noncritical code quickly becomes a dependency for something critical. This is one of the most important causes of unreliability or hard-to-debug software.

Go takes a unique path by treating the error as a first-citizen language feature. It assumes we want to write reliable software, making error handling explicit, easy, and uniform across libraries and interfaces. Let's see some examples in Example 2-4.

Example 2-4. Multiple function signatures with different return arguments

```
func noErrCanHappen() int { ❶
    // ...
    return 204
}

func doOrErr() error { ❷
    // ...
    if shouldFail() {
        return errors.New("ups, XYZ failed")
    }
    return nil
}

func intOrErr() (int, error) { ❸
    // ...
    if shouldFail() {
        return 0, errors.New("ups, XYZ2 failed")
    }
    return noErrCanHappen(), nil
}
```

❶ The critical aspect here is that functions and methods define the error flow as part of their signature. In this case, the `noErrCanHappen` function states that there is no way any error can happen during its invocation.

❷ By looking at the `doOrErr` function signature, we know some errors can happen. We don't know what type of error yet; we only know it is implementing a built-in `error` interface. We also know that there was no error if the error is nil.

❸ The fact that Go functions can return multiple arguments is leveraged when calculating some result in a "happy path." If the error can happen, it should be the last return argument (always). From the caller side, we should only touch the result if the error is nil.

It's worth noting that Go has an exception mechanism called `panics`, which are recoverable using the `recover()` built-in function. While useful or necessary for certain cases (e.g., initialization), you should never use `panics` for conventional error handling in your production code in practice. They are less efficient, hide failures, and overall surprise the programmers. Having errors as part of invocation allows the compiler and programmer to be prepared for error cases in the normal execution path. Example 2-5 shows how we can handle errors if they occur in our function execution path.

Example 2-5. Checking and handling errors

```
import "github.com/efficientgo/core/errors" ❶

func main() {
   ret := noErrCanHappen()
   if err := nestedDoOrErr(); err != nil { ❷
      // handle error
   }
   ret2, err := intOrErr()
   if err != nil {
      // handle error
   }
   // ...
}

func nestedDoOrErr() error {
   // ...
   if err := doOrErr(); err != nil {
      return errors.Wrap(err, "do") ❸
   }
   return nil
}
```

❶ Notice that we did not import the built-in `errors` package, but instead used the open source drop-in replacement `github.com/efficientgo/core/errors. core` module. This is my recommended replacement for the `errors` package and the popular, but archived, `github.com/pkg/errors`. It allows a bit more advanced logic, like wrapping errors you will see in step three.

❷ To tell if an error happened, we need to check if the `err` variable is nil or not. Then, if an error occurs, we can follow with error handling. Usually, it means logging it, exiting the program, incrementing metrics, or even explicitly ignoring it.

❸ Sometimes, it's appropriate to delegate error handling to the caller. For example, if the function can fail from many errors, consider wrapping it with a `errors.Wrap` function to add a short context of what is wrong. For example, with `github.com/efficientgo/core/errors`, we will have context and stack trace, which will be rendered if `%+v` is used later.

How to Wrap Errors?

Notice that I recommended `errors.Wrap` (or `errors.Wrapf`) instead of the built-in way of wrapping errors. Go defines the `%w` identifier for the `fmt.Errors` type of function that allows passing an error. Currently, I would not recommend `%w` because it's not type safe and as explicit as `Wrap`, causing nontrivial bugs in the past.

The one way of defining errors and handling them is one of Go's best features. Interestingly, it is one of the language disadvantages due to verbosity and certain boilerplate involved. It sometimes might feel repetitive, but tools allow you to mitigate the boilerplate.

Some Go IDEs define code templates. For example, in JetBrain's GoLand product, typing **`err`** and pressing the Tab key will generate a valid `if err != nil` statement. You can also collapse or expand error handling blocks for readability.

Another common complaint is that writing Go can feel very "pessimistic," because the errors that may never occur are visible in plain sight. The programmer has to decide what to do with them at every step, which takes mental energy and time. Yet, in my experience it's worth the work and makes programs much more predictable and easier to debug.

Never Ignore Errors!

Due to the verbosity of error handling, it's tempting to skip `err != nil` checks. Consider not doing it unless you know a function will never return an error (and in future versions!). If you don't know what to do with the error, consider passing it to the caller by default. If you must ignore the error, consider doing it explicitly with the `_ =` syntax. Also, always use linters, which will warn you about some portion of unchecked errors.

Are there any implications of the error handling for general Go code runtime efficiency? Yes! Unfortunately, it's much more significant than developers usually anticipate. In my experience, error paths are frequently an order of magnitude slower and more expensive to execute than happy paths. One of the reasons is we tend not to ignore error flows during our monitoring or benchmarking steps (mentioned in "Efficiency-Aware Development Flow" on page 102).

Another common reason is that the construction of errors often involves heavy string manipulation for creating human-readable messages. As a result, it can be costly, especially with lengthy debugging tags, which are touched on later in this book. Understanding these implications and ensuring consistent and efficient error handling are essential in any software, and we will take a detailed look at that in the following chapters.

Strong Ecosystem

A commonly stated strong point of Go is that its ecosystem is exceptionally mature for such a "young" language. While items listed in this section are not mandatory for solid programming dialects, they improve the whole development experience. This is also why the Go community is so large and still growing.

First, Go allows the programmer to focus on business logic without necessarily reimplementing or importing third-party libraries for basic functionalities like YAML decoding or cryptographic hashing algorithms. Go standard libraries are high quality, robust, ultra-backward compatible, and rich in features. They are well benchmarked, have solid APIs, and have good documentation. As a result, you can achieve most things without importing external packages. For example, running an HTTP server is dead simple, as visualized in Example 2-6.

Example 2-6. Minimal code for serving HTTP requests[15]

```
package main

import  "net/http"

func handle(w http.ResponseWriter, _ *http.Request) {
   w.Write([]byte("It kind of works!"))
}

func main() {
   http.ListenAndServe(":8080", http.HandlerFunc(handle))
}
```

In most cases, the efficiency of standard libraries is good enough or even better than third-party alternatives. For example, especially lower-level elements of packages, `net/http` for HTTP client and server code, or `crypto`, `math`, and `sort` parts (and

15 Such code is not recommended for production, but the only things that would need to change are avoiding using global variables and checking all errors.

more!), have a good amount of optimizations to serve most of the use cases. This allows developers to build more complex code on top while not worrying about the basics like `sorting` performance. Yet that's not always the case. Some libraries are meant for specific usage, and misusing them may result in significant resource waste. We will look at all the things you need to be aware of in Chapter 11.

Another highlight of the mature ecosystem is a basic, official in-browser Go editor called Go Playground (*https://oreil.ly/9Os3y*). It's a fantastic tool if you want to test something out quickly or share an interactive code example. It's also straightforward to extend, so the community often publishes variations of the Go Playground to try and share previously experimental language features like generics (*https://oreil.ly/f0qpm*) (which are now part of the primary language and explained in "Generics" on page 63).

Last but not least, the Go project defines its templating language, called Go templates (*https://oreil.ly/FdEZ8*). In some way, it's similar to Python's Jinja2 language (*https://oreil.ly/U6Em1*). While it sounds like a side feature of Go, it's beneficial in any dynamic text or HTML generation. It is also often used in popular tools like Helm (*https://helm.sh*) or Hugo (*https://gohugo.io*).

Unused Import or Variable Causes Build Error

The compilation will fail if you define a variable in Go but never read any value from it or don't pass it to another function. Similarly, it will fail if you added a package to the `import` statement but don't use that package in your file.

I see that Go developers have gotten used to this feature and love it, but it is surprising for newcomers. Failing on unused constructs can be frustrating if you want to play with the language quickly, e.g., create some variable without using it for debugging purposes.

There are, however, ways to handle these cases explicitly! You can see a few examples of dealing with these usage checks in Example 2-7.

Example 2-7. Various examples of unused and used variables

```
package main

func use(_ int) {}

func main() {
    var a int // error: a declared but not used ❶

    b := 1 // error: b declared but not used ❶

    var c int
    d := c // error: d declared but not used ❶
```

```
    e := 1
    use(e) ❷

    f := 1
    _ = f ❸
}
```

❶ Variables a, b, and c are not used, so they cause a compilation error.

❷ Variable e is used.

❸ Variable f is technically used for an explicit no identifier (_). Such an approach is useful if you explicitly want to tell the reader (and compiler) that you want to ignore the value.

Similarly, unused imports will fail the compilation process, so tools like `goimports` (mentioned in "Consistent Tooling" on page 45) automatically remove unused ones. Failing on unused variables and imports effectively ensures that code stays clear and relevant. Note that only internal function variables are checked. Elements like unused `struct` fields, methods, or types are not checked.

Unit Testing and Table Tests

Tests are a mandatory part of every application, small or big. In Go, tests are a natural part of the development process—easy to write, and focused on simplicity and readability. If we want to talk about efficient code, we need to have solid testing in place, allowing us to iterate over the program without worrying about regressions. Add a file with the *_test.go* suffix to introduce a unit test to your code within a package. You can write any Go code within that file, which won't be reachable from the production code. There are, however, four types of functions you can add that will be invoked for different testing parts. A certain signature distinguishes these types, notably function name prefixes: `Test`, `Fuzz`, `Example`, or `Benchmark`, and specific arguments.

Let's walk through the unit test type in Example 2-8. To make it more interesting, it's a table test. Examples and benchmarks are explained in "Code Documentation as a First Citizen" on page 55 and "Microbenchmarks" on page 275.

Example 2-8. Example unit table test

```
package max

import (
    "math"
    "testing"
```

```
    "github.com/efficientgo/core/testutil"
)

func TestMax(t *testing.T) { ❶
    for _, tcase := range []struct { ❷
        a, b     int
        expected int
    }{
        {a: 0, b: 0, expected: 0},
        {a: -1, b: 0, expected: 0},
        {a: 1, b: 0, expected: 1},
        {a: 0, b: -1, expected: 0},
        {a: 0, b: 1, expected: 1},
        {a: math.MinInt64, b: math.MaxInt64, expected: math.MaxInt64},
    } {
        t.Run("", func(t *testing.T) { ❸
            testutil.Equals(t, tcase.expected, max(tcase.a, tcase.b)) ❹
        })
    }
}
```

❶ If the function inside the *_test.go* file is named with the `Test` word and takes exactly `t *testing.T`, it is considered a "unit test." You can run them through the `go test` command.

❷ Usually, we want to test a specific function using multiple test cases (often edge cases) that define different input and expected output. This is where I would suggest using table tests. First, define your input and output, then run the same function in an easy-to-read loop.

❸ Optionally, you can invoke `t.Run`, which allows you to specify a subtest. Defining those on dynamic test cases like table tests is a good practice. It will enable you to navigate to the failing case quickly.

❹ The Go `testing.T` type gives useful methods like `Fail` or `Fatal` to abort and fail the unit test, or `Error` to continue running and check other potential errors. In our example, I propose using a simple helper called `testutil.Equals` from our open source core library (*https://oreil.ly/yAit9*), giving you a nice diff.[16]

Write tests often. It might surprise you, but writing unit tests for critical parts up front will help you implement desired features much faster. This is why I recommend

16 This assertion pattern is also typical in other third-party libraries like the popular `testify` package (*https://oreil.ly/I47fD*). However, I am not a fan of the `testify` package, because there are too many ways of doing the same thing.

following some reasonable form of test-driven development, covered in "Efficiency-Aware Development Flow" on page 102.

This information should give you a good overview of the language goals, strengths, and features before moving to more advanced features.

Advanced Language Elements

Let's now discuss the more advanced features of Go. Similar to the basics mentioned in the previous section, it's crucial to overview core language capabilities before discussing efficiency improvements.

Code Documentation as a First Citizen

Every project, at some point, needs solid API documentation. For library-type projects, the programmatic APIs are the main entry point. Robust interfaces with good descriptions allow developers to hide complexity, bring value, and avoid surprises. A code interface overview is essential for applications, too, allowing anyone to understand the codebase quickly. Reusing an application's Go packages in other projects is also not uncommon.

Instead of relying on the community to create many potentially fragmented and incompatible solutions, the Go project developed a tool called `godoc` (*https://oreil.ly/TQXxv*) from the start. It behaves similarly to Python's Docstring (*https://oreil.ly/UdkzS*) and Java's Javadoc (*https://oreil.ly/wlWGT*). `godoc` generates a consistent documentation HTML website directly from the code and its comments.

The amazing part is that you don't have many special conventions that would directly make the code comments less readable from the source code. To use this tool effectively, you need to remember five things. Let's go through them using Examples 2-9 and 2-10. The resulting HTML page, when `godoc` is invoked (*https://oreil.ly/EYJlx*), can be seen in Figure 2-1.

Example 2-9. Example snippet of block.go file with `godoc` compatible documentation

```
// Package block contains common functionality for interacting with TSDB blocks
// in the context of Thanos.
package block ❶

import ...

const (
   // MetaFilename is the known JSON filename for meta information. ❷
   MetaFilename = "meta.json"
)
```

```
// Download the downloads directory... ❷
// BUG(bwplotka): No known bugs, but if there was one, it would be outlined here. ❸
func Download(ctx context.Context, id ulid.ULID, dst string) error {
// ...

// cleanUp cleans the partially uploaded files. ❹
func cleanUp(ctx context.Context, id ulid.ULID) error {
// ...
```

❶ Rule 1: The optional package-level description must be placed on top of the package entry with no intervening blank line and start with the `Package <name>` prefix. If any source files have these entries, `godoc` will collect them all. If you have many files, the convention is to have the *doc.go* file with just the package-level documentation, package statement, and no other code.

❷ Rule 2: Any public construct should have a full sentence commentary, starting with the name of the construct (it's important!), right before its definition.

❸ Rule 3: Known bugs can be mentioned with `// BUG(who)` statements.

❹ Private constructs can have comments, but they will never be exposed in the documentation since they are private. Be consistent and start them with a construct name, too, for readability.

Example 2-10. Example snippet of block_test.go file with godoc compatible documentation

```
package block_test

import ...

func ExampleDownload() { ❶
    // ...

    // Output: ... ❷
}
```

❶ Rule 4: If you write a function named `Example<ConstructName>` in the test file, e.g., `block_test.go`, the `godoc` will generate an interactive code block with the desired examples. Note that the package name must have a *_test* suffix, too, representing a local testing package that tests the package without access to private fields. Since examples are part of the unit test, they will be actively run and compiled.

❷ Rule 5: If the example has the last comment starting with `// Output:`, the string after it will be asserted with the standard output after the example, allowing the example to stay reliable.

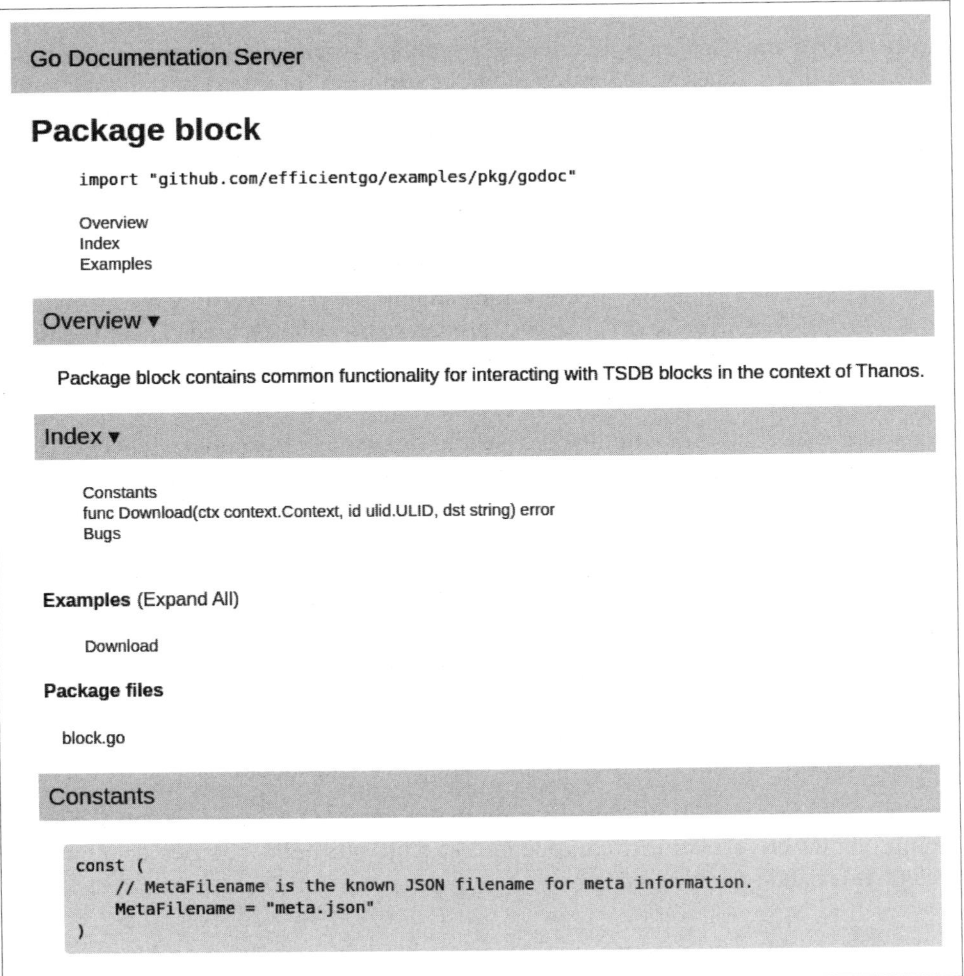

Figure 2-1. godoc output of Examples 2-9 and 2-10

I highly recommend sticking to those five simple rules. Not only because you can manually run `godoc` and generate your documentation web page, but the additional benefit is that these rules make your Go code comments structured and consistent. Everyone knows how to read them and where to find them.

I recommend using complete English sentences in all comments, even if the will not appear in `godoc`. It will help you keep your code commentary self-explanatory and explicit. After all, comments are for humans to read.

Furthermore, the Go team maintains a public documentation website (*https://pkg.go.dev*) that scrapes all requested public repositories for free. Thus, if your public code repository is compatible with `godoc`, it will be rendered correctly, and users can read the autogenerated documentation for every module or package version.

Backward Compatibility and Portability

Go has a strong take on backward compatibility guarantees. This means that core APIs, libraries, and language specifications should never break old code created for Go 1.0 (*https://oreil.ly/YOKfu*). This was proven to be well executed. There is a lot of trust in upgrading Go to the latest minor or patch versions. Upgrades are, in most cases, smooth and without significant bugs and surprises.

Regarding efficiency compatibility, it's hard to discuss any guarantees. There is (usually) no guarantee that the function that does two memory allocations now will not use hundreds in the next version of the Go project and any library. There have been surprises between versions in efficiency and speed characteristics. The community is working hard on improving the compilation and language runtime (more in "Go Runtime" on page 59 and Chapter 4). Since the hardware and operating systems are also developed, the Go team is experimenting with different optimizations and features to allow everyone to execute more efficiently. Of course, we don't speak about major performance regression here, as that is usually noticed and fixed in the release candidate period. Yet if we want our software to be deliberately fast and efficient, we need to be more vigilant and aware of the changes Go introduces.

Source code is compiled into binary code that is targeted to each platform. Yet Go tooling allows cross-platform compilation, so you can build binaries to almost all architectures and operating systems.

When you execute the Go binary, which was compiled for a different operating system (OS) or architecture, it can return cryptic error messages. For example, a common error is an Exec format error when you try running binary for Darwin (macOS) on Linux. You must recompile the code source for the correct architecture and OS if you see this.

Regarding portability, we can't skip mentioning the Go runtime and its characteristics.

Go Runtime

Many languages decided to solve portability across different hardware and operating systems by using virtual machines. Typical examples are Java Virtual Machine (JVM) (*https://oreil.ly/fhOmL*) for Java bytecode compatible languages (e.g., Java or Scala), and Common Language Runtime (CLR) (*https://oreil.ly/StGbU*) for .NET code, e.g., C#. Such a virtual machine allows for building languages without worrying about complex memory management logic (allocation and releasing), differences between hardware and operating systems, etc. JVM or CLR interprets the intermediate bytecode and transfers program instructions to the host. Unfortunately, while making it easier to create a programming language, they also introduce some overhead and many unknowns.[17] To mitigate the overhead, virtual machines often use complex optimizations like just-in-time (JIT) compilation (*https://oreil.ly/XXARz*) to process chunks of specific virtual machine bytecode to machine code on the fly.

Go does not need any "virtual machine." Our code and used libraries compile fully to machine code during compilation time. Thanks to standard library support of large operating systems and hardware, our code, if compiled against particular architecture, will run there with no issues.

Yet something is running in the background (concurrently) when our program starts. It's the Go runtime (*https://oreil.ly/mywcZ*) logic that, among other minor features of Go, is responsible for memory and concurrency management.

Object-Oriented Programming

Undoubtedly, object-oriented programming (OOP) got enormous traction over the last decades. It was invented around 1967 by Alan Kay, and it's still the most popular paradigm in programming.[18] OOP allows us to leverage advanced concepts like encapsulation, abstraction, polymorphisms, and inheritance (*https://oreil.ly/8hA0u*). In principle, it allows us to think about code as some objects with attributes (in Go fields) and behaviors (methods) telling each other what to do. Most OOP examples talk about high-level abstractions like an animal that exposes the `Walk()` method or a car that allows to `Ride()`, but in practice, objects are usually less abstract yet still helpful, encapsulated, and described by a class. There are no classes in Go, but there

---

17 Since programs, e.g., in Java, compile to Java bytecode, many things happen before the code is translated to actual machine-understandable code. The complexity of this process is too great to be understood by a mere mortal, so machine learning "AI" tools were created (*https://oreil.ly/baNvh*) to auto-tune JVM.

18 A survey in 2020 (*https://oreil.ly/WrtCH*) shows that among the top 10 used programming languages, 2 mandates object-oriented programming (Java, C#), 6 encourage it, and 2 do not implement OOP. I personally almost always favor object-oriented programming for algorithms that have to hold some context larger than three variables between data structures or functions.

are `struct` types equivalents. Example 2-11 shows how we can write OOP code in Go to compact multiple block objects into one.

Example 2-11. Example of the OOP in Go with `Group` that can behave like `Block`

```
type Block struct { ❶
    id          uuid.UUID
    start, end time.Time
    // ...
}

func (b Block) Duration() time.Duration { ❶
    return b.end.Sub(b.start)
}

type Group struct {
    Block ❷

    children []uuid.UUID
}

func (g *Group) Merge(b Block) { ❸
    if g.end.IsZero() || g.end.Before(b.end) {
        g.end = b.end
    }
    if g.start.IsZero() || g.start.After(b.start) {
        g.start = b.start
    }
    g.children = append(g.children, b.id)
}

func Compact(blocks ...Block) Block {
    sort.Sort(sortable(blocks)) ❹

    g := &Group{}
    g.id = uuid.New()
    for _, b := range blocks {
        g.Merge(b)
    }
    return g.Block ❺
}
```

❶ In Go, there is no separation between structures and classes, like in C++. In Go, on top of basic types like `integer`, `string`, etc., there is a `struct` type that can have methods (behaviors) and fields (attributes). We can use structures as a `class` equivalent to *encapsulate* more complex logic under a more straightforward interface. For example, the `Duration()` method on `Block` tells us the duration of the time range covered by the block.

❷ If we add some struct, e.g., `Block`, into another struct, e.g., `Group`, without any name, such a `Block struct` is considered embedded instead of being a field. Embedding allows Go developers to get the most valuable part of *inheritance*, borrowing the embedded structure fields and methods. In this case, `Group` will have `Block`'s fields and `Duration` method. This way, we can reuse a significant amount of code in our production codebases.

❸ There are two types of methods you can define in Go: using the "value receiver" (e.g., as in the `Duration()` method) or using the "pointer receiver" (with `*`). The so-called receiver is the variable after `func`, which represents the type we are adding a method to, in our case `Group`. We will mention this in "Values, Pointers, and Memory Blocks" on page 176, but the rule regarding which one to use is straightforward:

- Use the value receiver (no `func (g Group) SomeMethod()`) if your method does not modify the `Group` state. For the value receiver, every time we invoke it, the `g` will create a local copy of the `Group` object. It is equivalent to `func SomeMethod(g Group)`.
- Use the pointer receiver (e.g., `func (g *Group) SomeMethod()`) if your method is meant to modify the local receiver state or if any other method does that. It is equivalent to `func SomeMethod(g *Group)`. In our example, if the `Group.Merge()` method would be a value receiver, we will not persist `g.childen` changes or potentially inject `g.start` and `g.end` values. Additionally, for consistency, it's always recommended to have a type with all pointer receiver methods if at least one requires a pointer.

❹ To compact multiple blocks together, our algorithm requires a sorted list of blocks. We can use the standard library `sort.Sort` (*https://oreil.ly/N6ZWS*), which expects the `sort.Interface` interface. The `[]Block` slice does not implement this interface, so we convert it to our temporary `sortable` type, explained in Example 2-13.

❺ This is the only missing element for true inheritance. Go does not allow casting specific types into another type unless it's an alias or strict single-struct embedding (shown in Example 2-13). After that, you can only cast the interface into some type. That's why we need to specify embedded `struct` and `Block` explicitly. As a result, Go is often considered a language that does not support full inheritance.

What does Example 2-11 give us? First, the `Group` type can reuse `Block` functionality, and if done correctly, we can use `Group` as any other `Block`.

Embedding Multiple Types

You can embed as many unique structures as you want within one struct.

There is no priority for these—the compilation will fail if the compiler can't tell which method to use because two embedded types have the same SomeMethod() method. In such cases, use the type name to explicitly tell the compiler what should be used.

As mentioned in Example 2-11, Go also allows defining interfaces that tell what methods struct has to implement to match it. Note that there is no need to mark a specific struct explicitly that implements a particular interface, as in other languages like Java. It's enough just to implement the required methods. Let's see an example of sorting interface exposed by the standard library in Example 2-12.

Example 2-12. Sorting interface from the standard sort Go library

```
// A type, typically a collection, that satisfies sort.Interface can be
// sorted by the routines in this package. The methods require that the
// elements of the collection be enumerated by an integer index.
type Interface interface {
    // Len is the number of elements in the collection.
    Len() int
    // Less reports whether the element with
    // index i should sort before the element with index j.
    Less(i, j int) bool
    // Swap swaps the elements with indexes i and j.
    Swap(i, j int)
}
```

To use our type in the sort.Sort function, it has to implement all sort.Interface methods. Example 2-13 shows how sortable type does it.

Example 2-13. Example of the type that can be sorted using sort.Slice

```
type sortable []Block ❶

func (s sortable) Len() int             { return len(s) }
func (s sortable) Less(i, j int) bool { return s[i].start.Before(s[j].start) } ❷
func (s sortable) Swap(i, j int)        { s[i], s[j] = s[j], s[i] }

var _ sort.Interface = sortable{} ❸
```

❶ We can embed another type (e.g., a slice of Block elements) as the only thing in our sortable struct. This allows easy (but explicit) casting between []Block and sortable, as we used in the Compact method in Example 2-11.

❷ We can sort by increasing the `start` time using the `time.Time.Before(...)` (*https://oreil.ly/GQ2Ru*) method.

❸ We can assert our `sortable` type implements `sort.Interface` using this single-line statement, which fails compilation otherwise. I recommend using such statements whenever you want to ensure your type stays compatible with a particular interface in the future!

To sum up, `struct` methods, fields, and interfaces are an excellent yet simple way of writing both procedural composable and object-oriented code. In my experience, eventually it satisfies both low-level and high-level programming needs during our software development. While Go does not support all inheritance aspects (type to type casting), it provides enough to satisfy almost all OOP cases.

Generics

Since version 1.18, Go supports generics (*https://oreil.ly/qYyuQ*), one of the community's most desired features. Generics, also called parametric polymorphism (*https://oreil.ly/UIUAg*), allow type-safe implementations of the functionalities we want to reuse across different types.

The demand for generics in Go started quite big discussions in the Go team and community because of two main problems:

Two ways of doing the same thing

From the beginning, Go already supported type-safe reusable code via interfaces. You could see that in the preceding OOP example—the `sort.Sort` (*https://oreil.ly/X2NxR*) can be reusable by all types that implement a `sort.Interface` presented in Example 2-12. We can sort our custom `Block` type by implementing those methods in Example 2-13. Adding generics means we have two ways of doing a thing (*https://oreil.ly/dL8uE*) in many cases.

However, `interfaces` can be more troublesome for users of our code and slow at times due to some runtime overhead (*https://oreil.ly/8tSVf*).

Overhead

Implementing generics can have many negative consequences for the language. Depending on the implementation, it can impact different things. For example:

- We can just skip implementing them like in C, which slows programmers.
- We can use monomorphization (*https://oreil.ly/B062N*), which essentially copies the code for each type that will be used. This impacts compile time and binary size.

- We can use boxing like in Java, which is quite similar to the Go interface implementation. In this case, we impact execution time or memory usage.

> The generic dilemma is this: do you want slow programmers, slow compilers and bloated binaries, or slow execution times?
>
> —Russ Cox, "The Generic Dilemma" (*https://oreil.ly/WjjV4*)

After many proposals and debates, the final (extremely detailed!) design (*https://oreil.ly/k9cCR*) was accepted. Initially, I was very skeptical, but the accepted generic use turned out to be clear and reasonable. So far, the community also didn't jump ahead and abuse these mechanics as was feared. We tend to see generics used very rarely—only when needed, as it makes the code more complex to maintain.

For example, we could write a generic sort for all basic types like `int`, `float64`, or even `strings`, as presented in Example 2-14.

Example 2-14. Example implementation of the generic sort for basic types

```
// import "golang.org/x/exp/constraints" ❶

type genericSortableBasic[T constraints.Ordered] []T ❶

func (s genericSortableBasic[T]) Len() int           { return len(s) }
func (s genericSortableBasic[T]) Less(i, j int) bool { return s[i] < s[j] } ❷
func (s genericSortableBasic[T]) Swap(i, j int)      { s[i], s[j] = s[j], s[i] }

func genericSortBasic[T constraints.Ordered](slice []T) { ❸
    sort.Sort(genericSortableBasic[T](slice))
}

func Example() {
    toSort := []int{-20, 1, 10, 20}
   sort.Ints(toSort) ❹

   toSort2 := []int{-20, 1, 10, 20}
   genericSortBasic[int](toSort2) ❹
    // ...
}
```

❶ Thanks to generics (also called type parameters), we can implement a single type that will implement `sort.Interface` (see Example 2-13) for all basic types. We can provide custom constraints that look mostly like interfaces to limit the types that can be used as a type parameter. Here we use a type that represents `Integer | Float | ~string` constraints, so any type that supports comparison operators. We can put any other interface, like `any` to match all types. We can also use a special `comparable` keyword that will allow us to use the object of `T comparable` as a `map` key.

❷ Any element of s slice is now expected to be of type T with Ordered constraints, so the compiler will allow us to compare them for Less functionality.

❸ We can now implement a sort function for any basic type that will leverage sort.Sort implementation.

❹ We don't need to implement type-specific functions like sort.Ints. We can do genericSortBasic[<type>]([]<type>) as long as the slice is of the types that can be ordered!

This is great, but it only works for basic types. Unfortunately, we cannot override operators like < in Go (yet), so to implement generic sort for more complex types, we have to do a bit more work. For example, we could design our sort to expect each type to implement the func <typeA> Compare(<typeA>) int method.[19] If we add this method to the Block in Example 2-11, we can sort it easily, as presented in Example 2-15.

Example 2-15. Example implementation of the generic sort for certain types of objects

```
type Comparable[T any] interface { ❶
    Compare(T) int
}

type genericSortable[T Comparable[T]] []T ❷

func (s genericSortable[T]) Len() int            { return len(s) }
func (s genericSortable[T]) Less(i, j int) bool { return s[i].Compare(s[j]) > 0 } ❷
func (s genericSortable[T]) Swap(i, j int)      { s[i], s[j] = s[j], s[i] }

func genericSort[T Comparable[T]](slice []T) {
    sort.Sort(genericSortable[T](slice))
}

func (b Block) Compare(other Block) int { ❸
    // ...
}

func Example() {
    toSort := []Block{ /* ... */ }
    sort.Sort(sortable(toSort)) ❹

    toSort2 := []Block{ /* ... */ }
    genericSort[Block](toSort2) ❹
}
```

19 I prefer functions to methods (*https://oreil.ly/Et9CE*), as they're easier to use in most cases.

❶ Let's design our constraint. We expect every type to have a `Compare` method that accepts the same type. Because constraints and interfaces can also have type parameters, we can implement such requirements.

❷ We can now provide a type that implements a `sort.Interface` interface for such kinds of objects. Notice the nested `T` in `Comparable[T]`, as our interface also is generic!

❸ Now we can implement `Compare` for our `Block` type.

❹ Thanks to this, we don't need to implement a `sortable` type for every custom type we want to sort. As long as the type has the `Compare` method, we can use `genericSort`!

The accepted design shows advantages in cases where the user interface alone would be cumbersome. But what about the generics dilemma problem? The design allows any implementation (*https://oreil.ly/rZBtz*), so what trade-off was chosen at the end? We won't go into the details in this book, but Go uses the dictionaries and stenciling (*https://oreil.ly/poLls*) algorithm, which is between monomorphization and boxing.[20]

Generic Code Will Be Faster?

The specific implementation of generics in Go (which can change over time) means that the generic implementation, in theory, should be faster than interfaces but slower than implementing certain functionality for a specific type by hand. In practice, however, the potential difference is, in most cases, negligible, so use the most readable and easy-to-maintain option first.

In my experience, the difference might matter in the efficiency-critical code, but the results do not always follow the theory. For example, sometimes generic implementation is faster (*https://oreil.ly/9cEIb*), and sometimes using interfaces might be more efficient (*https://oreil.ly/tiOhS*). Conclusion? Always perform benchmarks (Chapter 8) to be sure!

To sum up, these facts are what I found crucial when teaching others programming in Go, based on my own experience with the language. Moreover, it will be helpful when diving deeper into the runtime performance of Go later in this book.

However, if you have never programmed in Go before, it's worth going through other materials like the tour of Go (*https://oreil.ly/J3HE3*) before jumping to the subsequent

20 The summary was well explained on the *PlanetScale* blog post (*https://oreil.ly/ksqO0*).

sections and chapters of this book. Make sure you try writing your own basic Go program, write a unit test, and use loops, switches, and concurrency mechanisms like channels and routines. Learn common types and standard library abstraction. As a person coming to a new language, you need to produce a program returning valid results before ensuring that it executes quickly and efficiently.

We learned about some basic and advanced characteristics of Go, so it's time to unwrap the efficiency aspects of the language. How easy is it to write good enough or high-performance code in Go?

Is Go "Fast"?

Recently, many companies have rewritten their products (e.g., from Ruby, Python, and Java) to Go.[21] Two repeatedly stated reasons for moving to Go or starting a new project in Go were readability and excellent performance. Readability comes from simplicity and consistency (e.g., single way of error handling as you remember from "Single Way of Handling Errors" on page 47), and it's where Go excels, but what about performance? Is Go fast compared to other languages like Python, Java, or C++?

In my opinion, this question is badly formed. Given time and room for complexities, any language can be as fast as your machine and operating system allow. That's because, in the end, the code we write is compiled into machine code that uses the exact CPU instructions. Also, most languages allow delegating execution to other processes, e.g., written in optimized Assembly. Unfortunately, sometimes all we use to decide if a language is "fast" are raw, semi-optimized short program benchmarks that compare execution time and memory usage across languages. While it tells us something, it effectively does not show practical aspects, e.g., how complex the programming for efficiency was.[22]

Instead, we should look at a programming language in terms of how hard and practical it is to write efficient code (not just fast), and how much readability and reliability such a process sacrifices. I believe the Go language has a superior balance between those elements while keeping it fast and trivial to write basic, functional code.

One of the reasons for being able to write efficient code more easily is the hermetic compilation stage, the relatively small amount of unknowns in the Go runtime (see

21 To name a few public changes, we've seen the Salesforce case (*https://oreil.ly/H3WsC*), AppsFlyer (*https://oreil.ly/iazde*), and Stream (*https://oreil.ly/NSJLD*).

22 For example, when we look at some benchmarks (*https://oreil.ly/s7qTj*), we see Go as sometimes faster, sometimes slower than Java. Yet if we look at CPU loads, every time Go or Java is faster, it's simply faster because, for example, the implementation allowed fewer CPU cycles to be wasted on memory access. You can achieve that in any programming language. The question is, how hard was it to achieve this? We don't usually measure how much time we spend to optimize code in each particular language, how easy it is to read or extend such code after optimizations, etc. Only those metrics might tell us which programming language is "faster."

"Go Runtime" on page 59), the easy-to-use concurrency framework, and the maturity of the debugging, benchmarking, and profiling tools (discussed in Chapters 8 and 9). Those Go characteristics did not appear from thin air. Not many know, but Go was designed on the shoulders of giants: C, Pascal, and CSP.

> In 1960, language experts from America and Europe teamed up to create Algol 60. In 1970, the Algol tree split into the C and the Pascal branch. ~40 years later, the two branches join again in Go.
>
> —Robert Griesemer, "The Evolution of Go" (*https://oreil.ly/a4V1e*)

As we can see in Figure 2-2, many of the names mentioned in Chapter 1 are grandfathers of Go. The great concurrency language CSP created by Sir Hoare, Pascal declarations and packages created by Wirth, and C basic syntax all contributed to how Go looks today.

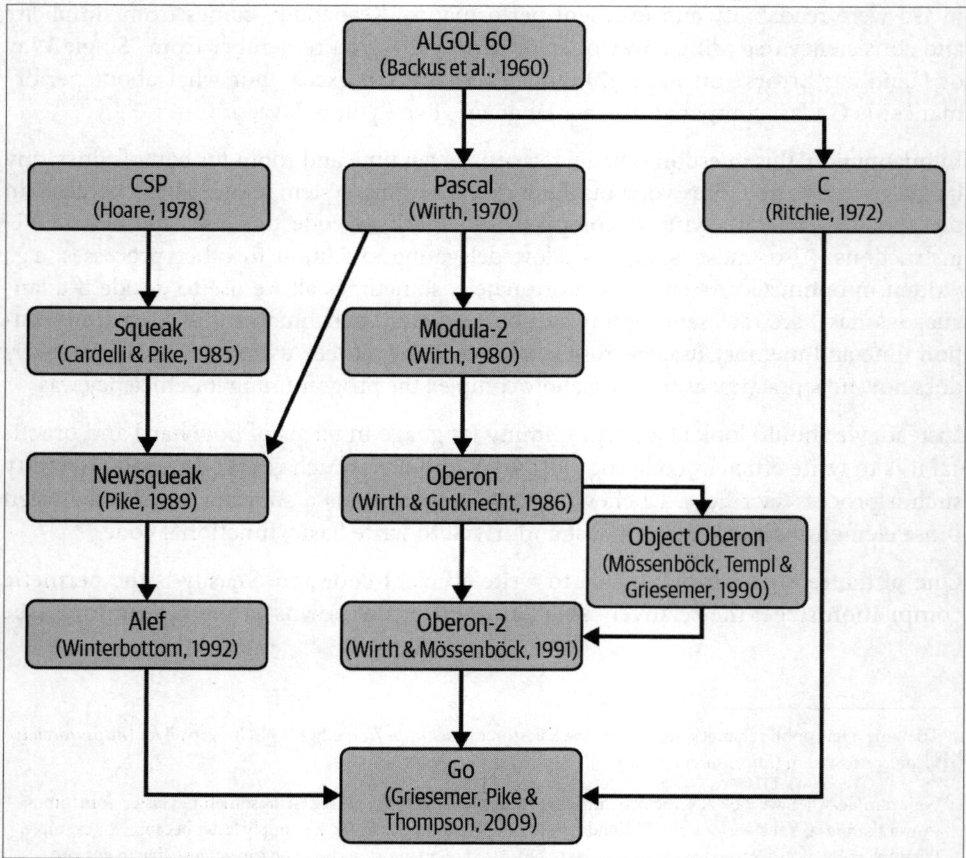

Figure 2-2. Go genealogy

But not everything can be perfect. In terms of efficiency, Go has its own Achilles' heel. As you will learn in "Go Memory Management" on page 172, memory usage can sometimes be hard to control. Allocations in our program can be surprising (especially for new users), and the garbage collections automatic memory release process has some overhead and eventual behavior. Especially for data-intensive applications, it takes effort to ensure memory or CPU efficiency, similar to machines with strictly limited RAM capacities (e.g., IoT).

Yet the decision to automate this process is highly beneficial, allowing the programmer to not worry about memory cleanup, which has proven to be even worse and sometimes catastrophic (e.g., deallocating memory twice). An excellent example of alternative mechanisms that other languages use is Rust. It implements a unique memory ownership model that replaces automatic global garbage collection. Unfortunately, while more efficient, it turns out that writing code in Rust is much more complicated than in Go. That's why we see higher adoption of Go. This reflects the Go team's ease-of-use trade-off in this element.

Fortunately, there are ways to mitigate the negative performance consequences of the garbage collection mechanism in Go and keep our software lean and efficient. We will go through those in the following chapters.

Summary

In my opinion, Go is an incredibly elegant and consistent language. Moreover, it offers many modern and innovative features that make programming more effective and reliable. Plus, the code is readable and maintainable by design.

This is a critical foundation for the efficiency improvements we will discuss later in this book. Like any other feature, optimizations always add complexity, so it's easier to modify simple code than to complicate already complex code. Simplicity, safety, and readability are paramount, even for efficient code. Make sure you know how to achieve that without thinking about efficiency first!

Many resources go into more details for elements I could spend only a subchapter on. If you are interested to learn more, there is nothing better than practice. If you need more experience with Go before we jump into optimizations, here is a short list of excellent resources:

- "Effective Go" (*https://oreil.ly/9auky*)
- "How to Write Go Code" (*https://oreil.ly/uS51g*)
- "A Tour of Go" (*https://oreil.ly/LpGBN*)
- "Practical Go Lessons" (*https://oreil.ly/VnFms*) by Maximilien Andile, available for free in the digital version

- Contributing to any open source project in Go, for example, through the CNCF mentoring initiatives (*https://oreil.ly/Y3D2Q*) we offer four or more times a year

The true power of the Go optimizations, benchmarking, and efficiency practices comes when used in practice, in everyday programming. Therefore, I want to empower you to marry efficiency with other good techniques around reliability or abstractions for practical use. While fully tailored logic sometimes has to be built for a critical path (as you will see in Chapter 10), the basic, often good enough, efficiency comes from understanding simple rules and language capabilities. That's why I focused on giving you a better overview of Go and its features in this chapter. With this knowledge, we can now move to Chapter 3, where we will learn how to start the journey to improve the efficiency and overall performance of our program's execution when we need to.

CHAPTER 3
Conquering Efficiency

It's action time! In Chapter 1, we learned that software efficiency matters. In Chapter 2, we studied the Go programming language—its basics and advanced features. Next, we discussed Go's capabilities of being easy to read and write. Finally, we mentioned that it could also be an effective language for writing efficient code.

Undoubtedly, achieving better efficiency in your program does not come without work. In some cases, the functionality you try to improve is already well optimized, so further optimization without system redesign might take a lot of time and only make a marginal difference. However, there might be other cases where the current implementation is heavily inefficient. Removing instances of wasted work can improve the program's efficiency in only a few hours of developer time. The true skill here as an engineer is to know, ideally after a short amount of research, which situation you are currently in:

- Do you need to improve anything on the performance side?
- If yes, is there a potential for the removal of wasted cycles?
- How much work is needed to reduce the latency of function X?
- Are there any suspicious overallocations?
- Should you stop overusing network bandwidth and sacrifice memory space instead?

This chapter will teach you the tools and methodologies to help you answer these questions effectively.

If you are struggling with these skills, don't worry! It's normal. The efficiency topic is not trivial. Despite the demand, this space is still not mastered by many, and even major software players sometimes make poor decisions. It's surprising how often what looks like high-quality software is shipped with fairly apparent inefficiencies. For instance, at the beginning of 2021, one user optimized the loading time of the popular game *Grand Theft Auto Online* from six minutes to two minutes (*https://oreil.ly/ast0m*) without access to the source code! As mentioned in Chapter 1, this game cost a staggering ~$140 million and a few years to make. Yet, it had an obvious efficiency bottleneck with a naive JSON parsing algorithm and deduplication logic that took most of the game loading time and worsened the game experience. This person's work is outstanding, but they used the same techniques you are about to learn. The only difference is that our job might be a bit easier—hopefully, you don't need to reverse engineer the binary written in C++ code on the way!

In the preceding example, the company behind the game missed the apparent waste of computation impacting the game's loading performance. It's unlikely that the company didn't have the resources to get an expert to optimize this part. Instead, it's a decision based on specific trade-offs, where the optimization wasn't worth the investment since there might have been higher-priority development tasks. In the end, one would say that an inefficiency like this didn't stop the success of the game. It did the job, yes, but for example, my friends and I were never fans of the game because of the loading time. I would argue that without this silly "waste," success might have been even bigger.

Laziness or Deliberate Efficiency Descoping?

There are other amusing examples of situations where a certain aspect of software efficiency could be descoped given certain circumstances. For instance, there is the amusing story about missile software developers (*https://oreil.ly/mJ8Mi*) who decided to accept certain memory leaks since the missile would be destroyed at the end of the application run. Similarly, we hear the story about "deliberate" memory leaks in low-latency trading software (*https://oreil.ly/PgzHQ*) that is expected to run only for very short durations.

You could say that the examples where the efficiency work was avoided and nothing tragically bad happened were pragmatic approaches. In the end, extra knowledge and work needed to fix leaks or slowdowns were avoided. Potentially yes, but what if these decisions were not data driven? We don't know, but these decisions might have been made out of laziness and ignorance without any valid data points that the fix would indeed take too much effort. What if developers in each example didn't fully understand the small effort needed? What if they didn't know how to optimize the problematic parts of the software? Would they make better decisions otherwise? Take less risk? I would argue yes.

In this chapter, I will introduce the topic of optimizations, starting with explaining the definition and initial approach in "Beyond Waste, Optimization Is a Zero-Sum Game". In the next section, "Optimization Challenges" on page 79, we will summarize the challenges we have to overcome while attempting to improve the efficiency of our software.

In "Understand Your Goals" on page 80, we will try to tame our software's tendency and temptation to maximize optimization effort by setting clear efficiency goals. We need only to be fast or efficient "enough." This is why setting the correct performance requirements from the start is so important. Next, in "Resource-Aware Efficiency Requirements" on page 86, I will propose a template and pragmatic process anyone can follow. Finally, those efficiency requirements will be useful in "Got an Efficiency Problem? Keep Calm!" on page 94, where I will teach you a professional flow for handling performance issues you or someone else has reported. You will learn that the optimization process could be your last resort.

In "Optimization Design Levels" on page 98, I will explain how to divide and isolate your optimization effort for easier conquering. Finally, in "Efficiency-Aware Development Flow" on page 102, we will combine all the pieces into a unified optimization process I always use and want to recommend to you: reliable flow, which applies to any software or design level.

There is a lot of learning ahead of us, so let's start understanding what optimization means.

Beyond Waste, Optimization Is a Zero-Sum Game

It is not a secret that one of many weapons in our arsenal to overcome efficiency issues is an effort called "optimization." But what does optimization mean, exactly? What's the best way to think about it and master it?

Optimization is not exclusively reserved for software efficiency topics. We also tend to optimize many things in our life, sometimes unconsciously. For example, if we cook a lot, we probably have salt in a well-accessible place. If our goal is to gain weight, we eat more calories. If we travel in the early morning, we pack and prepare the day before. If we commute, we tend to use that time by listening to audiobooks. If our commute to the office is painful, we consider moving closer to a better transportation system. All of these are optimization techniques that are meant to improve our life toward a specific goal. Sometimes we need a significant change. On the other hand, minor incremental improvements are often enough as they are magnified through repetition for a more substantial impact.

In engineering, the word "optimization" has its roots in mathematics (*https://oreil.ly/a1lou*), which means finding the best solution from all possible solutions for a problem constrained by a set of rules. Typically in computer science, however, we use the word "optimization" to describe an act of improving the system or program execution for a specific aspect. For instance, we can optimize our program to load a file faster or decrease peak memory utilization while serving a request on a web server.

We Can Optimize for Anything

Generally, optimization does not necessarily need to improve our program's efficiency characteristics if that is not our goal. For example, if we aim to improve security, maintainability, or code size, we can optimize for that too. Yet, in this book, when we talk about optimizations, they will be on an efficiency background (improving resource consumption or speed).

The goal of efficiency optimization should be to modify code (generally without changing its functionality[1]) so that its execution is either overall more efficient or at least more efficient in the categories we care about (and worse in others).

The important part is that, from a high-level view, we can perform the optimization by doing either of two things (or both):

- We can eliminate "wasted" resource consumption.
- We can trade one resource consumption for another or deliberately sacrifice other software qualities (so-called trade-off).

Let me explain the difference between these two by describing the first type of change—reducing so-called waste.

Reasonable Optimizations

Our program consists of a code—a set of instructions that operates on some data and uses various resources on our machines (CPU, memory, disk, power, etc.). We write this code so our program can perform the requested functionality. But everything involved in the process is rarely perfect (or integrated perfectly): our programmed code, compiler, operating systems, and even hardware. As a result, we sometimes introduce "waste." Wasted resource consumption represents a relatively unnecessary operation in our programs that takes precious time, memory, or CPU time, etc. Such waste might have been introduced as a deliberate simplification, by accident, tech debt, oversight, or just unawareness of better approaches. For example:

1 There might be exceptions. There might be domains where it's acceptable to approximate results. Sometimes we can (and should) also drop nice-to-have features if they block the critical efficiency characteristics we want.

- We might have accidentally left some debugging code that introduces massive latency in the heavily used function (e.g., `fmt.Println` statements).
- We performed an unnecessary, expensive check because the caller has already verified the input.
- We forgot to stop certain goroutines (a concurrency paradigm we will explain in detail in "Go Runtime Scheduler" on page 138), which are no longer required, yet still running, which wastes our memory and CPU time.[2]
- We used a nonoptimized function from a third-party library, when an optimized one exists in a different, well-maintained library that does the same thing faster.
- We saved the same piece of data a couple of times on disk, while it could be just reused and stored once.
- Our algorithm might have performed checks too many times when it could have done less for free (e.g., naive search versus binary search on sorted data).

The operation performed by our program or consumption of specific resources is a "waste" if, by eliminating it, we don't sacrifice anything else. And "anything" here means anything we particularly care for, such as extra CPU time, other resource consumption, or nonefficiency-related qualities like readability, flexibility, or portability. Such elimination makes our software, overall, more efficient. Looking closer, you might be surprised at how much waste every program has. It just waits for us to notice it and take it back!

Our program's optimization by reducing "waste" is a simple yet effective technique. In this book, we will call it a reasonable optimization, and I suggest doing it every time you notice such waste, even if you don't have time to benchmark it afterward. Yes. You heard me right. It should be part of coding hygiene. Note that to treat it as "reasonable" optimization, it has to be obvious. As the developer, you need to be sure that:

- Such optimization eliminates some additional work of the program.
- It does not sacrifice any other meaningful software quality or functionality, especially readability.

Look for the things that might be "obviously" unnecessary. Eliminating such unnecessary work is easily obtainable and does no harm (otherwise, it's not waste).

2 Situations where resources are not cleaned after each periodic functionality due to leftover concurrent routine are often referred to as memory leaks.

> **Be Mindful of Readability**
>
> The first thing that usually gets impacted by any code modification is readability. If reducing some obvious waste meaningfully reduces readability, or you need to spend a few hours experimenting on readable abstractions for it, it is not a reasonable optimization.
>
> That's fine. We can deal with that later, and we will talk about it in "Deliberate Optimizations" on page 77. If it impacts readability, we need data to prove it's worth it.

Cutting "waste" is also an effective mental model. Like humans who are rewarded for being intelligently lazy (*https://oreil.ly/u8IDm*), we also want to maximize the value our program brings with minimum runtime work.

One would say that reasonable optimization is an example of the anti-pattern often called "premature optimization" that many have been warned against (*https://oreil.ly/drziD*). And I cannot agree more that reducing obvious waste like this is a premature optimization since we don't assess and measure its impact. But I would argue that if we are sure that such premature optimization deals no harm, other than a little extra work, let's acknowledge that it is premature optimization but is reasonable, still do it, and move on.

If we go back to our commute to work example, if we notice we have a few stones in our shoes, of course we pick them out so we can walk without pain. We don't need to assess, measure, or compare if removing the stones improved our commute time or not. Getting rid of stones will help us somehow, and it's not harmful to do so (we don't need to take stones with us every time we go)! :)

> If you are dealing with something which is the noise, you don't deal with that right away because the payoff of investing time and energy is very small. But if you are walking through your codebase and you notice an opportunity for notable improvement (say 10% or 12%), of course, you reach down and pick it up.
>
> —Scott Meyers, "Things That Matter" (*https://oreil.ly/T9VFz*)

Initially, when you are new to programming or a particular language, you might not know which operations are unnecessary waste or if eliminating the potential waste will harm your program. That's fine. The "obviousness" comes from practice, so don't guess here. If you are guessing, it means the optimization is not obvious. You will learn what's reasonable with experience, and we will practice this together in Chapters 10 and 11.

Reasonable optimizations yield consistent performance improvements and often simplify or make our code more readable. However, we might want to take a more deliberate approach for bigger efficiency impacts, where the result might be less obvious, as explained in the next section.

Deliberate Optimizations

Beyond waste, we have operations that are critically important for our functionality. In this case, we can say we have a zero-sum game.[3] This means we have a situation where we cannot eliminate a certain operation that uses resource A (e.g., memory) without using more resource B (e.g., CPU time) or other quality (e.g., readability, portability, or correctness).

The optimizations that are not obvious or require us to make a certain trade-off can be called *deliberate*[4] since we have to spend a little bit more time on them. We can understand the trade-off, measure or assess it, and decide to keep it or throw it away.

Deliberate optimizations are not worse in any way. On the contrary, they often significantly impact the latency or resource consumption you want to cut. For example, if our request is too slow on a web server, we can consider optimizing latency by introducing a cache. Caching will allow us to save the result from expensive computation for requests asking for the same data. In addition, it saves CPU time and the need to introduce complex parallelization logic. Yet we will sacrifice memory or disk usage during the server's lifetime and potentially introduce some code complexity. As a result, deliberate optimization might not improve the program's overall efficiency, but it can improve the efficiency of a particular resource usage that we care about at the moment. Depending on the situation, the sacrifice might be worth it.

However, the implication of having certain sacrifices means we have to perform such optimization in a separate development phase isolated from the functionality one, as explained in "Efficiency-Aware Development Flow" on page 102. The reason for this is simple. First, we have to be sure that we understand what we sacrifice and whether the impact is not too big. Unfortunately, humans are quite bad at estimating such impacts.

For example, a common way to reduce network bandwidth and disk usage is to compress the data before sending it or storing it. However, simultaneously it requires us to decompress (decode) when receiving or reading the data. The potential balance of the resources used by our software before and after introducing compression can be seen in Figure 3-1.

---

3 Zero-sum game comes from game and economic theory. It describes a situation where one player can only win X if other players in total lost exactly X.

4 I got inspired for dividing optimizations on reasonable and deliberate by the community-driven go-perfbook (*https://oreil.ly/RuxfU*) led by Damian Gryski. In his book, he also mentioned the "dangerous" optimization category. I don't see a value in splitting classes further since there is a fuzzy borderline between deliberate and dangerous that depends on the situation and personal taste.

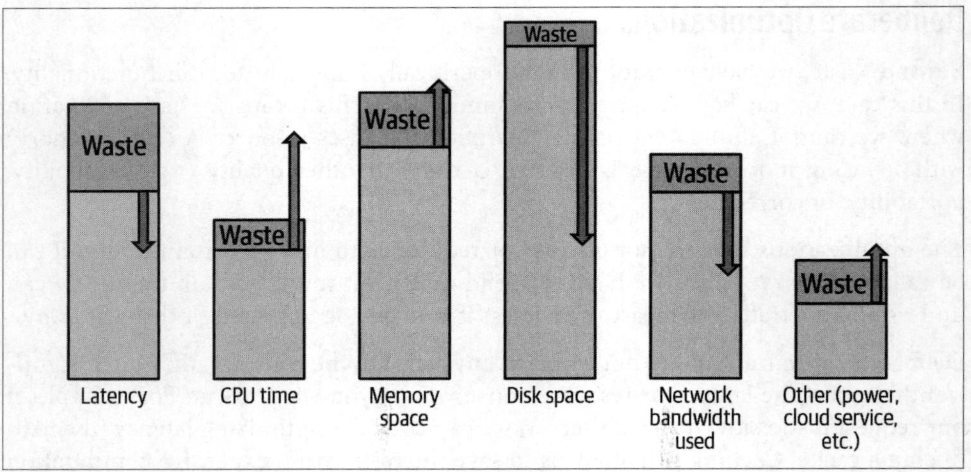

Figure 3-1. Potential impact on latency and resource usage if we compress the data before sending it over the network and saving it on disk

The exact numbers will vary, but the CPU resource will potentially be used more after compression addition. Instead of a simple data write operation, we must go through all bytes and compress them. It takes some time, even for the best lossless compression algorithms (e.g., `snappy` or `gzip`). Still, a smaller amount of messages to send over the network and disk writes might improve the total latency of such an operation. All of the compression algorithms require some extra buffers, so additional memory usage is also expected.

To sum up, there are strong implications for categorizing optimization reasonably and deliberately. If we see a potential efficiency improvement, we must be aware of its unintended consequences. There might be cases where it's reasonable and easy to obtain optimization. For example, we might have peeled some unnecessary operations from our program for free. But more often than not, making our software efficient in every aspect is impossible, or we impact other software qualities. This is when we get into a zero-sum game, and we must take a deliberate look at these problems. In this book and practice, you will learn what situations you are in and how to predict these consequences.

Before we bring the two types of optimizations into our development flow, let's discuss the efficiency optimization challenges we must be aware of. We will go through the most important ones in the next section.

Optimization Challenges

I wouldn't need to write this book if optimizing our software was easy. It's not. The process can be time-consuming and prone to mistakes. This is why many developers tend to ignore this topic or learn it later in their careers. But don't feel demotivated! Everyone can be an effective and pragmatic efficiency-aware developer after some practice. Knowing about the optimization obstacles should give us a good indication of what we should focus on to improve. Let's go through some fundamental problems:

Programmers are bad at estimating what part is responsible for the performance problem.
: We are really bad at guessing which part of the program consumes the most resources and how much. However, it's essential to find these problems because, generally, the Pareto Principle (*https://oreil.ly/eZIl5*) applies. It states that 80% of the time or resources consumed by our program come only from 20% of the operations it performs. Since any optimization is time-consuming, we want to focus on that critical 20% of operations, not some noise. Fortunately, there are tools and methods for estimating this, which we will touch on in Chapter 9.

Programmers are notoriously bad at estimating exact resource consumption.
: Similarly, we often make wrong assumptions on whether certain optimizations should help. Our guesses get better with experience (and hopefully after reading this book). Yet, it's best to *never trust your judgment*, and always measure and verify all numbers after deliberate optimizations (discussed in depth in Chapter 7). There are just too many layers in software executions with many unknowns and variables.

Maintaining efficiency over time is hard.
: The complex software execution layers mentioned previously are constantly changing (new versions of operating systems, hardware, firmware, etc.), not to mention the program's evolution and future developers who might touch your code. We might have spent weeks optimizing one part, but it could be irrelevant if we don't guard against regressions. There are ways to automate or at least structure the benchmarking and verification process for the efficiency of our program, because things change every day, as discussed in Chapter 6.

Reliable verification of current performance is very difficult.
: As we will learn in "Efficiency-Aware Development Flow" on page 102, the solution to the aforementioned challenges is to benchmark, measure, and validate the efficiency. Unfortunately, these are difficult to perform and prone to errors. There are many reasons: inability to simulate the production environment closely enough, external factors like noisy neighbors, lack of warm-up phase, wrong data sets, or microbenchmark accidental compiler optimizations.

This is why we will spend some time on this topic in "Reliability of Experiments" on page 256.

Optimizing can easily impact other software qualities.
: Solid software is great at many qualities: functionality, compatibility, usability, reliability, security, maintainability, portability, and efficiency. Each of these characteristics is nontrivial to get right, so they cause some cost to the development process. The importance of each can differ depending on your use cases. However, there are safe minimums of each software quality to be maintained for your program to be useful. This might be challenging when you add more features and optimization.

Specifically, in Go we don't have strict control over memory management.
: As we learned in "Go Runtime" on page 59, Go is garbage-collected language. While it's lifesaving for the simplicity of our code, memory safety, and developer velocity, it has downsides that can be seen when we want to be memory efficient. There are ways to improve our Go code to use less memory, but things can get tricky since the memory release model is eventual. Usually, the solution is simply to allocate less. We will go through memory management in "Do We Have a Memory Problem?" on page 152.

When is our program efficient "enough"?
: In the end, all optimizations are never fully free. They require a bigger or smaller effort from the developer. Both reasonable and deliberate optimizations require prior knowledge and time spent on implementation, experimentations, testing, and benchmarking. Given that, we need to find justification for this effort. Otherwise, we can spend this time somewhere else. Should we optimize away this waste? Should we trade the consumption of resource X for resource Y? Is such conversion useful for us? The answer might be "no." And if "yes," how much efficiency improvement is enough?

Regarding the last point, this is why it's extremely important to know your goals. What things, resources, and qualities do you (or your boss) care about during the development? It can vary depending on what you build. In the next section, I will propose a pragmatic way of stating performance requirements for a piece of software.

Understand Your Goals

> Before you proceed toward such lofty goals [program efficiency optimization], you should examine your reasons for doing so. Optimization is one of many desirable goals in software engineering and is often antagonistic to other important goals such as stability, maintainability, and portability. At its most cursory level (efficient implementation, clean non-redundant interfaces), optimization is beneficial and should always be applied. But at its most intrusive (inline assembly, pre-compiled/self-modified code,

loop unrolling, bit-fielding, superscalar and vectorizing) it can be an unending source of time-consuming implementation and bug hunting. Be cautious and wary of the cost of optimizing your code.

—Paul Hsieh, "Programming Optimization" (*https://oreil.ly/PQ4pk*)

By our definition, efficiency optimization improves our program resource consumption or latency. It's highly addictive to challenge ourselves and explore how fast our program can be.[5] First, however, we need to understand that optimization aims to not make our program perfectly efficient or "optimal" (as that might be simply impossible or feasible) but rather suboptimal enough. But what does "enough" mean for us? When do you stop? What if there isn't a need to even start optimizing?

One answer is to optimize when stakeholders (or users) ask for better efficiency in the software we develop until they are happy. But unfortunately, this is usually very difficult for a few reasons:

XY problem (https://oreil.ly/AolRQ).
: Stakeholders often ask for better efficiency, whereas a better solution is elsewhere. For example, many people complain about the heavy memory usage of the metric system if they try to monitor unique events. Instead, the potential solution might be to use logging or tracing systems for such data instead of making the metric system faster.[6] As a result, we can't always trust the initial user requests, especially around efficiency.

Efficiency is not a zero-sum game.
: Ideally, we need to see the big picture of all efficiency goals. As we learned in "Deliberate Optimizations" on page 77, one optimization for latency might cause more memory usage or impact other resources, so we can't react to every user complaint about efficiency without thinking. Of course, it helps when software is generally lean and efficient, but most likely we can't produce a single software that satisfies both the user who needs a latency-sensitive real-time event-capturing solution and the user who needs ultra-low memory used during such an operation.

---

5 No one said challenging ourselves is bad in certain situations. If you have time, playing with initiatives like Advent of Code (*https://oreil.ly/zT0Bl*) is a great way to learn or even compete! This is, however, different than the situation where we are paid to develop functional software effectively.

6 I experienced this a lot while maintaining the Prometheus project (*https://prometheus.io*), where we were constantly facing situations where users tried to ingest unique events into Prometheus. The problem is that we designed Prometheus as an efficient metric monitoring solution with a bespoke time-series database that assumed storing aggregated samples over time. If the ingested series were labeled with unique values, Prometheus slowly but surely began to use many resources (we call it a high-cardinality situation).

Stakeholders might not understand the optimization cost.

Everything costs, especially optimization effort and maintaining highly optimized code. Technically speaking, only physics laws limit us on how optimized software can be.[7] At some point, however, the benefit we gain from optimization versus the cost of finding and developing such optimization is impractical. Let's expand on the last point.

Figure 3-2 shows a typical correlation between the efficiency of the software and different costs.

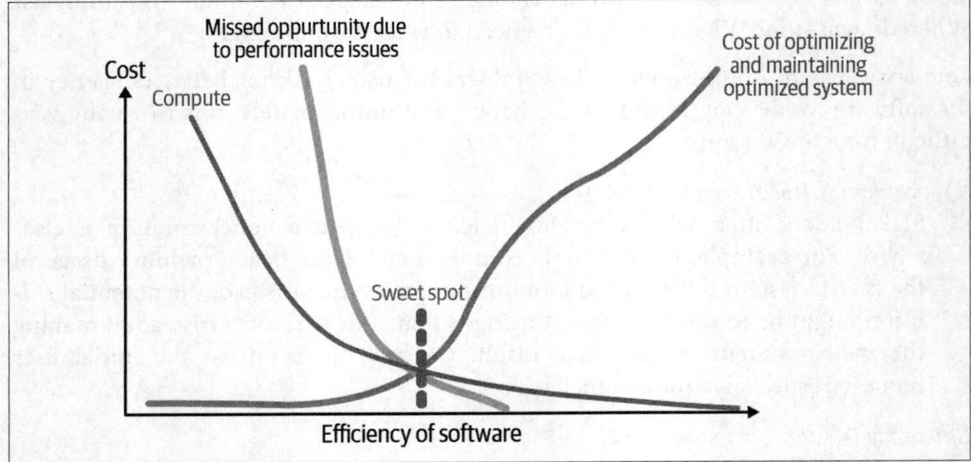

Figure 3-2. Beyond the "sweet spot," the cost of gaining higher efficiency might be extremely high

Figure 3-2 explains why at some "sweet spot" point, it might not be feasible to invest more time and resources in our software efficiency. Beyond some point, the cost of optimizing and developing optimized code can quickly surpass the benefits we get from leaner software, like computational cost and opportunities. We might need to spend exponentially more of the expensive developer time, and need to introduce clever, nonportable tricks, dedicated machine code, dedicated operating systems, or even specialized hardware.

In many cases, optimizations beyond the sweet spot aren't worth it, and it might be better to design a different system or use other flows to avoid such work. Unfortunately, there is also no single answer to where the sweet spot is. Typically, the longer

7 Just imagine, with all the resources in the world, we could try optimizing the software execution to the limits of physics. And once we are there, we could spend decades on research that pushes boundaries with things beyond the current physics we know. But, practically speaking, we might never find the "true" limit in our lifetime.

the lifetime planned for the software, the larger its deployment is, and the more investment is worth putting into it. On the other hand, if you plan to use your program only a few short times, your sweet spot might be at the beginning of this diagram, with very poor efficiency.

The problem is that users and stakeholders will not be aware of this. While ideally, product owners help us find that out, it's often the developer's role to advise the level of those different costs, using tools we will learn in Chapters 6 and 7.

However, whatever numbers we agree on, the best idea to solve the "when is enough" problem and have clear efficiency requirements is to write them down. In the next section, I will explain why. In "Resource-Aware Efficiency Requirements" on page 86, I will introduce the lightweight formula for them. Then in "Acquiring and Assessing Efficiency Goals" on page 89, we will discuss how to acquire and assess those efficiency requirements.

Efficiency Requirements Should Be Formalized

As you probably already know, every software development starts with the functional requirements gathering stage (FR stage). An architect, product manager, or yourself has to go through potential stakeholders, interview them, gather use cases and, ideally, write them down in some functional requirements document. The development team and stakeholders then review and negotiate functionality details in this document. The FR document describes what input your program should accept, and what behavior and output a user expects. It also mentions prerequisites, like what operating systems the application is meant to be running on. Ideally, you get formal approval on the FR document, and it becomes your "contract" between both parties. Having this is extremely important, especially when you are compensated for building the software:

- FR tells developers what they should focus on. It tells you what inputs should be valid and what things a user can configure. It dictates what you should focus on. Are you spending your time on something stakeholders paid for?
- It's easier to integrate with software with a clear FR. For example, stakeholders might want to design or order further system pieces that will be compatible with your software. They can start doing this before your software is even finished!
- FR enforces clear communication. Ideally, the FR is written and formal. This is helpful, as people tend to forget things, and it's easy to miscommunicate. That's why you write it all down and ask stakeholders for review. Maybe you misheard something?

You do formal functional requirements for bigger systems and features. For a smaller piece of software, you tend to write them up for some issue in your backlog, e.g.,

GitHub or GitLab issues, and then document them. Even for tiny scripts or little programs, set some goals and prerequisites—maybe a specific environment (e.g., Python version) and some dependencies (GPU on the machine). When you want others to use it effectively, you have to mention your software's functional requirements and goals.

Defining and agreeing on functional requirements is well adopted in the software industry. Even if a bit bureaucratic, developers tend to like those specifications because it makes their life easier—requirements are then more stable and specific.

Probably you know where I am going with this. Surprisingly, we often neglect to define similar requirements focused on the more nonfunctional aspects of the software we are expected to build, for example, describing a required efficiency and speed of the desired functionality.[8]

Such efficiency requirements are typically part of the nonfunctional requirement (NFR) (*https://oreil.ly/AQWLm*) documentation or specification. Its gathering process ideally should be similar to the FR process, but for all other qualities requested, software should have: portability, maintainability, extensibility, accessibility, operability, fault tolerance and reliability, compliance, documentation, execution efficiency, and so on. The list is long.

The NFR name can be in some way misleading since many qualities, including efficiency, massively impact our software functionality. As we learned in Chapter 1, efficiency and speed are critical for user experience.

In reality, NFRs are not very popular to use during software development, based on my experience and research. I found multiple reasons:

- Conventional NFR specification is considered bureaucratic and full of boilerplate. Especially if the mentioned qualities are not quantifiable and not specific, NFR for every software will look obvious and more or less similar. Of course, all software should be readable, maintainable, as fast as possible using minimum resources, and usable. This is not helpful.
- There are no easy-to-use, open, and accessible standards for this process. The most popular ISO/IEC 25010:2011 standard (*https://oreil.ly/IzqJo*) costs around $200 to read. It has a staggering 34 pages, and hasn't been changed since the last revision in 2017.

8 I was never explicitly asked to create a nonfunctional specification, and the same with people around me (*https://oreil.ly/Ui2tu*).

- NFRs are usually too complex to be applicable in practice. For example, the ISO/IEC 25010 standard previously mentioned specifies 13 product characteristics with 42 subcharacteristics in total (*https://oreil.ly/0MMcb*). It is hard to understand and takes too much time to gather and walk through.
- As we will learn in "Optimization Design Levels" on page 98, our software's speed and execution efficiency depend on more factors than our code. The typical developer usually can impact the efficiency by optimizing algorithms, code, and compiler. It's then up to the operator or admin to install that software, fit it into a bigger system, configure it, and provide the operating system and hardware for that workload. When developers are not in the domain of running their software on "production," it's hard for them to talk about runtime efficiency.

The SRE Domain

Site Reliability Engineering (SRE) (*https://sre.google*) introduced by Google is a role focused on marrying these two domains: software development and operators/administrators. Such engineers have experience running and building their software on a large scale. With more hands-on experience, it's easier to talk about efficiency requirements.

- Last but not least, we are humans and full of emotions. Because it's hard to estimate the efficiency of our software, especially in advance, it's not uncommon to feel humiliated when setting efficiency or speed goals. This is why we sometimes unconsciously refrain from agreeing to quantifiable performance goals. It can be uncomfortable, and that's normal.

OK, scratch that, we aren't going there. We need something more pragmatic and easier to work with. Something that will state our rough goals for efficiency and speed of the requested software and will be a starting point for some contracts between consumers and the development team. Having such efficiency requirements on top of functional ones up front is enormously helpful because:

We know exactly how fast or resource efficient our software has to be.
: For instance, let's say we agree that a certain operation should use 1 GB of memory, 2 CPU seconds, and take 2 minutes at maximum. If our tests show that it takes 2 GB of memory and 1 CPU second for 1 minute, then there is no point in optimizing latency.

We know if we have room for a trade-off or not.
: In the preceding example, we can precalculate or compress things to improve memory efficiency. We still have 1 CPU second to spare, and we can be slower for 1 minute.

Without official requirements, users will implicitly assume some efficiency expectations.
For example, maybe our program was accidentally very fast for a certain input. Users can assume this is by design, and they will depend on the fact in the future, or for other parts of the systems. This can lead to poor user experience and surprises.[9]

It's easier to use your software in a bigger system.
More often than not, your software will be a dependency on another piece of software and form a bigger system. Even a basic efficiency requirements document can tell system architects what to expect from the component. It can help enormously with further system performance assessments and capacity planning tasks.

It's easier to provide operational support.
When users do not know what performance to expect from your software, you will have difficulty supporting it over time. There will be many back-and-forths with the user on what is acceptable efficiency and what's not. Instead, with clear efficiency requirements, it is easier to tell if your software was underutilized or not, and as a result, the issue might be on the user side.

Let's summarize our situation. We know efficiency requirements can be enormously useful. On the other hand, we also know they can be tedious and full of boilerplate. So let's explore some options and see if we can find some balance between the requirement gathering effort and the value it brings.

Resource-Aware Efficiency Requirements

No one has defined a good standard process for creating efficiency requirements, so let's try to define one (*https://oreil.ly/DCzpu*)! Of course, we want it to be as lightweight a process as possible, but let's start with the ideal situation. What is the perfect set of information someone could put into some Resource-Aware Efficiency Requirements (RAER) document? Something that will be more specific and actionable than "I want this program to run adequately snappy."

In Example 3-1, you can see an example of a data-driven, minimal RAER for a single operation in some software.

9 Funnily enough, with enough program users, even with a formal performance and reliability contract, all your system's observable behaviors will depend on somebody. This is known as Hyrum's Law (*https://oreil.ly/UcrQo*).

Example 3-1. The example RAER entry

```
Program: "The Ruler"
Operation: "Fetching alerting rules for one tenant from the storage using HTTP."
Dataset: "100 tenants having 1000 alerting rules each."

Maximum Latency: "2s  for 90th percentile"
CPU Cores Limit: "2"
Memory Limit: "500 MB"
Disk Space Limit: "1 GB"
...
```

Ideally, this RAER is a set of records with efficiency requirements for certain operations. In principle, a single record should have information like:

- The operation, API, method, or function it relates to.
- The size and shape dataset we operate on, e.g., input or data stored (if any).
- Maximum latency of the operation.
- The resource consumption budget for this operation on that dataset, e.g., memory, disk, network bandwidth, etc.

Now, there is bad news and good news. The bad news is that, strictly speaking, such records are unrealistic to gather for all small operations. This is because:

- There are potentially hundreds of different operations that run during the software execution.
- There is an almost infinite number of dataset shapes and sizes (e.g., imagine an SQL query being an input, and stored SQL data being a dataset: we have a near-infinite amount of option permutations).
- Modern hardware with an operating system has thousands of elements that can be "consumed" when we execute our software. Overall, CPU seconds and memory are common, but what about the space and bandwidth of individual CPU caches, memory bus bandwidth, number of TCP sockets taken, file descriptors used, and thousands of other elements? Do we have to specify all that can be used?

The good news is that we don't need to provide all the small details. This is similar to how we deal with functional requirements. Do we focus on all possible user stories and details? No, just the most important ones. Do we define all possible permutations of valid inputs and expected outputs? No, we only define a couple of basic characteristics around boundaries (e.g., information has to be a positive integer). Let's look at how we can simplify the level of details of the RAER entry:

- Focus on the most utilized and expensive operations our software does first. These will impact the software resource usage the most. We will discuss benchmarking and profiling that will help you with this later in this book.
- We don't need to outline requirements for all tiny resources that might be consumed. Start with those that have the highest impact and matter the most. Usually, it means specific requirements toward CPU time, memory space, and storage (e.g., disk space). From there, we can iterate and add other resources that will matter in the future. Maybe our software needs some unique, expensive, and hard-to-find resources that are worth mentioning (e.g., GPU). Maybe a certain consumption poses a limit to overall scalability, e.g., we could fit more processes on a single machine if our operation would use fewer TCP sockets or disk IOPS. Add them only if they matter.
- Similar to what we do in unit tests when validating functionality, we can focus only on important categories of inputs and datasets. If we pick edge cases, we have a high chance of providing resource requirements for the worst- and best-case datasets. That is an enormous win already.
- Alternatively, there is a way to define the relation of input (or dataset) to the allowed resource consumption. We can then describe this relation in the form of mathematical functions, which we usually call *complexity* (discussed in "Asymptotic Complexity with Big O Notation" on page 243). Even with some approximation, it's quite an effective method. Our RAER for the operation `/rules` in Example 3-1 could then be described, as seen in Example 3-2.

Example 3-2. The example RAER entry with complexities or throughput instead of absolute numbers

```
Program: "The Ruler"
Operation: "Fetching alerting rules for one tenant from the storage using HTTP."
Dataset: "X tenants having Y alerting rules each."

Maximum Latency: "2*Y ms for 90th percentile"
CPU Cores Limit: "2"
Memory Limit: "X + 0.4 * Y MB"
Disk Space Limit: "0.1 * X GB"
...
```

Overall, I would even propose to include the RAER in the functional requirement (FR) document mentioned previously. Put it in another section called "Efficiency Requirements." After all, without rational speed and efficiency, our software can't be called fully functional, can it?

To sum up, in this section we defined the Resource-Aware Efficiency Requirements specification that gives us approximations of the needs and expected performance toward our software efficiency. It will be extremely helpful for the further development and optimization techniques we learn in this book. Therefore, I want to encourage you to understand the performance you aim for, ideally before you start developing your software and optimizing or adding more features to it.

Let's explain how we can possess or create such RAERs ourselves for the system, application, or function we aim to provide.

Acquiring and Assessing Efficiency Goals

Ideally, when you come to work on any software project, you have something like a RAER already specified. In bigger organizations, you might have dedicated people like project or product managers who will gather such efficiency requirements on top of functional requirements. They should also make sure the requirements are possible to fulfill. If they don't gather the RAER, don't hesitate to ask them to provide such information. It's often their job to give it.

Unfortunately, in most cases, there are no specific efficiency requirements, especially in smaller companies, community-driven projects, or, obviously, your personal projects. In those cases, we need to acquire the efficiency goals ourselves. How do we start?

This task is, again, similar to functional goals. We need to bring value to users, so ideally, we need to ask them what they need in terms of speed and running costs. So we go to the stakeholders or customers and ask what they need in terms of efficiency and speed, what they are willing to pay for, and what the constraints are on their side (e.g., the cluster has only four servers or the GPU has only 512 MB of internal memory). Similarly, with features, good product managers and developers will try to translate user performance needs into efficiency goals, which is not trivial if the stakeholders are not from the engineering space. For example, the "I want this application to run fast" statement has to be translated into specifics.

If the stakeholder can't give the latency numbers they might expect from your software, just pick a number. It can be high for a start, which is great for you, but it will make your life easier later. Perhaps this will trigger discussions on the stakeholder side on the implications of that number.

Very often, there are multiple personas of the system users too. For example, let's imagine our company will run our software as a service for the customer, and the service has already defined a price. In this case, the user cares about the speed and correctness, and our company will care about the efficiency of the software, as this translates to how much net profit the running service will have (or loss if the computation cost of running our software is too large). In this typical software as a service (SaaS) example, we have not one but two sources of input for our RAER.

Dogfooding

Very often, for smaller coding libraries, tools, and our infrastructure software, we are both developers and users. In this case, setting RAERs from the user's perspective is much easier. That is only one of the reasons why using the software you create is a good practice (*https://oreil.ly/xBgef*). This approach is often called "eating your own dog food" (dogfooding).

Unfortunately, even if a user is willing to define the RAER, the reality is not so perfect. Here comes the difficult part. Are we sure that what was proposed from the user perspective is doable within the expected amount of time? We know the demand, but we must validate it with the supply we can provide regarding our team skill set, technological possibilities, and time needed. Usually, even if some RAER is given, we need to perform our own diligence and define or assess the RAER from an achievability perspective. This book will teach you all that is required to accomplish this task.

In the meantime, let's go through one example of the RAER definition process.

Example of Defining RAER

Defining and assessing complex RAERs can get complicated. However, starting with potentially trivial yet clear requirements is reasonable if you have to do it from scratch.

Setting these requirements boils down to the user perspective. We need to find the minimum requirements that make your software valuable in its context. For example, let's say we need to create software that applies image enhancements on top of a set of images in JPEG format. In RAER, we can now treat such image transforming as an *operation*, and the set of image files and chosen enhancement as our *input*.

The second item in our RAER is the latency of our operation. It is better to have it as fast as possible from a user perspective. Yet our experience should tell us that there are limits on how quickly we can apply the enhancement to images (especially if large and many). But how can we find a reasonable latency number requirement that would work for potential users and make it possible for our software?

It's not easy to agree on a single number, especially when we are new to the efficient world. For example, we could potentially guess that 2 hours for a single image process might be too long, and 20 nanoseconds is not achievable, but it's hard to find the middle ground here. Yet as mentioned in "Efficiency Requirements Should Be Formalized" on page 83, I would encourage you to try defining one number, as it would make your software much easier to assess!

Defining Efficiency Requirements Is Like Negotiating Salary

Agreeing to someone's compensation for their work is similar to finding the requirement sweet spot for our program's latency or resource usage. The candidate wants the salary to be the highest possible. As an employer, you don't want to overpay. It's also hard to assess the value the person will be providing and how to set meaningful goals for such work. What works in salary negotiating works when defining RAER: don't set too high expectations, look at other competitors, negotiate, and have trial periods!

One way to define RAER details like latency or resource consumption is to check the competition. Competitors are already stuck in some kind of limits and framework for stating their efficiency guarantees. You don't need to set those as your numbers, but they can give you some clue of what's possible or what customers want.

While useful, checking competition is often not enough. Eventually, we have to estimate what's roughly possible with the system and algorithm we have in mind and the modern hardware. We can start by defining the initial naive algorithm. We can assume our first algorithm won't be the most efficient, but it will give us a good start on what's achievable with little effort. For example, let's assume for our problem that we want to read an image in JPEG format from disk (SSD), decode it to memory, apply enhancement, encode it back, and write it to disk.

With the algorithm, we can start discussing its potential efficiency. However, as you will learn in "Optimization Design Levels" on page 98 and "Reliability of Experiments" on page 256, efficiency depends on many factors! It's tough to measure it on an existing system, not to mention forecasting it just from the unimplemented algorithm.

This is where the complexity analysis with napkin math comes into play!

Napkin Math

Sometimes referred to as back-of-the-envelope calculation, *napkin math* is a technique of making rough calculations and estimations based on simple, theoretical assumptions. For example, we could assume latency for certain operations in computers, e.g., a sequential read of 8 KB from SSD is taking approximately 10 µs while writing 1 ms.[10] With that, we could calculate how long it takes to read and write 4 MB of sequential data. Then we can go from there and calculate overall latency if we make a few reads in our system, etc.

Napkin math is only an estimate, so we need to treat it with a grain of salt. Sometimes it can be intimidating to do since it all feels abstract. Yet such quick calculation is always a fantastic test on whether our guesses and initial system ideas are correct. It gives early feedback worth our time, especially around common efficiency requirements like latency, memory, or CPU usage.

We will discuss both complexity analysis and napkin math in detail in "Complexity Analysis" on page 240, but let's quickly define the initial RAER for our example JPEG enhancement problem space.

Complexity allows us to represent efficiency as the function of the latency (or resource usage) to the input. What's our input for the RAER discussion? Assume the worst case first. Find the slowest part of your system and what input can trigger that. In our example, we can imagine that the largest image we allow in our input (e.g., 8K resolution) is the slowest to process. The requirement of processing a set of images makes things a bit tricky. For now, we can assume the worst case and start negotiating with that. The worst case is that images are different, and we don't use concurrency. This means our latency will potentially be a function of $x * N$, where x is the latency of the biggest image, and N is the number of images in the set.

Given the worst-case input of an 8K image in JPEG format, we can try to estimate the complexities. The size of the input depends on the number of unique colors, but most of the images I found were around 4 MB, so let's have this number represent our average input size. Using data from Appendix A, we can calculate that such input will take at least 5 ms to read and 0.5 s to save on a disk. Similarly, encoding and decoding from JPEG format likely means at least looping through and allocating up to 7680 × 4320 pixels (around 33 million) in memory. Looking at the `image/jpeg` standard Go library (*https://oreil.ly/3Fnbz*), each pixel is represented by three `uint8` numbers (*https://oreil.ly/JmgZf*) to represent color in YCbCr format (*https://oreil.ly/lWiTf*).

10 We use napkin math more often in this book and during optimizations, so I prepared a small cheat sheet for latency assumptions in Appendix A.

That means approx 100 million unsigned 8-byte integers. We can then find out both the potential runtime and space complexities:

Runtime
: We need to fetch each element from memory (~5 ns for a sequential read from RAM) twice (one for decode, one for encode), which means 2 * 100 million * 5 ns, so 1 second. As a result of this quick math, we now know that without applying any enhancements or more tricky algorithms, such an operation for the single image will be no faster than 1s + 0.5s, so 1.5 seconds.

  Since napkin math is only an estimate, plus we did not account for the actual enhancing operation, it would be safe to assume we are wrong up to three times. This means we could use 5 seconds as the initial latency requirement for a single image to be safe, so 5 * *N* seconds for *N* images.

Space
: For the naive algorithm that reads the whole image to memory, storing that image will probably be the operation that allocates the most memory. With the mentioned three `uint8` numbers per pixel, we have 33 million * 3 * 8 bytes, so a maximum of 755 MB of memory usage.

We assumed typical cases and unoptimized algorithms, so we expect to be able to improve those initial numbers. But it might as well be fine for the user to wait 50 seconds for 10 images and use 1 GB of memory on each image. Knowing those numbers allows descoping efficiency work when possible!

To be more confident of the calculations we did, or if you are stuck in napkin math calculations, we could perform a quick benchmark[11] for the critical, slowest operation in our system. So I wrote a single benchmark for reading, decoding, encoding, and saving 8K images using the standard Go `jpeg` library. Example 3-3 shows the summarization of the benchmark results.

Example 3-3. Go microbenchmark results of reading, decoding, encoding, and saving an 8K JPEG file

```
name        time/op
DecEnc-12   1.56s ±2%
name        alloc/op
DecEnc-12   226MB ± 0%
name        allocs/op
DecEnc-12    18.8 ±3%
```

11 We will discuss benchmarks in detail in Chapter 7.

It turns out that our runtime calculations were quite accurate. It takes 1.56 seconds on average to perform a basic operation on an 8K image! However, the allocated memory is over three times better than we thought. Closer inspection of the `YCbCr struct's comment` (*https://oreil.ly/lm3T4*) reveals that this type stores on `Y` sample per pixel, but each `Cb` and `Cr` sample can span over one or more pixels, which might explain the difference.

Acquiring and assessing RAERs seems complex, but I recommend doing the exercise and getting those numbers before any serious development. Then, with benchmarking and napkin math, we can quickly understand if the RAERs are achievable with the rough algorithm we have in mind. The same process can also be used to tell if there is room for more easy-to-achieve optimization, as described in "Optimization Design Levels" on page 98.

With the ability to obtain, define, and assess your RAER, we can finally attempt to conquer some efficiency issues! In the next section, we will discuss steps I would recommend to handle such sometimes stressful situations professionally.

Got an Efficiency Problem? Keep Calm!

First of all, don't panic! We all have been there. We wrote a piece of code and tested it on our machine, which worked great. Then, proud of it, we released it to others, and immediately someone reported performance issues. Maybe it can't run fast enough on other people's machines. Perhaps it uses an unexpected amount of RAM with other users' datasets.

When facing efficiency issues in the program we build, manage, or are responsible for, we have several choices. But before you make any decisions, there is one critical thing you have to do. When issues happen, clear your mind from negative emotions about yourself or the team you worked with. It's very common to blame yourself or others for mistakes. It is only natural to feel an uncomfortable sense of guilt when someone complains about your work. However, everyone (including us) must understand that the topic of efficiency is challenging. On top of that, inefficient or buggy code happens every day, even for the most experienced developers. Therefore, there should be no shame in making mistakes.

Why do I write about emotions in a programming book? Because psychological safety is an important reason why developers take the wrong approach toward code efficiency. Procrastinating, feeling stuck, and being afraid to try new things or scratch bad ideas are only some of the negative consequences. From my own experience, if we start blaming ourselves or others, we won't solve any problems. Instead, we kill innovation and productivity, and introduce anxiety, toxicity, and stress. Those feelings can further prevent you from making a professional, reasonable decision on how to proceed with the reported efficiency issues or any other problems.

Blameless Culture Matters

Highlighting a blameless attitude is especially important during the "postmortem" process, which the Site Reliability Engineers perform after incidents. For example, sometimes costly mistakes are triggered by a single person. While we don't want to discourage this person or punish them, it is crucial to understand the cause of the incident to prevent it. Furthermore, the blameless approach enables us to be honest about facts while respecting others, so everyone feels safe to escalate issues without fear.

We should stop worrying too much, and with a clear mind, we should follow a systematic, almost robotic process (yes, ideally all of this is automated someday!). Let's face it, practically speaking, not every performance issue has to be followed by optimization. The potential flow for the developer I propose is presented in Figure 3-3. Note that the optimization step is not on the list yet!

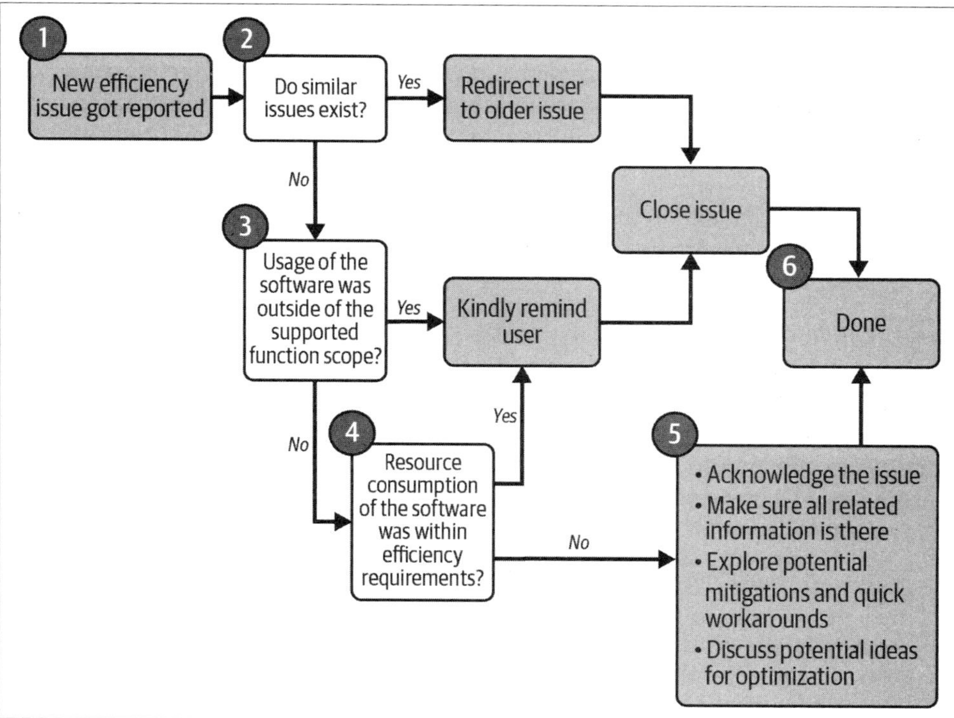

Figure 3-3. Recommended flow for efficiency issue triaging

Here, we outline six steps to do when an efficiency issue is reported:

Step 1: An efficiency issue was reported on our bug tracker.
: The whole process starts when someone reports an efficiency issue for the software we are responsible for. If more than one issue was reported, always begin the process shown in Figure 3-3 for every single issue (divide and conquer).

  Note that going through this process and putting things through a bug tracker should be your habit, even for small personal projects. How else would you remember in detail all the things you want to improve?

Step 2: Check for duplicates.
: This might be trivial, but try to be organized. Combine multiple issues for a single, focused conversation. Save time. Unfortunately, we are not yet at the stage where automation (e.g., artificial intelligence) can reliably find duplicates for us.

Step 3: Validate the circumstances against functional requirements.
: In this step, we have to ensure that the efficiency issue reporter used supported functionality. We design software for specific use cases defined in functional requirements. Due to the high demand for solving various unique yet sometimes similar use cases, users often try to "abuse" our software to do something it was never meant to do. Sometimes they are lucky, and things work. Sometimes it ends with crashes, unexpected resource usage, or slowdowns.[12]

  Similarly, we should do the same if the agreed prerequisites are not matched. For example, the unsupported, malformed request was sent, or the software was deployed on a machine without the required GPU resource.

Step 4: Validate the situation against RAERs.
: Some expectations toward speed and efficiency cannot or do not need to be satisfied. This is where the formal efficiency requirements specification discussed in "Resource-Aware Efficiency Requirements" on page 86 is invaluable. If the reported observation (e.g., response latency for the valid request) is still within the agreed-on software performance numbers, we should communicate that fact and move on.[13]

  Similarly, when the issue author deployed our software with an HDD disk where SSD was required, or the program was running on a machine with lower CPU cores than stated in the formal agreement, we should politely close such a bug report.

12 For example, see the instance of the XY problem mentioned in "Understand Your Goals" on page 80.

13 The reporter of the issue can obviously negotiate a change in the specification with the product owner if they think it's important enough or they want to pay additionally, etc.

Functional or Efficiency Requirements Can Change!

There might also be cases where the functional or efficiency specification did not predict certain corner cases. As a result, the specification might need to be revised to match reality. Requirements and demands evolve, and so should performance specifications and expectations.

Step 5: Acknowledge the issue, note it for prioritization, and move on.

Yes, you read it right. After you check the impact and all the previous steps, it's often acceptable (and even recommended!) to do almost nothing about the reported problem at the current moment. There might be more important things that need our attention—maybe an important, overdue feature or another efficiency issue in a different part of the code.

The world is not perfect. We can't solve everything. Exercise your assertiveness. Notice that this is not the same as ignoring the problem. We still have to acknowledge that there is an issue and ask follow-up questions that will help find the bottleneck and optimize it at a later date. Make sure to ask for the exact software version they are running. Try to provide a workaround or hints on what's happening so the user can help you find the root cause. Discuss ideas of what could be wrong. Write it all down in the issue. This will help you or another developer have a great starting point later. Communicate clearly that you will prioritize this issue with the team in the next prioritization session for the potential optimization effort.

Step 6: Done, issue was triaged.

Congratulations, the issue is handled. It's either closed or open. If it's open after all those steps, we can now consider its urgency and discuss the next steps with the team. Once we plan to tackle a specific issue, the efficiency flow in "Efficiency-Aware Development Flow" on page 102 will tell you how to do it effectively. Fear not. It might be easier than you think!

This Flow Is Applicable for Both SaaS and Externally Installed Software

The same flow is applicable for the software that is installed and executed by the user on their laptop, smartphone, or servers (sometimes called "on-premise" installation), as well as when it's managed by our company "as a service" (software as a service—SaaS). We developers should still try to triage all issues systematically.

We divided optimizations into reasonable and deliberate. Let's not hesitate and make the next division. To simplify and isolate the problem of software efficiency optimizations, we can divide it into levels, which we can then design and optimize in isolation. We will discuss those in the next section.

Optimization Design Levels

Let's take our previous real-life example of the long commute to work every day (we will use this example a couple of times in this chapter!). If such a commute makes you unhappy because it takes a considerable effort and is too long, it might make sense to optimize it. There are, however, so many levels we can do this on:

- We can start small, by buying more comfortable shoes for walking distances.
- We could buy an electric scooter or a car if that helps.
- We could plan the journey so it takes less time or distance to travel.
- We could buy an ebook reader and invest in a book-reading hobby to not waste time.
- Finally, we could move closer to the workplace or even change jobs.

We could do one such optimization in those separate "levels" or all, but each optimization takes some investment, trade-off (buying a car costs money), and effort. Ideally, we want to minimize the effort while maximizing value and making a difference.

There is another crucial aspect of those levels: optimizations from one level can be impacted or devalued if we do optimization on a higher level. For instance, let's say we did many optimizations to our commute on one level. We bought a better car, organized car sharing to save money on fuel, changed our work time to avoid traffic, etc. Imagine we would now decide to optimize on a higher level: move to an apartment within walking distance of our workplace. In such a case, any effort and investment in previous optimizations are now less valuable (if not fully wasted). This is the same in the engineering field. We should be aware of where we spend our optimization effort and when.

When studying computer science, one of the students' first encounters with optimization is learning theory about algorithms and data structures. They explore how to optimize programs using different algorithms with better time or space complexities (explained in "Asymptotic Complexity with Big O Notation" on page 243). While changing the algorithm we use in our code is an important optimization technique, we have many more areas and variables we can optimize to improve our software efficiency. To appropriately talk about the performance, there are more levels that software depends on.

Figure 3-4 presents the levels that take a significant part in software execution. This list of levels is inspired by Jon Louis Bentley's list made in 1982,[14] and it's still very accurate.

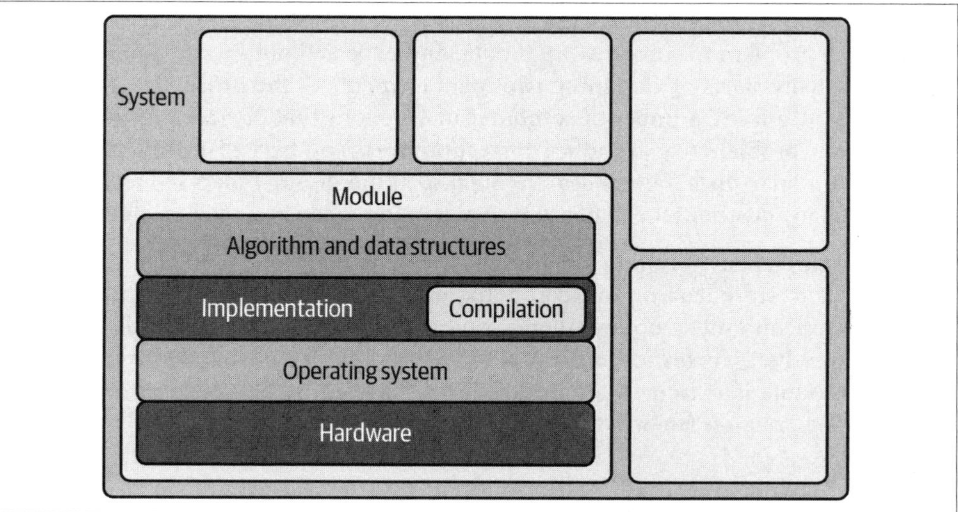

Figure 3-4. Levels that take part in software execution. We can provide optimization in each of these in isolation.

This book outlines five optimization design levels, each with its optimization approaches and verification strategies. So let's dig into them, from the highest to the lowest:

System level

In most cases, our software is part of some bigger system. Maybe it's one of many distributed processes or a thread in the bigger monolith application. In all cases, the system is structured around multiple modules. A module is a small software component that encapsulates certain functionality behind the method, interface, or other APIs (e.g., network API or file format) to be interchanged and modified more easily.

Each Go application, even the smallest, is an executable module that imports the code from other modules. As a result, your software depends on other components. Optimizing at the system level means changing what modules are used, how they are linked together, who calls which component, and how often. We could say we are designing algorithms that work across modules and APIs, which are our data structures.

14 Jon Louis Bentley, *Writing Efficient Programs* (Prentice Hall, 1982).

It is nontrivial work that requires multiple-team efforts and good architecture design up front. But, on the other hand, it often brings enormous efficiency improvements.

Intramodule algorithm and data structure level
: Given a problem to solve, its input data, and expected output, the module developer usually starts by designing two main elements of the procedure. First is the *algorithm*, a finite number of computer instructions that operate on data and can solve our problem (e.g., produce correct output). You have probably heard about many popular ones: binary search, quicksort, merge sort, map-reduce, and others, but any custom set of steps your program does can be called an algorithm.

  The second element is *data structures*, often implied by a chosen algorithm. They allow us to store data on our computer, e.g., input, output, or intermittent data. There are unlimited options here, too: arrays, hash maps, linked lists, stacks, queues, others, mixes, or custom ones. A solid choice of the algorithms within your module is extremely important. They have to be revised for your specific goals (e.g., request latency) and the input characteristics.

Implementation (code) level
: Algorithms in the module do not exist until they are written in code, compilable to machine code. Developers have huge control here. We can have an inefficient algorithm implemented efficiently, which fulfils our RAERs. On the other hand, we can have an amazing, efficient algorithm implemented poorly that causes unintended system slowdowns. Optimizing at the code level means taking a program written in a higher-level language (e.g., Go) that implements a specific algorithm, and producing a more efficient program in any aspect we want (e.g., latency) that uses the same algorithm and yields the same, correct output.

  Typically, we optimize on both algorithm and code levels together. In other cases, settling on one algorithm and focusing only on code optimizations is easier. You will see both approaches in Chapters 10 and 11.

> Some previous materials consider the compilation step as an individual level. I would argue that code-level optimization techniques have to embody compiler-level ones. There is a deep synergy between your implementation and how the compiler will translate it to machine code. As developers, we have to understand this relationship. We will explore Go compiler implications more in "Understanding Go Compiler" on page 118.

Operating system level
: These days, our software is never executed directly on the machine hardware and never runs alone. Instead, we run operating systems that split each software

execution into processes (then threads), schedule them on CPU cores, and provide other essential services, like memory and IO management, device access, and more. On top of that, we have additional virtualization layers (virtual machines, containers) that we can put in the operating system bucket, especially in cloud-native environments.

All those layers pose some overhead that can be optimized by those who control the operating system development and configuration. In this book, I assume that Go developers can rarely impact this level. Yet, we can gain a lot by understanding the challenges and usage patterns that will help us achieve efficiency on other, higher levels. We will go through them in Chapter 4, mainly focusing on Unix operating systems and popular virtualization techniques. I assume in this book that device drivers and firmware also fit into this category.

Hardware level

Finally, at some point, a set of instructions translated from our code is executed by the computer CPU units, with internal caches that are connected to other essential parts in the motherboard: RAM, local disks, network interfaces, input and output devices, and more. Usually, as developers or operators, we can abstract away from this complexity (which also varies across hardware products) thanks to the operating system level mentioned before. Yet the performance of our applications is limited by hardware constraints. Some of them might be surprising. For example, were you aware of the existence of NUMA nodes for multicore machines and how they can affect our performance (*https://oreil.ly/r1slU*)? Did you know that memory buses between CPU and memory nodes have limited bandwidth? It's an extensive topic that may impact our software efficiency optimization processes. We will explore this topic briefly in Chapters 4 and 5, together with the mechanisms Go employs to tackle these issues.

What are the practical benefits of dividing our problem space into levels? First of all, studies[15] show that when it comes to application speed, it is often possible to achieve speedups with factors of 10 to 20 at any of the mentioned levels, if not more. This is also similar to my experience.

The good news is that this implies the possibility of focusing our optimizations on just one level to gain the desired system efficiency.[16] However, suppose you optimized your implementation 10 to 20 times on one level. In that case, it might be hard to

15 Raj Reddy and Allen Newell's "Multiplicative Speedup of Systems" (in *Perspectives on Computer Science*, A.K. Jones, ed., Academic Press) elaborates on potential speedups of a factor of about 10 for each software design level. What's even more exciting is the fact that for hierarchical systems, the speedups from different levels multiplies, which offers massive potential for performance boost when optimizing.

16 This is a quite powerful thought. For example, imagine you have your application returning a result in 10 m. Reducing it to 1 m by optimizing on one level (e.g., an algorithm) is a game changer.

optimize this level further without significant sacrifices in development time, readability, and maintainability (our sweet spot from Figure 3-2). So you might have to look at another level to gain more.

The bad news is that you might be unable to change certain levels. For example, as programmers, we generally don't have the power to easily change the compiler, operating system, or hardware. Similarly, system administrators won't be able to change the algorithm the software is using. Instead, they can replace systems and configure or tune them.

Beware of the Optimization Biases!

It is sometimes funny (and scary!) how different engineering groups within a single company come up with highly distinct solutions to the same efficiency problems.

If the group has more system administrators or DevOps engineers, the solution is often to switch to another system, software, or operating system or try to "tune" them. In contrast, the software engineering group will mostly iterate on the same codebase, optimizing system, algorithm, or code levels.

This bias comes from the experience of changing each level, but it can have negative impacts. For example, switching the whole system, e.g., from RabbitMQ (*https://oreil.ly/ZVYo1*) to Kafka (*https://oreil.ly/wPpUD*), is a considerable effort. If you are doing this only because RabbitMQ "feels slow" without trying to contribute, perhaps a simple code-level optimization might be excessive. Or another way around, trying to optimize the efficiency of the system designed for different purposes on the code level might not be sufficient.

We discussed what optimization is, and we mentioned how to set performance goals, handle efficiency issues, and the design levels we operate in. Now it's time to hook everything together and combine this knowledge into the complete development cycle.

Efficiency-Aware Development Flow

> The primary concerns of the programmer during the early part of a program's life should be the overall organization of the programming project and producing correct and maintainable code. Furthermore, in many contexts, the cleanly designed program is often efficient enough for the application at hand.
>
> —Jon Louis Bentley, *Writing Efficient Programs*

Hopefully, at this point, you are aware that we have to think about performance, ideally from the early development stages. But there are risks—we don't develop code for it to be just efficient. We write programs for specific functionality that match the functional requirements we set or get from stakeholders. Our job is to get this work done effectively, so a pragmatic approach is necessary. How might developing a working but efficient code look from a high-level point of view?

We can simplify the development process into nine steps, as presented in Figure 3-5. For lack of a better term, let's call it the *TFBO* flow—test, fix, benchmark, and optimize.

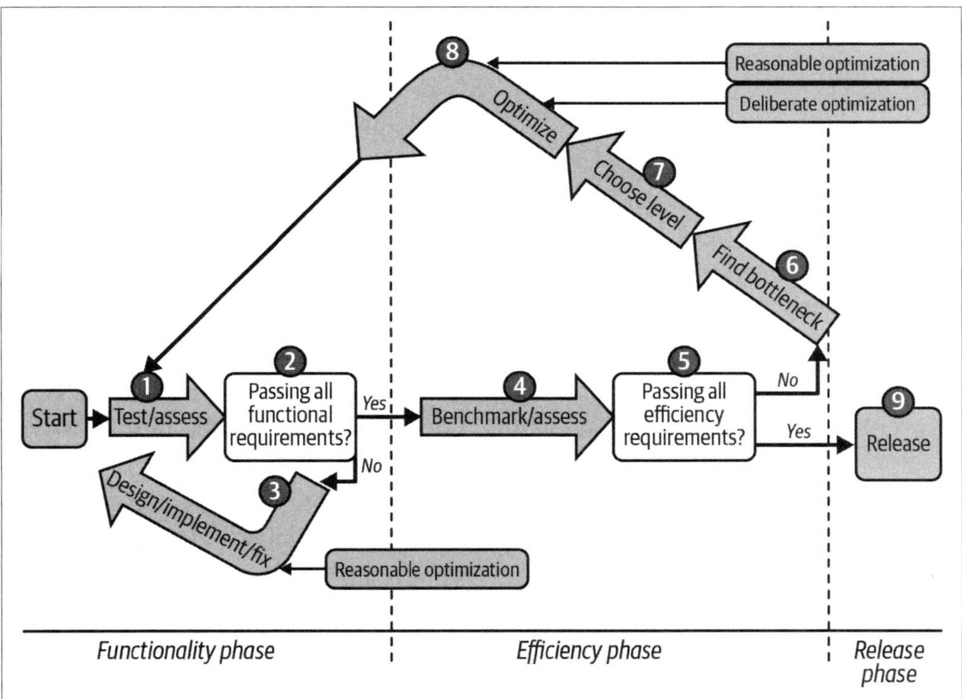

Figure 3-5. Efficiency-aware development flow

The process is systematic and highly iterative. Requirements, dependencies, and environments are changing, so we have to work in smaller chunks too. The TFBO process can feel a little strict, but trust me, mindful and effective software development requires some discipline. It applies to cases when you create new software from scratch, add a feature, or change the code. TFBO should work for software written in any language, not only Go. It is also applicable for all levels mentioned in "Optimization Design Levels" on page 98. Let's go through the nine TFBO steps.

Functionality Phase

> It is far, far easier to make a correct program fast than it is to make a fast program correct.
>
> —H. Sutter and A. Alexandrescu, *C++ Coding Standards: 101 Rules, Guidelines, and Best Practices* (*https://oreil.ly/hq0zw*) (Addison-Wesley, 2004)

Always start with functionality first. Whether we aim to start a new program, add new functionality, or just optimize an existing program, we should always begin with the design or implementation of the functionality. Make it work, make it simple, readable, maintainable, secure, etc., according to goals we have set, ideally in written form. Especially when you are starting your journey as a software engineer, focus on one thing at a time. With practice, we can add more reasonable optimizations early on.

1. Test functionality first

It might feel counterintuitive for some, but you should almost always start with a verification framework for the expected functionality. The more automated it is, the better. This also applies when you have a blank page and start developing a new program. This development paradigm is called test-driven development (TDD). It is mainly focused on code reliability and feature delivery velocity efficiency. In a strict form, on the code level, it mandates a specific flow:

1. Write a test (or extend an existing one) that expects the feature to be implemented.
2. Make sure to run all tests and see the new tests failing for expected reasons. If you don't see the failure or other failures, fix those tests first.
3. Iterate with the smallest possible changes until all tests pass and the code is clean.

TDD eliminates many unknowns. Imagine if we would not follow TDD. For example, we add a feature, and we write a test. It's easy to make a mistake that always passes the test even without our feature. Similarly, let's say we add the test after implementation, which passes, but other previously added tests fail. Most likely, we did not run a test before the implementation, so we don't know if everything worked before. TDD ensures you don't run into those questions at the end of your work, enormously improving reliability. It also reduces implementation time, allowing safe code modifications and giving you feedback early.

Furthermore, what if the functionality we wanted to implement is already done and we didn't notice? Writing a test first would reveal that quickly, saving us time. Spoiler alert: we will use the same principles for benchmark-driven optimization in step 4 later!

The TDD can be easily understood as a code-level practice, but what if you design or optimize algorithms and systems? The answer is that the flow remains the same, but our testing strategy must be applied on a different level, e.g., validating system design.

Let's say we implemented a test or performed an assessment on what is currently designed or implemented. What's next?

2. Do we pass the functional tests?

With the results from step 1, our work is much easier—we can perform data-driven decisions on what to do next! First, we should compare tests or assessment results with our agreed functional requirements. Is the current implementation or design fulfilling the specification? Great, we can jump to step 4. However, if tests fail or the functionality assessment shows some functionality gap, it's time to go to step 3 and fix this situation.

The problem is when you don't have those functional requirements stated anywhere. As discussed in "Efficiency Requirements Should Be Formalized" on page 83, this is why asking for functional requirements or defining them on your own is so important. Even the simplest bullet-point list of goals, written in the project README, is better than nothing.

Now, let's explore what to do if the current state of our software doesn't pass functional verification.

3. If the tests fail, we have to fix, implement, or design the missing parts

Depending on the design level we are at, in this step, we should design, implement, or fix the functional parts to close the gap between the current state and the functional expectation. As we discussed in "Reasonable Optimizations" on page 74, no optimizations other than the obvious, reasonable optimizations are allowed here. Focus on readability, design of modules, and simplicity. For example, don't bother thinking if it's more optimal to pass an argument by pointer or value or if parsing integers here will be too slow unless it's obvious. Just do whatever makes sense from a functional and readability standpoint. We don't validate efficiency yet, so let's forget about deliberate optimizations for now.

As you might have noticed in Figure 3-5, steps 1, 2, and 3 compose a small loop. This gives us an early feedback loop whenever we change things in our code or design. Step 3 is like us steering the direction of our boat called "software" when sailing over the ocean. We know where we want to go and understand how to look at the sun or stars in the right direction. Yet without precise feedback tools like GPS, we can end up sailing to the wrong place and only realizing it after weeks have gone by. This is why it's beneficial to validate our sailing position in short intervals for early feedback!

This is the same for our code. We don't want to work for months only to learn that we didn't get closer to what we expected from the software. Leverage the functionality phase loop by making a small iteration of code or design change, going to step 1 (run tests), step 2, and going back to step 3 to do another little correction.[17] This is the most effective development cycle engineers have found over the years. All modern methodologies like extreme programming (*https://oreil.ly/rhx8W*), Scrum, Kanban, and other Agile (*https://oreil.ly/sKZUA*) techniques are built on a small iterations premise.

After potentially hundreds of iterations, we might have software or design that fulfills, in step 2, the functional requirements we have set for ourselves for this development session. Finally, it's time to ensure our software is fast and efficient enough! Let's look at that in the next section.

Efficiency Phase

Once we are happy with the functional aspects of our software, it's time to ensure it matches the expected resource consumption and speed.

Splitting phases and isolating them from each other seems like a burden at first glance, but it will organize your developer workflow better. It gives us deep focus, ruling our early unknowns and mistakes, and helps us avoid expensive focus context switches.

Let's start our efficiency phase by performing the initial (baseline) efficiency validation in step 4. Then, who knows, maybe our software is efficient enough without any changes!

4. Efficiency assessment

Here we employ a similar strategy to step 1 of the functionality phase, but toward efficiency space. We can define an equivalent of the TDD method explained in step 1. Let's call it benchmark-driven optimization (BDO). In practice, step 4 looks like this process at the code level:

1. Write benchmarks (or extend existing ones) for all the operations from the efficiency requirements we want to compare against. Do it even if you know that the current implementation is not efficient yet. We will need that work later. It is not trivial, and we will discuss this aspect in detail in Chapter 8.

17 Ideally, we would have functionality checks for every code stroke or event of the saved code file. The earlier the feedback loop, the better. The main blocker for this is the time required to perform all tests and their reliability.

2. Ideally, run all the benchmarks to ensure your changes did not impact unrelated operations. In practice, this takes too much time, so focus on one part of the program (e.g., one operation) you want to check and run benchmarks only for that. Save the results for later. This will be our baseline.

Similar to step 1, the higher-level assessment might require different tools. Equipped with results from benchmarks or assessments, let's go to step 5.

5. Are we within RAERs?

In this step, we must compare the results from step 4 with the RAERs we gathered. For example, is our latency within the acceptable norm for the current implementation? Is the amount of resources our operation consumes within what we agreed? If yes, then no optimization is needed!

Again, similar to step 2, we have to establish requirements or rough goals for efficiency. Otherwise, we have zero ideas if the numbers we see are acceptable or not. Again, refer to "Acquiring and Assessing Efficiency Goals" on page 89 on how to define RAERs.

With this comparison, we should have a clear answer. Are we within acceptable thresholds? If yes, we can jump straight to the release process in step 9. If not, there is exciting optimization logic ahead of us in steps 6, 7, and 8. Let's walk through those now.

6. Find the main bottleneck

Here we must address the first challenge mentioned in "Optimization Challenges" on page 79. We are typically bad at guessing which part of the operation causes the biggest bottleneck; unfortunately, that's where our optimization should focus first.

The word *bottleneck* describes a place where most consumption of specific resources or software comes from. It might be a significant number of disk reads, deadlock, memory leak, or a function executed millions of times during a single operation. A single program usually has only a few of these bottlenecks. To perform effective optimization, we must first understand the bottleneck's consequences.

As part of this process, we need first to understand the underlying root cause of the problem we found in step 5. We will discuss the best tools for this job in Chapter 9.

Let's say we found the set of functions executed the most or another part of a program that consumes the most resources. What's next?

7. Choice of level

In step 7, we must choose how we want to tackle the optimization. Should we make the code more efficient? Perhaps we could improve the algorithm? Or maybe optimize on the system level? In extreme cases, we might also want to optimize the operating system or hardware!

The choice depends on what's more pragmatic at the moment and where we are in our efficiency spectrum in Figure 3-1. The important part is to stick to single-level optimization at one optimization iteration. Similar to the functionality phase, make short iterations and small corrections.

Once we know the level we want to make more efficient or faster, we are ready to perform optimization!

8. Optimize!

This is what everyone was waiting for. Finally, after all that effort, we know:

- What place in the code or design to optimize for the most impact.
- What to optimize for—what resource consumption is too large.
- How much sacrifice we can make on other resources because we have RAER. There will be trade-offs.
- On what level we are optimizing.

These elements make the optimization process much easier and often even make it possible to begin with. Now we focus on the mental model we introduced in "Beyond Waste, Optimization Is a Zero-Sum Game" on page 73. We are looking for *waste*. We are looking for places where we can do *less work*. There are always things that can be eliminated, either for free or by doing other work using another resource. I will introduce some patterns in Chapter 11 and show examples in Chapter 10.

Let's say we found some ideas for improvement. This is when you should implement it or design it (depending on the level). But what's next? We cannot just release our optimization like this simply because:

- We don't know that we did not introduce functional issues (bugs).
- We don't know if we improved any performance.

This is why we have to perform the full cycle now (no exceptions!). It's critical to go to step 1 and test the optimized code or design. If there are problems, we must fix them or revert optimization (steps 2 and 3).

It is tempting to ignore the functional testing phase when iterating on optimizations. For example, what can go wrong if you only reduce one allocation by reusing some memory?

I often caught myself doing this, and it was a painful mistake. Unfortunately, when you find that your code cannot pass tests after a few iterations of optimizations, it is hard to find what caused it. Usually, you have to revert all and start from scratch. Therefore, I encourage you to run a scoped unit test every time after the optimization attempt.

Once we gain confidence that our optimization did not break any basic functionality, it's crucial to check if our optimization improved the situation we want to improve. It's important to run *the same* benchmark, ensuring that nothing changes except the optimization you did (step 4). This allows us to reduce unknowns and iterate on our optimization in small parts.

With the results from this recent step 4, compare it with the baseline made in the initial visit to step 4. This crucial step will tell us if we optimized anything or introduced performance regression. Again, don't assume anything. Let the data speak for itself! Go has amazing tools for that, which we will discuss in Chapter 8.

If the new optimization doesn't have a better efficiency result, we simply try different ideas again until it works out. If the optimization has better results, we save our work and go to step 5 to check if it's enough. If not, we have to make another iteration. It's often useful to build another optimization on what we already did. Maybe there is something more to improve!

We repeat this cycle, and after a few (or hundreds), we hopefully have acceptable results in step 5. In this case, we can move to step 9 and enjoy our work!

9. Release and enjoy!

Great job! You went through the full iteration of the efficiency-aware development flow. Your software is now fairly safe to be released and deployed in the wild. The process might feel bureaucratic, but it's easy to build an instinct for it and follow it naturally. Of course, you might already be using this flow without noticing!

Summary

As we learned in this chapter, conquering efficiency is not trivial. However, certain patterns exist that help to navigate this process systematically and effectively. For example, the TFBO flow was immensely helpful for me to keep my efficiency-aware development pragmatic and effective.

Some of the frameworks incorporated in the TFBO, like test-driven development and benchmark-driven optimizations, might seem tedious initially. However, similar to the saying, "Give me six hours to chop a tree, I will spend four hours sharpening an axe" (*https://oreil.ly/qNPId*), you will notice that spending time on a proper test and benchmark will save you tons of effort in the long term!

The main takeaways are that we can divide optimizations into reasonable and deliberate ones. Then, to be mindful of the trade-offs and our effort, we discussed defining RAER so we can assess our software toward a formal goal everyone understands. Next, we mentioned what to do when an efficiency problem occurs and what optimizations levels there are. Finally, we discussed TFBO flow, which guides us through the practical development process.

To sum up, finding optimization can be considered a problem-solving skill. Noticing waste is not easy, and it comes with a lot of practice. This is somewhat similar to being good at programming interviews. In the end, what helps is the experience of seeing past patterns that were not efficient enough and how they were improved. Through this book, we will exercise those skills and uncover many tools that can help us in this journey.

Yet before that, there are important things to learn about modern computer architecture. We can learn typical optimization patterns by examples, but the optimizations do not generalize very well (*https://oreil.ly/eNkOY*). We won't be able to find them effectively and apply them in unique contexts without understanding the mechanisms that make those optimizations effective. In the next chapter, we will discuss how Go interacts with the key resources in typical computer architecture.

CHAPTER 4

How Go Uses the CPU Resource (or Two)

> One of the most useful abstractions we can make is to treat properties of our hardware and infrastructure systems as resources. CPU, memory, data storage, and the network are similar to resources in the natural world: they are finite, they are physical objects in the real world, and they must be distributed and shared between various key players in the ecosystem.
>
> —Susan J. Fowler, *Production-Ready Microservices* (*https://oreil.ly/8xO1v*) (O'Reilly, 2016)

As you learned in "Behind Performance" on page 3, software efficiency depends on how our program uses the hardware resources. If the same functionality uses fewer resources, our efficiency increases and the requirements and net cost of running such a program decrease. For example, if we use less CPU time (CPU "resource") or fewer resources with slower access time (e.g., disk), we usually reduce the latency of our software.

This might sound simple, but in modern computers, these resources interact with each other in a complex, nontrivial way. Furthermore, more than one process is using these resources, so our program does not use them directly. Instead, these resources are managed for us by an operating system. If that wasn't complex enough, especially in cloud environments, we often "virtualize" the hardware further so it can be shared across many individual systems in an isolated way. That means there are methods for "hosts" to give access to part of a single CPU or disk to a "guest" operating system that thinks it's all the hardware that exists. In the end, operating systems and virtualization mechanisms create layers between our program and the actual physical devices that store or compute our data.

To understand how to write efficient code or improve our program's efficiency effectively, we have to learn the characteristics, purpose, and limits of the typical computer resources like CPU, different types of storage, and network. There is no shortcut here. Furthermore, we can't ignore understanding how these physical components are managed by the operating system and typical virtualization layers.

In this chapter, we will examine our program execution from the point of view of the CPU. We will discuss how Go uses CPUs for single and multiple core tasking.

We won't discuss all types of computer architectures with all mechanisms of all existing operating systems, as this would be impossible to fit in one book, never mind one chapter. So instead, this chapter will focus on a typical x86-64 CPU architecture with Intel or AMD, ARM CPUs, and the modern Linux operating system. This should get you started and give you a jumping-off point if you ever run your program on other, unique types of hardware or operating systems.

We will start with exploring CPU in a modern computer architecture to understand how modern computers are designed, mainly focusing on the CPU, or processor. Then I will introduce the Assembly language, which will help us understand how the CPU core executes instructions. After that, we will dig into the Go compiler to build awareness of what happens when we do a `go build`. Furthermore, we will jump into the CPU and memory wall problem, showing you why modern CPU hardware is complex. This problem directly impacts writing efficient code on these ultracritical paths. Finally, we will enter the realm of multitasking by explaining how the operating system scheduler tries to distribute thousands of executing programs on outnumbered CPU cores and how the Go runtime scheduler leverages that to implement an efficient concurrency framework for us to use. We will finish with the summary on when to use concurrency.

Mechanical Sympathy

Initially, this chapter might get overwhelming, especially if you are new to low-level programming. Yet, awareness of what is happening will help us understand the optimizations, so focus on understanding high-level patterns and characteristics of each resource (e.g., how the Go scheduler works). We don't need to know how to write machine code manually or how to, blindfolded, manufacture the computer.

Instead, let's treat this with curiosity about how things work under the computer case in general. In other words, we need to have mechanical sympathy (*https://oreil.ly/Co2IM*).

To understand how the CPU architecture works, we need to explain how modern computers operate. So let's dive into that in the next section.

CPU in a Modern Computer Architecture

All we do while programming in Go is construct a set of statements that tells the computer what to do, step-by-step. Given predefined language constructs like variables, loops, control mechanisms, arithmetic, and I/O operations, we can implement any algorithms that interact with data stored in different mediums. This is why Go, like many other popular programming languages, can be called imperative—as developers, we have to describe how the program will operate. This is also how hardware is designed nowadays—it is imperative too. It waits for program instructions, optional input data, and the desired place for output.

Programming wasn't always so simple. Before general-purpose machines, engineers had to design fixed program hardware to achieve requested functionality, e.g., a desk calculator. Adding a feature, fixing a bug, or optimizing required changing the circuits and manufacturing new devices. Probably not the easiest time to be a "programmer"!

Fortunately, around the 1950s, a few inventors worldwide figured out the opportunity for the universal machine that could be programmed using a set of predefined instructions stored in memory. One of the first people to document this idea was a great mathematician, John von Neumann, and his team.

> It is evident that the machine must be capable of storing in some manner not only the digital information needed in a given computation ..., the intermediate results of the computation (which may be wanted for varying lengths of time), but also the instructions which govern the actual routine to be performed on the numerical data. ... For an all-purpose machine, it must be possible to instruct the device to carry out whatsoever computation that can be formulated in numerical terms.
>
> —Arthur W. Burks, Herman H. Goldstine, and John von Neumann, *Preliminary Discussion of the Logical Design of an Electronic Computing Instrument* (Institute for Advanced Study, 1946)

What's noteworthy is that most modern general-purpose computers (e.g., PCs, laptops, and servers) are based on John von Neumann's design. This assumes that program instructions can be stored and fetched similar to storing and reading program data (instruction input and output). We fetch both the instruction to be performed (e.g., `add`) and data (e.g., addition operands) by reading bytes from a certain memory address in the main memory (or caches). While it doesn't sound like a novel idea now, it established how general-purpose machines work. We call this Von Neumann computer architecture, and you can see its modern, evolved variation in Figure 4-1.[1]

1 To be technically strict, modern computers nowadays have distinct caches for program instructions and data, while both are stored the same on the main memory. This is the so-called modified Harvard architecture. At the optimization levels we aim for in this book, we can safely skip this level of detail.

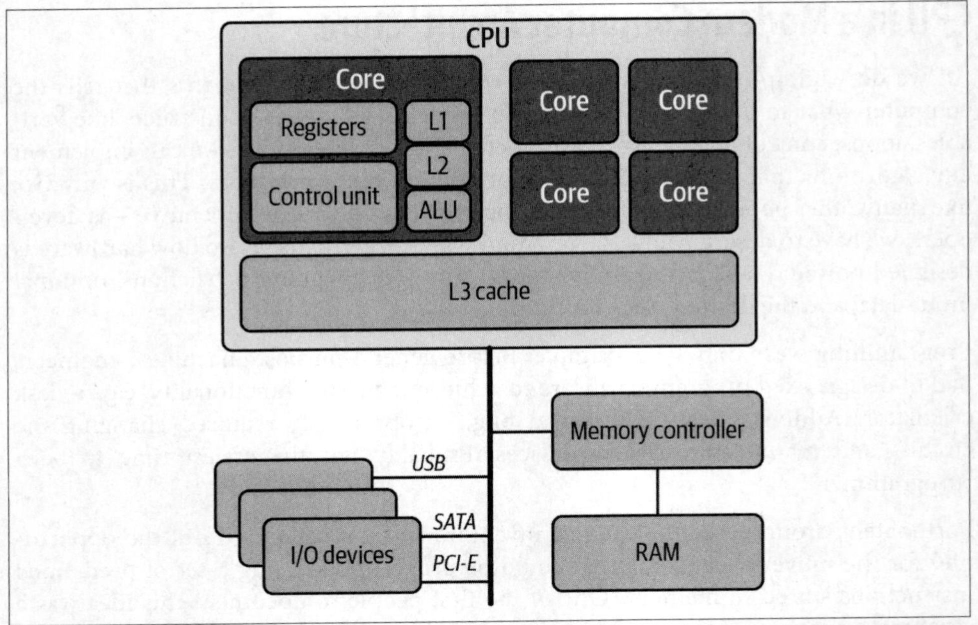

Figure 4-1. High-level computer architecture with a single multicore CPU and uniform memory access (UMA)

At the heart of modern architecture, we see a CPU consisting of multiple cores (four to six physical cores are the norm in the 2020s PCs). Each core can execute desired instructions with certain data saved in random-access memory (RAM) or any other memory layers like registers or L-caches (discussed later).

The RAM explained in Chapter 5 performs the duty of the main, fast, volatile memory that can store our data and program code as long as the computer is powered. In addition, the memory controller makes sure RAM is supplied with a constant power flow to keep the information on RAM chips. Last, the CPU can interact with various external or internal input/output (I/O) devices. From a high-level view, an I/O device means anything that accepts sending or receiving a stream of bytes, for example, mouse, keyboard, speaker, monitor, HDD or SSD disk, network interface, GPU, and thousands more.

Roughly speaking, CPU, RAM, and popular I/O devices like disks and network interfaces are the essential parts of computer architecture. This is what we use as "resources" in our RAERs mentioned in "Efficiency Requirements Should Be Formalized" on page 83 and what we are usually optimizing for in our software development.

In this chapter, we will focus on the brain of our general-purpose machines—the CPU. When should we care about CPU resources? Typically, from an efficiency standpoint, we should start looking at our Go process CPU resource usage when either of the following occurs:

- Our machine can't do other tasks because our process uses all the available CPU resource computing capacity.
- Our process runs unexpectedly slow, while we see higher CPU consumption.

There are many techniques to troubleshoot these symptoms, but we must first understand the CPU's internal working and program execution basics. This is the key to efficient Go programming. Furthermore, it explains the numerous optimization techniques that might surprise us initially. For example, do you know why in Go (and other languages), we should avoid using linked lists like structures if we plan to iterate over them a lot, despite their theoretical advantages like quick insertion and deletion?

Before we learn why, we must understand how the CPU core executes our programs. Surprisingly, I found that the best way to explain this is by learning how the Assembly language works. Trust me on this; it might be easier than you think!

Assembly

The CPU core, indirectly, can execute programs we write. For example, consider the simple Go code in Example 4-1.

Example 4-1. Simple function that reads numbers from a file and returns the total sum

```
func Sum(fileName string) (ret int64, _ error) {
   b, err := os.ReadFile(fileName)
   if err != nil {
      return 0, err
   }

   for _, line := range bytes.Split(b, []byte("\n")) {
      num, err := strconv.ParseInt(string(line), 10, 64)
      if err != nil {
         return 0, err
      }

      ret += num ❶
   }

   return ret, nil
}
```

❶ The main arithmetic operation in this function adds a parsed number from the file into a `ret` integer variable representing the total sum.

While such language is far from, let's say, spoken English, unfortunately, it is still too complex and incomprehensible for the CPU. It is not "machine-readable" code. Thankfully every programming language has a dedicated tool called a compiler[2] that (among other things discussed in "Understanding Go Compiler" on page 118) translates our higher-level code to machine code. You might be familiar with a `go build` command that invokes a default Go compiler.

The machine code is a sequence of instructions written in binary format (famous zeros and ones). In principle, each instruction is represented by a number (`opcode`) followed by optional operands in the form of a constant value or address in the main memory. We can also refer to a few CPU core registers, which are tiny "slots" directly on the CPU chip that can be used to store intermediate results. For example, on AMD64 CPU, we have sixteen 64-bit general-purpose registers referred to as RAX, RBX, RDX, RBP, RSI, RDI, RSP, and R8-R15.

While translating to machine code, the compiler often adds additional code like extra memory safety bound checks. It automatically changes our code for known efficiency patterns for a given architecture. Sometimes this might not be what we expect. This is why inspecting the resulting machine code when troubleshooting some efficiency problems is sometimes useful. Another advanced example of humans needing to read machine code is when we need to reverse engineer programs without source code.

Unfortunately, machine code is impossible to read for humans unless you are a genius. However, there is a great tool we can use in such situations. We can compile Example 4-1 code to Assembly language (*https://oreil.ly/3xZAs*) instead of machine code. We can also disassemble the compiled machine code to Assembly. The Assembly language represents the lowest code level that can be practically read and (in theory) written by human developers. It also represents well what will be interpreted by the CPU when converted to machine code.

It is worth mentioning that we can disassemble compiled code into various Assembly dialects. For example:

- To Intel syntax (*https://oreil.ly/alpt4*) using the standard Linux tool `objdump -d -M intel <binary>` (*https://oreil.ly/kZO3j*)
- To AT&T syntax (*https://oreil.ly/k6bKs*) using the similar command `objdump -d -M att <binary>` (*https://oreil.ly/cmAW9*)

2 For scripted (interpreted) languages, there is no complete code compilation. Instead, there is an interpreter that compiles the code statement by statement. Another unique type of language is represented by a family of languages that use Java Virtual Machine (JVM). Such a machine can dynamically switch from interpreting to just-in-time (JIT) compilation for runtime optimizations.

- To Go "pseudo" assembly language (*https://oreil.ly/lT07J*) using Go tooling `go tool objdump -s <binary>` (*https://oreil.ly/5I9t2*)

All three of these dialects are used in the various tools, and their syntax varies. To have an easier time, always ensure what syntax your disassembly tool uses. The Go Assembly is a dialect that tries to be as portable as possible, so it might not exactly represent the machine code. Yet it is usually consistent and close enough for our purposes. It can show all compilation optimization discussed in "Understanding Go Compiler" on page 118. This is why Go Assembly is what we will use throughout this book.

Do I Need to Understand Assembly?

You don't need to know how to program in Assembly to write efficient Go code. Yet a rough understanding of Assembly and the decompilation process are essential tools that can often reveal hidden, lower-level computation waste. Practically speaking, it's useful primarily for advanced optimizations when we have already applied all of the more straightforward optimizations. Assembly is also beneficial for understanding the changes the compiler applies to our code when translating to machine code. Sometimes these might surprise us! Finally, it also tells us how the CPU works.

In Example 4-2 we can see a tiny, disassembled part of the compiled Example 4-1 (using `go tool objsdump -s`) that represents `ret += num` statement.[3]

Example 4-2. Addition part of code in Go Assembly language decompiled from the compiled Example 4-1

```
// go tool objdump -s sum.test
ret += num
0x4f9b6d      488b742450     MOVQ 0x50(SP), SI  ❶
0x4f9b72      4801c6        ADDQ AX, SI  ❷
```

❶ The first line represents a quadword (64 bit) MOV instruction (*https://oreil.ly/SDE5R*) that tells the CPU to copy the 64-bit value from memory under the address stored in register `SP` plus 80 bytes and put that into the `SI` register.[4] The compiler decided that `SI` will store the initial value of the return argument in our function, so the `ret` integer variable for the `ret+=num` operation.

3 Similar output to Example 4-2 can be obtained by compiling the source code to Assembly using `go build -gcflags -S <source>`.

4 Note that in the Go Assembly register, names are abstracted for portability. Since we will compile to 64-bit architecture, `SP` and `SI` will mean RSP and RSI registers.

❷ As a second instruction, we tell the CPU to add a quadword value from the `AX` register to the `SI` register. The compiler used the `AX` register to store the `num` integer variable, which we parsed from the `string` in previous instructions (outside of this snippet).

The preceding example shows `MOVQ` and `ADDQ` instructions. To make things more complex, each distinct CPU implementation allows a different set of instructions, with different memory addressing, etc. The industry created the Instruction Set Architecture (ISA) (*https://oreil.ly/eTzST*) to specify a strict, portable interface between software and hardware. Thanks to the ISA, we can compile our program, for example, to machine code compatible with the ISA for x86 architecture and run it on any x86 CPU.[5] The ISA defines data types, registers, main memory management, fixed set of instructions, unique identification, input/output model, etc. There are various ISAs (*https://oreil.ly/TLxJn*) for different types of CPUs. For example, both 32-bit and 64-bit Intel and AMD processors use x86 ISA, and ARM uses its ARM ISA (for example, new Apple M chips use ARMv8.6-A (*https://oreil.ly/NZqT1*)).

As far as Go developers are concerned, the ISA defines a set of instructions and registers our compiled machine code can use. To produce a portable program, a compiler can transform our Go code into machine code compatible with a specific ISA (architecture) and the type of the desired operating system. In the next section, let's look at how the default Go compiler works. On the way, we will uncover mechanisms to help the Go compiler produce efficient and fast machine code.

Understanding Go Compiler

The topic of building effective compilers can fill a few books. In this book, however, we will try to understand the Go compiler basics that we, as Go developers interested in efficient code, have to be aware of. Generally, many things are involved in executing the Go code we write on the typical operating system, not only compilation. First, we need to compile it using a compiler, and then we have to use a linker to link different object files together, including potentially shared libraries. These compile and link procedures, often called *building*, produce the executable ("binary") that the operating system can execute. During the initial start, called *loading*, other shared libraries can be dynamically loaded too (e.g., Go plug-ins).

There are many code-building methods for Go code, designed for different target environments. For example, Tiny Go (*https://oreil.ly/c2C5E*) is optimized to produce binaries for microcontrollers, gopherjs (*https://oreil.ly/D83Jq*) produces JavaScript for

5 There can be incompatibilities, but mostly with special-purpose instructions like cryptographic or SIMD instructions, which can be checked at runtime if they are available before execution.

in-browser execution, and android (*https://oreil.ly/83Wm1*) produces programs executable on Android operating systems. However, this book will focus on the default and most popular Go compiler and linking mechanism available in the `go build` command. The compiler itself is written in Go (initially in C). The rough documentation and source code can be found here (*https://oreil.ly/qcrLt*).

The `go build` can build our code into many different outputs. We can build executables that require system libraries to be dynamically linked on startup. We can build shared libraries or even C-compatible shared libraries. Yet the most common and recommended way of using Go is to build executables with all dependencies statically linked in. It offers a much better experience where invocation of our binary does not need any system dependency of a specific version in a certain directory. It is a default build mode for code with a starting `main` function that can also be explicitly invoked using `go build -buildmode=exe`.

The `go build` command invokes both compilation and linking. While the linking phase also performs certain optimizations and checks, the compiler probably performs the most complex duty. The Go compiler focuses on a single package at once. It compiles package source code into the native code that the target architecture and operating systems support. On top of that, it validates, optimizes that code, and prepares important metadata for debugging purposes. We need to "collaborate" with the compiler (and operating system and hardware) to write efficient Go and not work against it.

> I tell everyone, if you're not sure how to do something, ask the question around what is the most idiomatic way to do this in Go. Because many of those answers are already tuned to being sympathetic with the operating system of the hardware.
>
> —Bill Kennedy, "Bill Kennedy on Mechanical Sympathy" (*https://oreil.ly/X3XzI*)

To make things more interesting, `go build` also offers a special cross-compilation mode if you want to compile a mix of Go code that uses functions implemented in C, C++, or even Fortran! This is possible if you enable a mode called `cgo` (*https://oreil.ly/Xjh9U*), which uses a mix of C (or C++) compiler and Go compiler. Unfortunately, `cgo` is not recommended (*https://oreil.ly/QojX3*), and it should be avoided if possible. It makes the build process slow, the performance of passing data between C and Go is questionable, and non-`cgo` compilation is already powerful enough to cross-compile binaries for different architectures and operating systems. Luckily, most of the libraries are either pure Go or are using pieces of Assembly that can be included in the Go binary without `cgo`.

To understand the impact of the compiler on our code, see the stages the Go compiler performs in Figure 4-2. While `go build` includes such compilation, we can trigger just the compilation (without linking) alone using `go tool compile`.

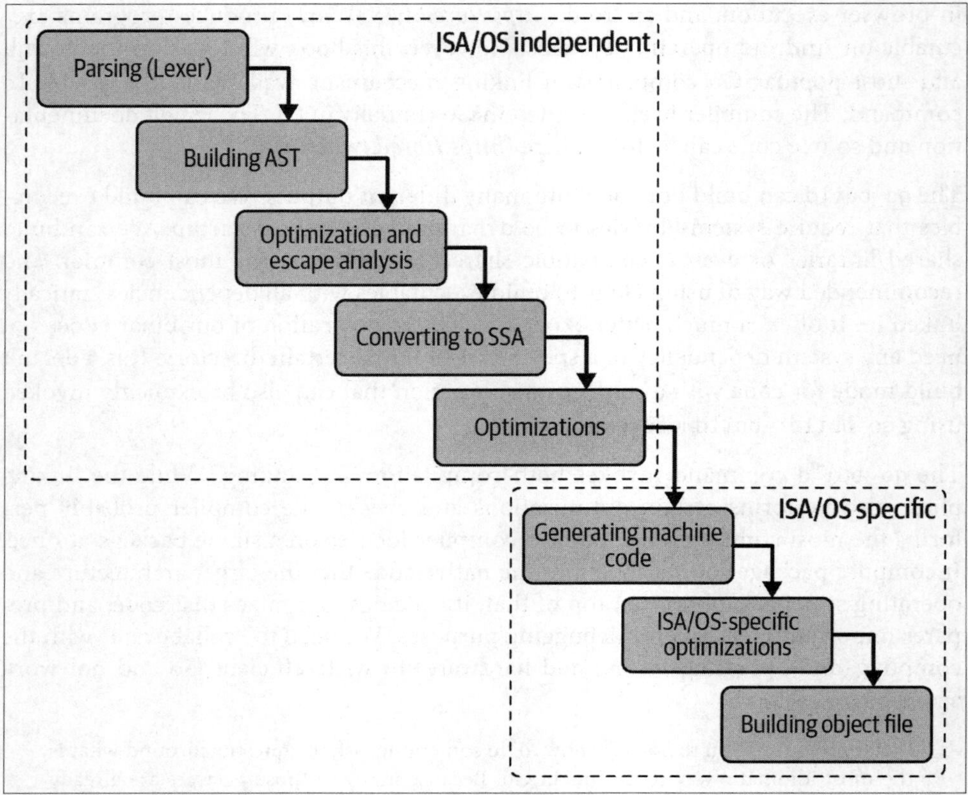

Figure 4-2. Stages performed by the Go compiler on each Go package

As mentioned previously, the whole process resides around the packages you use in your Go program. Each package is compiled in separation, allowing parallel compilation and separation of concerns. The compilation flow presented in Figure 4-2 works as follows:

1. The Go source code is first tokenized and parsed. The syntax is checked. The syntax tree references files and file positions to produce meaningful error and debugging information.
2. An abstract syntax tree (AST) is built. Such a tree notion is a common abstraction that allows developers to create algorithms that easily transform or check parsed statements. While in AST form, code is initially type-checked. Declared but not used items are detected.
3. The first pass of optimization is performed. For example, the initial dead code is eliminated, so the binary size can be smaller and less code needs to be compiled. Then, escape analysis (mentioned in "Go Memory Management" on page 172) is

performed to decide which variables can be placed on the stack and which have to be allocated on the heap. On top of that, in this stage, function inlining occurs for simple and small functions.

Function Inlining

Functions[6] in programming language allow us to create abstractions, hide complexities, and reduce repeated code. Yet the cost of calling execution is nonzero. For example, a function with a single argument call needs ~10 extra CPU instructions (*https://oreil.ly/4OPbI*).[7] So, while the cost is fixed and typically at the level of nanoseconds, it can matter if we have thousands of these calls in the hot path and the function body is small enough that this execution call matters.

There are also other benefits of inlining. For example, the compiler can apply other optimizations more effectively in code with fewer functions and does not need to use heap or large stack memory (with copy) to pass arguments between function scopes. Heap and stack are explained in "Go Memory Management" on page 172.

The compiler automatically substitutes some function calls with the exact copy of its body. This is called *inlining* or *inline expansion* (*https://oreil.ly/JGde3*). The logic is quite smart. For instance, from Go 1.9, the compiler can inline both leaf and mid-stack functions (*https://oreil.ly/CX2v0*).

Manual Inlining Is Rarely Needed

It is tempting for beginner engineers to micro-optimize by manually inlining some of their functions. However, while developers had to do it in the early days of programming, this functionality is a fundamental duty of the compiler, which usually knows better when and how to inline a function. Use that fact by focusing on your code readability and maintainability first regarding the choice of functions. Inline manually only as a last resort, and always measure.

6 Note that the structure methods, from a compiler perspective, are just functions, with the first argument being that structure, so the same inlining technique applies here.

7 A function call needs more CPU instructions since the program has to pass argument variables and return parameters through the stack, keep the current function's state, rewind the stack after the function call, add the new frame stack, etc.

4. After early optimizations on the AST, the tree is converted to the Static Single Assignment (SSA) form. This low-level, more explicit representation makes it easier to perform further optimization passes using a set of rules. For example, with the help of the SSA, the compiler can easily find places of unnecessary variable assignments.[8]
5. The compiler applies further machine-independent optimization rules. So, for example, statements like `y := 0*x` will be simplified to `y :=0`. The complete list of rules is enormous (*https://oreil.ly/QTljA*) and only confirms how complex this space is. Furthermore, some code pieces can be replaced by an intrinsic function (*https://oreil.ly/FMjT0*)—heavily optimized equivalent code (e.g., in raw Assembly).
6. Based on `GOARCH` and `GOOS` environment variables, the compiler invokes the `genssa` function that converts SSA to the machine code for the desired architecture (ISA) and operating system.
7. Further ISA- and operating system–specific optimizations are applied.
8. Package machine code that is not dead is built into a single object file (with the *.o* suffix) and debug information.

The final "object file" is compressed into a `tar` file called a Go *archive*, usually with *.a* file suffix.[9] Such archive files for each package can be used by Go linker (or other linkers) to combine all into a single executable, commonly called a *binary file*. Depending on the operating system, such a file follows a certain format, telling the system how to execute and use it. Typically for Linux, it will be an Executable and Linkable Format (*https://oreil.ly/jnicX*) (ELF). On Windows, it might be Portable Executable (*https://oreil.ly/SdohW*) (PE).

The machine code is not the only part of such a binary file. It also carries the program's static data, like global variables and constants. The executable file also contains a lot of debugging information that can take a considerable amount of binary size, like a simple symbols table, basic type information (for reflection), and PC-to-line mapping (*https://oreil.ly/akAR2*) (address of the instruction to the line in the source code where the command was). That extra information enables valuable debugging tools to link machine code to the source code. Many debugging tools use it, for example, "Profiling in Go" on page 331 and the aforementioned `objdump` tool.

---

8 Go tooling allows us to check the state of our program through each optimization in the SSA form thanks to the GOSSAFUNC environment variable. It's as easy as building our program with GOSSAFUNC=<function to see> go build and opening the resulting *ssa.html* file. You can read more about it here (*https://oreil.ly/32Zbd*).

9 You can unpack it with the tar <archive> or go tool pack e <archive> command. Go archive typically contains the object file and package metadata in the *__.PKGDEF* file.

For compatibility with debugging software like Delve or GDB, the DWARF table is also attached to the binary file.[10]

On top of the already long list of responsibilities, the Go compiler must perform extra steps to ensure Go memory safety (*https://oreil.ly/kkCRb*). For instance, the compiler can often tell during compile time that some commands will use a memory space that is safe to use (contains an expected data structure and is reserved for our program). However, there are cases when this cannot be determined during compilation, so additional checks have to be done at runtime, e.g., extra bound checks or nil checks.

We will discuss this in more detail in "Go Memory Management" on page 172, but for our conversation about CPU, we need to acknowledge that such checks can take our valuable CPU time. While the Go compiler tries to eliminate these checks when unnecessary (e.g., in the bound check elimination stage during SSA optimizations), there might be cases where we need to write code in a way that helps the compiler eliminate some checks.[11]

There are many different configuration options for the Go build process. The first large batch of options can be passed through `go build -ldflags="<flags>"`, which represents linker command options (*https://oreil.ly/g8dvv*) (the `ld` prefix traditionally stands for Linux linker (*https://oreil.ly/uJEda*)). For example:

- We can omit the DWARF table, thus reducing the binary size using `-ldflags="-w"` (recommended for production build if you don't use debuggers there).
- We can further reduce the size with `-ldflags= "-s -w"`, removing the DWARF and symbols tables with other debug information. I would not recommend the latter option, as non-DWARF elements allow important runtime routines, like gathering profiles.

Similarly, `go build -gcflags="<flags>"` represents Go compiler options (*https://oreil.ly/rRtRs*) (`gc` stands for `Go Compiler`; don't confuse it with GC, which means garbage collection, as explained in "Garbage Collection" on page 185). For example:

- `-gcflags="-S"` prints Go Assembly from the source code.
- `-gcflags="-N"` disables all compiler optimizations.

10 However, there are discussions to remove (*https://oreil.ly/xoijc*) it from the default building process.

11 Bound check elimination (*https://oreil.ly/E7FJI*) is not explained in this book, as it's a rare optimization idea.

- -gcflags="-m=<number>" builds the code while printing the main optimization decisions, where the number represents the level of detail. See Example 4-3 for the automatic compiler optimizations made on our Sum function in Example 4-1.

Example 4-3. Output of go build -gcflags="-m=1" sum.go on Example 4-1 code

```
# command-line-arguments
./sum.go:10:27: inlining call to os.ReadFile ❶
./sum.go:15:34: inlining call to bytes.Split ❶
./sum.go:9:10: leaking param: fileName ❷
./sum.go:15:44: ([]byte)("\n") does not escape ❸
./sum.go:16:38: string(line) escapes to heap ❹
```

❶ os.ReadFile and bytes.Split are short enough, so the compiler can copy the whole body of the Sum function.

❷ The fileName argument is "leaking," meaning this function keeps its parameter alive after it returns (it can still be on stack, though).

❸ Memory for []byte("\n") will be allocated on the stack. Messages like this help debug escape analysis. Learn more about it here (*https://oreil.ly/zBCyO*).

❹ Memory for string(line) will be allocated in a more expensive heap.

The compiler will print more details with an increased -m number. For example, -m=3 will explain why certain decisions were made. This option is handy when we expect certain optimization (inlining or keeping variables on the stack) to occur, but we still see an overhead while benchmarking in our TFBO cycle ("Efficiency-Aware Development Flow" on page 102).

The Go compiler implementation is highly tested and mature, but there are limitless ways of writing the same functionality. There might be edge cases when our implementation confuses the compiler, so it does not apply certain naive implementations. Benchmarking if there is a problem, profiling the code, and confirming with the -m option help. More detailed optimizations can also be printed using further options. For example, -gcflags="-d=ssa/check_bce/debug=1" prints all bound check elimination optimizations.

The Simpler the Code, the More Effective Compiler Optimizations Will Be

Too-clever code is hard to read and makes it difficult to maintain programmed functionality. But it also can confuse the compiler that tries to match patterns with their optimized equivalents. Using idiomatic code, keeping your functions and loops straightforward, increases the chances that the compiler applies the optimizations so you don't need to!

Knowing compiler internals helps, especially when it comes to more advanced optimizations tricks, which among other things, help compilers optimize our code. Unfortunately, it also means our optimizations might be a bit fragile regarding portability between different compiler versions. The Go team reserves rights to change compiler implementation and flags since they are not part of any specification. This might mean that the way you wrote a function that allows automatic inline by the compiler might not trigger inline in the next version of the Go compiler. This is why it's even more important to benchmark and closely observe the efficiency of your program when you switch to a different Go version.

To sum up, the compilation process has a crucial role in offloading programmers from pretty tedious work. Without compiler optimizations, we would need to write more code to get to the same efficiency level while sacrificing readability and portability. Instead, if you focus on making your code simple, you can trust that the Go compiler will do a good enough job. If you need to increase efficiency for a particular hot path, it might be beneficial to double-check if the compiler did what you expected. For example, it might be that the compiler did not match our code with common optimization; there is some extra memory safety check that the compiler could further eliminate or function that could be inlined but was not. In very extreme cases, there might be even a value to write a dedicated assembly code and import it from the Go code.[12]

The Go building process constructs fully executable machine code from our Go source code. The operating system loads machine code to memory and writes the first instruction address to the program counter (PC) register when it needs to be executed. From there, the CPU core can compute each instruction one by one. At first glance, it might mean that the CPU has a relatively simple job to do. But unfortunately, a memory wall problem causes CPU makers to continuously work on additional hardware optimizations that change how these instructions are executed. Understanding these mechanisms will allow us to control the efficiency and speed of our Go programs even better. Let's uncover this problem in the next section.

---

12 This is very often used in standard libraries for critical code.

CPU and Memory Wall Problem

To understand the memory wall and its consequences, let's dive briefly into CPU core internals. The details and implementation of the CPU core change over time for better efficiency (usually getting more complex), but the fundamentals stay the same. In principle, a Control Unit, shown in Figure 4-1, manages reads from memory through various L-caches (from smallest and fastest), decodes program instructions, coordinates their execution in the Arithmetic Logic Unit (ALU), and handles interruptions.

An important fact is that the CPU works in cycles. Most CPUs in one cycle can perform one instruction on one set of tiny data. This pattern is called the Single Instruction Single Data (SISD) in characteristics mentioned in Flynn's taxonomy (*https://oreil.ly/oQu0M*), and it's the key aspect of the von Neumann architecture. Some CPUs also allow Single Instruction Multiple Data (SIMD)[13] processing with special instructions like SSE, which allows the same arithmetic operation on four floating numbers in one cycle. Unfortunately, these instructions are not straightforward to use in Go and are therefore quite rarely seen.

Meanwhile, registers are the fastest local storage available to the CPU core. Because they are small circuits wired directly into the ALU, it takes only one CPU cycle to read their data. Unfortunately, there are also only a few of them (depending on the CPU, typically 16 for general use), and their size is usually not larger than 64 bits. This means they are used as short-time variables in our program lifetime. Some of the registers can be used for our machine code. Others are reserved for CPU use. For example, the PC register (*https://oreil.ly/TvHVd*) holds the address of the next instruction that the CPU should fetch, decode, and execute.

Computation is all about the data. As we learned in Chapter 1, there is lots of data nowadays, scattered around different storage mediums—uncomparably more than what's available to store in a single CPU register. Moreover, a single CPU cycle is faster than accessing data from the main memory (RAM)—on average, one hundred times faster, as we read from our rough napkin math of latencies in Appendix A that we will use throughout this book. As discussed in the misconception "Hardware Is Getting Faster and Cheaper" on page 17, technology allows us to create CPU cores with dynamic clock speed, yet the maximum is always around 4 GHz. Funny enough, the fact we can't make faster CPU cores is not the most important problem since our

13 On top of SISD and SIMD, Flynn's taxonomy also specifies MISD, which describes performing multiple instructions on the same data, and MIMD, which describes full parallelism. MISD is rare and only happens when reliability is important. For example, four flight control computers perform exactly the same computations for quadruple error checks in every NASA space shuttle. MIMD, on the other hand, is more common thanks to multicore or even multi-CPU designs.

CPU cores are already…too fast! It's a fact we cannot make faster memory, which causes the main efficiency issues in CPUs nowadays.

> We can execute something in the ballpark of 36 billion instructions every second. Unfortunately, most of that time is spent waiting for data. About 50% of the time in almost every application. In some applications upwards of 75% of the time is spent waiting for data rather than executing instructions. If this horrifies you, good. It should.
>
> — Chandler Carruth, "Efficiency with Algorithms, Performance with Data Structures" (*https://oreil.ly/I55mm*)

The aforementioned problem is often referred to as a "memory wall" problem (*https://oreil.ly/l5zgk*). As a result of this problem, we risk wasting dozens, if not hundreds, of CPU cycles per single instruction, since fetching that instruction and data (and then saving the results) takes ages.

This problem is so prominent that it has triggered recent discussions about revisiting von Neumann's architecture (*https://oreil.ly/xqbNU*) as machine learning (ML) workloads (e.g., neural networks) for artificial intelligence (AI) use become more popular. These workloads are especially affected by the memory wall problem because most of the time is spent performing complex matrix math calculations, which require traversing large amounts of memory.[14]

The memory wall problem effectively limits how fast our programs do their job. It also impacts the overall energy efficiency that matters for mobile applications. Nevertheless, it is the best common general-purpose hardware nowadays. Industry mitigated many of these problems by developing a few main CPU optimizations we will discuss below: the hierarchical cache system, pipelining, out-of-order execution, and hyperthreading. These directly impact our low-level Go code efficiency, especially in terms of how fast our program will be executed.

Hierachical Cache System

All modern CPUs include local, fast, small caches for often-used data. L1, L2, L3 (and sometimes L4) caches are on-chip static random-access memory (SRAM) circuits. SRAM uses different technology for storing data faster than our main memory RAM but is much more expensive to use and produce in large capacities (main memory is explained in "Physical Memory" on page 153). Therefore, L-caches are touched first when the CPU needs to fetch instruction or data for an instruction from the main

14 This is why we see specialized chips (called Neural Processing Units, or NPUs) appearing in the commodity devices—for example, Tensor Processing Unit (TPU) in Google phones, A14 Bionic chip in iPhones, and dedicated NPU in the M1 chip in Apple laptops.

memory (RAM). The way the CPU is using L-caches is presented in Figure 4-3.[15] In the example, we will use a simple CPU instruction `MOVQ`, explained in Example 4-2.

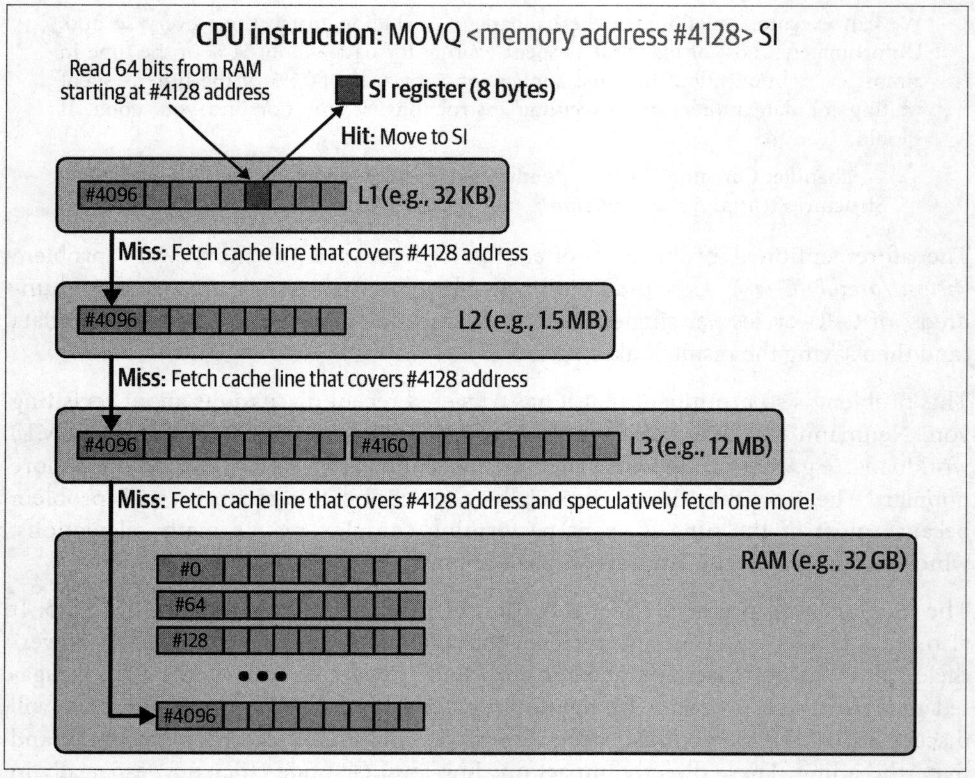

Figure 4-3. The "look up" cache method performed by the CPU to read bytes from the main memory through L-caches

To copy 64 bits (`MOVQ` command) from a specific memory address to register `SI`, we must access the data that normally resides in the main memory. Since reading from RAM is slow, it uses L-caches to check for data first. The CPU will ask the L1 cache for these bytes on the first try. If the data is not there (cache miss), it visits a larger L2 cache, then the largest cache L3, then eventually main memory (RAM). In any of these misses, the CPU will try to fetch the complete "cache line" (typically 64 bytes, so eight times the size of the register), save it in all caches, and only use these specific bytes.

15 Sizes of caches can vary. Example sizes are taken from my laptop. You can check the sizes of your CPU caches in Linux by using the `sudo dmidecode -t cache` command.

Reading more bytes at once (cache line) is useful as it takes the same latency as reading a single byte (explained in "Physical Memory" on page 153). Statistically, it is also likely that the next operation needs bytes next to the previously accessed area. L-caches partially mitigate the memory latency problem and reduce the overall amount of data to be transferred, preserving memory bandwidth.

The first direct consequence of having L-caches in our CPUs is that the smaller and more aligned the data structure we define, the better the efficiency. Such a structure will have more chances to fit fully in lower-level caches and avoid expensive cache misses. The second result is that instructions on sequential data will be faster since cache lines typically contain multiple items stored next to each other.

Pipelining and Out-of-Order Execution

If the data were magically accessible in zero time, we would have a perfect situation where every CPU core cycle performs a meaningful instruction, executing instructions as fast as CPU core speed allows. Since this is not the case, modern CPUs try to keep every part of the CPU core busy using cascading pipelining. In principle, the CPU core can perform many stages required for instruction execution at once in one cycle. This means we can exploit Instruction-Level Parallelism (ILP) to execute, for example, five independent instructions in five CPU cycles, giving us that sweet average of one instruction per cycle (IPC).[16] For example, in an initial 5-stage pipeline system (*https://oreil.ly/ccBg2*) (modern CPUs have 14–24 stages!), a single CPU core computes 5 instructions at the same time within a cycle, as presented in Figure 4-4.

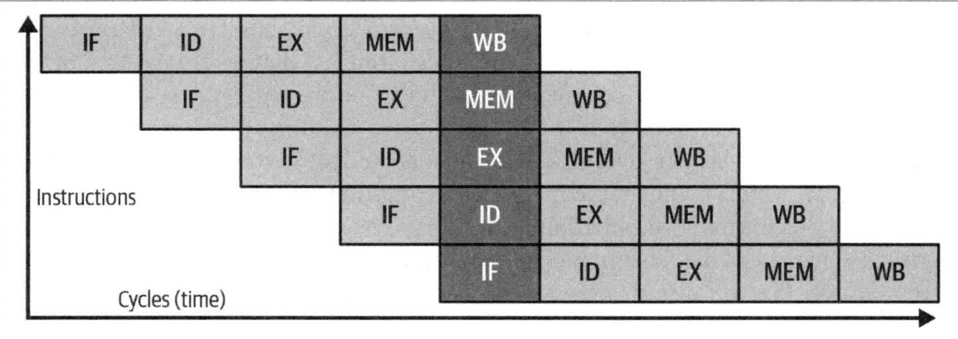

Figure 4-4. Example five-stage pipeline

16 If a CPU can in total perform up to one instruction per cycle (IPC ⇐ 1), we call it a scalar CPU. Most modern CPU cores have IPC ⇐ 1, but one CPU has more than one core, which makes IPC > 1. This makes these CPUs superscalar. IPC has quickly become a performance metric for CPUs.

The classic five-stage pipeline consists of five operations:

`IF`
: Fetch the instruction to execute.

`ID`
: Decode the instruction.

`EX`
: Start the execution of the instruction.

`MEM`
: Fetch the operands for the execution.

`WB`
: Write back the result of the operation (if any).

To make it even more complex, as we discussed in the L-caches section, it is rarely the case that even the fetch of the data (e.g., the `MEM` stage) takes only one cycle. To mitigate this, the CPU core also employs a technique called out-of-order execution (*https://oreil.ly/ccBg2*). In this method, the CPU attempts to schedule instructions in an order governed by the availability of the input data and execution unit (if possible) rather than by their original order in the program. For our purposes, it is enough to think about it as a complex, more dynamic pipeline that utilizes internal queues for more efficient CPU execution.

The resulting pipelined and out-of-order CPU execution is complex, but the preceding simplified explanation should be all we need to understand two critical consequences for us as developers. The first, trivial one is that every switch of the instruction stream has a huge cost (e.g., in latency),[17] because the pipeline has to reset and start from scratch, on top of the obvious cache trashing. We haven't yet mentioned the operating system overhead that must be added on top. We often call this a *context switch*, which is inevitable in modern computers since the typical operating systems use preemptive task scheduling. In these systems, the execution flow of the single CPU core can be preempted many times a second, which might matter in extreme cases. We will discuss how to influence such behavior in "Operating System Scheduler" on page 134.

17 Huge cost is not an overstatement. Latency of context switch depends on many factors, but it was measured that in the best case, direct latency (including operating system switch latency) is around 1,350 nanoseconds—2,200 nanoseconds if it has to migrate to a different core. This is only a direct latency, from the end of one thread to the start of another. The total latency that would include the indirect cost in the form of cache and pipeline warm-up could be as high as 10,000 nanoseconds (and this is what we see in Table A-1). During this time, we could compute something like 40,000 instructions.

The second consequence is that the more predictive our code is, the better. This is because pipelining requires the CPU cores to perform complex *branch predictions* to find instructions that will be executed after the current one. If our code is full of branches like `if` statements, `switch` cases, or jump statements like `continue`, finding even two instructions to execute simultaneously might be impossible, simply because one instruction might decide on what instruction will be done next. This is called data dependency. Modern CPU core implementation goes even further by performing speculative execution. Since it does not know which instruction is next, it picks the most likely one and assumes that such a branch will be chosen. Unnecessary executions on the wrong branches are better than wasted CPU cycles doing nothing. Therefore, many branchless coding techniques have emerged, which help the CPU predict branches and might result in faster code. Some methods are applied automatically by the Go compiler (*https://oreil.ly/VqKzx*), but sometimes, manual improvements have to be added.

Generally speaking, the simpler the code, with fewer nested conditionals and loops, the better for the branch predictor. This is why we often hear that the code that "leans to the left" is faster.

> In my experience [I saw] repeatedly that code that wants to be fast, go to the left of the page. So if you [write] like a loop and the if, and the for and a switch, it's not going to be fast. By the way, the Linux kernel, do you know what the coding standard is? Eight characters tab, 80 characters line width. You can't write bad code in the Linux kernel. You can't write slow code there. ... The moment you have too many ifs and decision points ... in your code, the efficiency is out of the window.
>
> —Andrei Alexandrescu, "Speed Is Found in the Minds of People" (*https://oreil.ly/6mERC*)

The existence of branch predictors and speculative approaches in the CPU has another consequence. It causes contiguous memory data structures to perform much better in pipelined CPU architecture with L-caches.

Contiguous Memory Structure Matters

Practically speaking, on modern CPUs, developers in most cases should prefer contiguous memory data structures like arrays instead of linked lists in their programs. This is because a typical linked-like list implementation (e.g., a tree) uses memory pointers to the next, past, child, or parent elements. This means that when iterating over such a structure, the CPU core can't tell what data and what instruction we will do next until we visit the node and check that pointer. This effectively limits the speculation capabilities, causing inefficient CPU usage.

Hyper-Threading

Hyper-Threading is Intel's proprietary name for the CPU optimization technique called *simultaneous multithreading* (SMT) (*https://oreil.ly/L5va6*).[18] Other CPU makers implement SMT too. This method allows a single CPU core to operate in a mode visible to programs and operating systems as two logical CPU cores.[19] SMT prompts the operating system to schedule two threads onto the same physical CPU core. While a single physical core will never execute more than one instruction at a time, more instructions in the queue help make the CPU core busy during idle times. Given the memory access wait times, this can utilize a single CPU core more without impacting the latency of the process execution. In addition, extra registers in SMT enable CPUs to allow for faster context switches between multiple threads running on a single physical core.

SMT has to be supported and integrated with the operating system. You should see twice as many cores as physical ones in your machine when enabled. To understand if your CPU supports Hyper-Threading, check the "thread(s) per core" information in the specifications. For example, using the `lscpu` Linux command in Example 4-4, my CPU has two threads, meaning Hyper-Threading is available.

Example 4-4. Output of the `lscpu` command on my Linux laptop

```
Architecture:                    x86_64
CPU op-mode(s):                  32-bit, 64-bit
Byte Order:                      Little Endian
Address sizes:                   39 bits physical, 48 bits virtual
CPU(s):                          12
On-line CPU(s) list:             0-11
Thread(s) per core:              2 ❶
Core(s) per socket:              6
Socket(s):                       1
NUMA node(s):                    1
Vendor ID:                       GenuineIntel
CPU family:                      6
Model:                           158
Model name:                      Intel(R) Core(TM) i7-9850H CPU @ 2.60GHz
CPU MHz:                         2600.000
CPU max MHz:                     4600.0000
CPU min MHz:                     800.0000
```

18 In some sources, this technique is also called CPU threading (aka hardware threads). I will avoid this terminology in this book due to possible confusion with operating system threads.

19 Do not confuse Hyper-Threading logical cores with virtual CPUs (vCPUs) referenced when we use virtualizations like virtual machines. Guest operating systems use the machine's physical or logical CPUs depending on host choice, but in both cases, they are called vCPUs.

❶ My CPU supports SMT, and it's enabled on my Linux installation.

The SMT is usually enabled by default but can be turned to on demand on newer kernels. This poses one consequence when running our Go programs. We can usually choose if we should enable or disable this mechanism for our processes. But should we? In most cases, it is better to keep it enabled for our Go programs as it allows us to fully utilize physical cores when running multiple different tasks on a single computer. Yet, in some extreme cases, it might be worth dedicating full physical core to a single process to ensure the highest quality of service. Generally, a benchmark on each specific hardware should tell us.

To sum up, all the aforementioned CPU optimizations and the corresponding programming techniques utilizing that knowledge tend to be used only at the very end of the optimization cycle and only when we want to squeeze out the last dozen nanoseconds on the critical path.

Three Principles of Writing CPU-Efficient Code on Critical Path

The three basic rules that will yield CPU-friendly code are as follows:

- Use algorithms that do less work.
- Focus on writing low-complexity code that will be easier to optimize for the compiler and CPU branch predictors. Ideally, separate "hot" from "cold" code.
- Favor contiguous memory data structures when you plan to iterate or traverse over them a lot.

With this brief understanding of CPU hardware dynamics, let's dive deeper into the essential software types that allow us to run thousands of programs simultaneously on shared hardware—schedulers.

Schedulers

Scheduling generally means allocating necessary, usually limited, resources for a certain process to finish. For example, assembling car parts must be tightly scheduled in a certain place at a certain time in a car factory to avoid downtime. We might also need to schedule a meeting among certain attendees with only certain time slots of the day free.

In modern computers or clusters of servers, we have thousands of programs that have to be running on shared resources like CPU, memory, network, disks, etc. That's why the industry developed many types of scheduling software (commonly called *schedulers*) focused on allocating these programs to free resources on many levels.

In this section, we will discuss CPU scheduling. Starting from the bottom level, we have an operating system that schedules arbitrary programs on a limited number of physical CPUs. Operating system mechanisms should tell us how multiple programs running simultaneously can impact our CPU resources and, in effect, our own Go program execution latency. It will also help us understand how a developer can utilize multiple CPU cores simultaneously, in parallel or concurrently, to achieve faster execution.

Operating System Scheduler

As with compilers, there are many different operating systems (OSes), each with different task scheduling and resource management logic. While most of the systems operate on similar abstractions (e.g., threads, processes with priorities), we will focus on the Linux operating system in this book. Its core, called the kernel, has many important functionalities, like managing memory, devices, network access, security, and more. It also ensures program execution using a configurable component called a scheduler.

> As a central part of resource management, the OS thread scheduler must maintain the following, simple, invariant: make sure that ready threads are scheduled on available cores.
>
> —J.P. Lozi et al., "The Linux Scheduler: A Decade of Wasted Cores" (*https://oreil.ly/bfiEW*)

The smallest scheduling unit for the Linux scheduler is called an OS thread (*https://oreil.ly/Lp2Sk*). The thread (sometimes also referred to as a *task* or *lightweight process*) contains an independent set of machine code in the form of CPU instructions designed to run sequentially. While threads can maintain their execution state, stack, and register set, they cannot run out of context.

Each thread runs as a part of the process. The process represents a program in execution and can be identified by its Process Identification Number (PID). When we tell Linux OS to execute our compiled program, a new process is created (for example, when a `fork` (*https://oreil.ly/IPKYU*) system call is used).

The process creation includes the assignment of a new PID, the creation of the initial thread with its machine code (our `func main()` in the Go code) and stack, files for standard outputs and input, and tons of other data (e.g., list of open file descriptors, statistics, limits, attributes, mounted items, groups, etc.). On top of that, a new memory address space is created, which has to be protected from other processes. All of that information is maintained under the dedicated directory */proc/<PID>* for the duration of the program execution.

Threads can create new threads (e.g., using the `clone` (*https://oreil.ly/6qSg3*) syscall) that will have independent machine code sequences but will share the same memory address space. Threads can also create new processes (e.g., using `fork` (*https://oreil.ly/idB06*)) that will run in isolation and execute the desired program. Threads maintain their execution state: Running, Ready, and Blocked. Possible transformations of these states are presented in Figure 4-5.

Thread state tells the scheduler what the thread is doing at the moment:

Running
: Thread is assigned to the CPU core and is doing its job.

Blocked
: Thread is waiting on some event that potentially takes longer than a context switch. For example, a thread reads from a network connection and is waiting for a packet or its turn on the mutex lock. This is an opportunity for the scheduler to step in and allow other threads to run.

Ready
: Thread is ready for execution but is waiting for its turn.

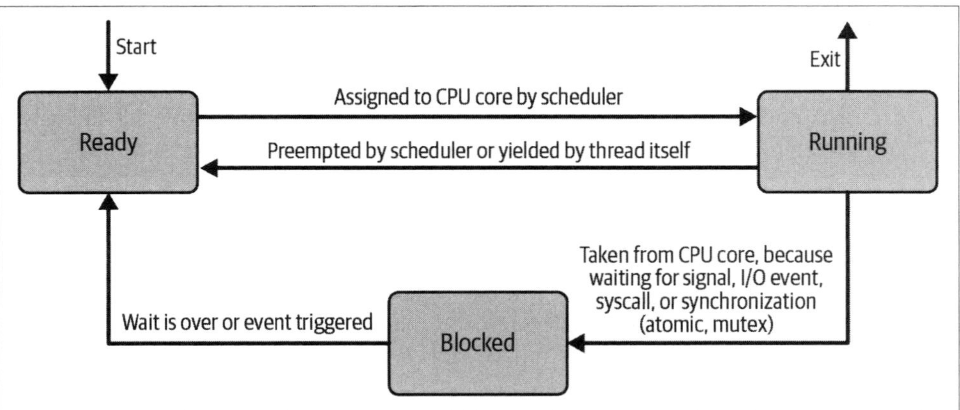

Figure 4-5. Thread states as seen by the Linux OS scheduler

As you might already notice, the Linux scheduler does a preemptive type of thread scheduling. Preemptive means the scheduler can freeze a thread execution at any time. In modern OS, we always have more threads to be executed than available CPU cores, so the scheduler must run multiple "ready" threads on a single CPU core. The thread is preempted every time it waits for an I/O request or other events. The thread can also tell the operating system to yield itself (e.g., using the `sched_yield` (*https://oreil.ly/QfnCs*) syscall). When preempted, it enters a "blocked" state, and another thread can take its place in the meantime.

The naive scheduling algorithm could wait for the thread to preempt itself. This would work great for I/O bound threads, which are often in the "Blocked" state—for example, interactive systems with graphical interfaces or lightweight web servers working with network calls. But what if the thread is CPU bound, which means it spends most of its time using only CPU and memory—for example, doing some computation-heavy jobs like linear search, multiplying matrixes, or brute-forcing a hashed password? In such cases, the CPU core could be busy on one task for minutes, which will starve all other threads in the system. For example, imagine not being able to type in your browser or resize a window for a minute—it would look like a long system freeze!

This primary Linux scheduler implementation addresses that problem. It is called a Completely Fair Scheduler (CFS), and it assigns threads in short turns. Each thread is given a certain slice of the CPU time, typically something between 1 ms and 20 ms, which creates the illusion that threads are running simultaneously. It especially helps desktop systems, which must be responsive to human interactions. There are a few other important consequences of that design:

- The more threads that want to be executed, the less time they will have in each turn. However, this can result in lower productive utilization of the CPU core, which starts to spend more time on expensive context switches.
- On the overloaded machine, each thread has shorter turns on the CPU core and can also end up having fewer turns per second. While none of the threads is completely starved (blocked), their execution can significantly slow down.

CPU Overloading

Writing CPU-efficient code means our program wastes significantly fewer CPU cycles. Of course, this is always great, but the efficient implementation might be still doing its job very slowly if the CPU is overloaded.

An overloaded CPU or system means too many threads are competing for the available CPU cores. As a result, the machine might be overscheduled, or a process or two spawns too many threads to perform some heavy task (we call this situation a *noisy neighbor*). If an overloaded CPU situation occurs, checking the machine CPU utilization metric should show us CPU cores running at 100% capacity. Every thread will be executed slower in such a case, resulting in a frozen system, timeouts, and lack of responsiveness.

- It is hard to rely on pure program execution latency (sometimes referred to as *wall time* or *wall-clock time*) to estimate our program CPU efficiency. This is because modern OS schedulers are preemptive, and the program often waits for other I/O or synchronizations. As a result, it's pretty hard to reliably check if, after a fix, our program utilizes the CPU better than the previous implementation. This is why the industry defined an important metric to gather how long our program's process (all its threads) spent in the "Running" state on all CPU cores. We usually call it CPU time and we will discuss it in "CPU Usage" on page 229.

CPU Time on an Overloaded Machine

Measuring CPU time is a great way to check our program's CPU efficiency. However, be careful when looking at the CPU time from some narrow window of process execution time. For example, lower CPU time might mean our process was not using much CPU during that moment, but it might also represent an overloaded CPU.

Overall, sharing processes on the same system has its problems. That's why in virtualized environments, we tend to reserve these resources. For example, we can limit CPU use of one process to 200 milliseconds of CPU time per second, so 20% of one CPU core.

- The final consequence of the CFS design is that it is too fair to ensure dedicated CPU time for a single thread. The Linux scheduler has priorities, a user-configurable "niceness" flag, and different scheduling policies. Modern Linux OS even has a scheduling policy that uses a special real-time scheduler in place of CFS for threads that need to be executed in the first order.[20]

  Unfortunately, even with a real-time scheduler, a Linux system cannot ensure that higher-priority threads will have all the CPU time they need, as it will still try to ensure that low-priority threads are not starved. Furthermore, because both CFS and real-time counterparts are preemptive, they are not deterministic and predictive. As a result, any task with hard real-time requirements (e.g., millisecond trading or airplane software) can't be guaranteed enough execution time before its deadline. This is why some companies develop their own schedulers or systems for strict real-time programs (*https://oreil.ly/oVsCz*) like Zephyr OS (*https://oreil.ly/hV7ym*).

Despite the somewhat complex characteristics of the CFS scheduler, it remains the most popular thread orchestration system available in modern Linux systems. In

20 There are lots of good materials (*https://oreil.ly/8OPW3*) about tuning up the operating system. Many virtualization mechanisms, like containers with orchestrating systems like Kubernetes, also have their notion of priorities and affinities (pinning processes to specific cores or machines). In this book, we focus on writing efficient code, but we must be aware that execution environment tuning has an important role in ensuring quick and reliable program executions.

2016 the CFS was also overhauled for multicore machines and NUMA architectures, based on findings from a famous research paper (*https://oreil.ly/kUEiQ*). As a result, threads are now smartly distributed across idle cores while ensuring migrations are not done too often and not among threads sharing the same resources.

With a basic understanding of the OS scheduler, let's dive into why the Go scheduler exists and how it enables developers to program multiple tasks to run concurrently on single or multiple CPU cores.

Go Runtime Scheduler

The Go concurrency framework is built on the premise that it's hard for a single flow of CPU instructions (e.g., function) to utilize all CPU cycles due to the I/O-bound nature of the typical workflow. While OS thread abstraction mitigates this by multiplexing threads into a set of CPU cores, the Go language brings another layer—a *goroutine*—that multiplexes functions on top of a set of threads. The idea for goroutines (*https://oreil.ly/TClXu*) is similar to coroutines (*https://oreil.ly/t7oXZ*), but since it is not the same (goroutines can be preempted) and since it's in Go language, it has the *go* prefix. Similar to the OS thread, when the goroutine is blocked on a system call or I/O, the Go scheduler (not OS!) can quickly switch to a different goroutine, which will resume on the same thread (or a different one if needed).

> Essentially, Go has turned I/O-bound work [on the application level] into CPU-bound work at the OS level. Since all the context switching is happening at the application level, we don't lose the same 12K instructions (on average) per context switch that we were losing when using threads. In Go, those same context switches are costing you 200 nanoseconds or 2.4K instructions. The scheduler is also helping with gains on cache-line efficiencies and NUMA. This is why we don't need more threads than we have virtual cores.
>
> —William Kennedy, "Scheduling in Go: Part II—Go Scheduler" (*https://oreil.ly/Z4sRA*)

As a result, we have in Go very cheap execution "threads" in the user space (a new goroutine only allocates a few kilobytes for the initial, local stack), which reduce the number of competing threads in our machine and allow hundreds of goroutines in our program without extensive overhead. Just one OS thread per CPU core should be enough to get all the work in our goroutines done.[21] This enables many readability patterns—like event loops, map-reduce, pipes, iterators, and more—without involving more expensive kernel multithreading.

21 Details around Go runtime implementing Go scheduling are pretty impressive (*https://oreil.ly/G9bFb*). Essentially, Go does everything to keep the OS thread busy (spinning the OS thread) so it's not moving to a blocking state as long as possible. If needed, it can steal goroutines from other threads, poll networks, etc., to ensure we keep the CPU busy so the OS does not preempt the Go process.

Using Go concurrency in the form of goroutines is an excellent way to:

- Represent complex asynchronous abstractions (e.g., events)
- Utilize our CPU to the fullest for I/O-bound tasks
- Create a multithreaded application that can utilize multiple CPUs to execute faster

Starting another goroutine is very easy in Go. It is built in the language via a `go <func>()` syntax. Example 4-5 shows a function that starts two goroutines and finishes its work.

Example 4-5. A function that starts two goroutines

```
func anotherFunction(arg1 string) { /*...*/ }

func function() {
   // ... ❶

   go func() {
      // ... ❷
   }()

   go anotherFunction("argument1") ❸

   return ❹
}
```

❶ The scope of the current goroutine.

❷ The scope of a new goroutine that will run concurrently any moment now.

❸ `anotherFunction` will start running concurrently any moment now.

❹ When `function` terminates, the two goroutines we started can still run.

It's important to remember that all goroutines have a flat hierarchy between each other. Technically, there is no difference when goroutine `A` started `B` or `B` started `A`. In both cases, both `A` and `B` goroutines are equal, and they don't know about each other.[22] They also cannot stop each other unless we implement explicit communication or synchronization and "ask" the goroutine to shut down. The only exception is

22 In practice, there are ways to get this information using debug tracing. However, we should not rely on the program knowing which goroutine is a parent goroutine for normal execution flow.

the main goroutine that starts with the `main()` function. If the main goroutine finishes, the whole program terminates, killing all other goroutines forcefully.

Regarding communication, goroutines, similarly to OS threads, have access to the same memory space within the process. This means that we can pass data between goroutines using shared memory. However, this is not so trivial because almost no operation in Go is atomic. Concurrent writing (or writing and reading) from the same memory can cause data races, leading to nondeterministic behavior or even data corruption. To solve this, we need to use synchronization techniques like explicit atomic function (as presented in Example 4-6) or mutex (as shown in Example 4-7), so in other words, a lock.

Example 4-6. Safe multigoroutine communication through dedicated atomic addition

```
func sharingWithAtomic() (sum int64) {
   var wg sync.WaitGroup ❶

   concurrentFn := func() {
      atomic.AddInt64(&sum, randInt64())
      wg.Done()
   }
   wg.Add(3)
   go concurrentFn()
   go concurrentFn()
   go concurrentFn()

   wg.Wait()
   return sum
}
```

❶ Notice that while we use atomic to synchronize additions between `concurrentFn` goroutines, we use additional `sync.WaitGroup` (another form of locking) to wait for all these goroutines to finish. We do the same in Example 4-7.

Example 4-7. Safe multigoroutine communication through mutex (lock)

```
func sharingWithMutex() (sum int64) {
   var wg sync.WaitGroup
   var mu sync.Mutex

   concurrentFn := func() {
      mu.Lock()
      sum += randInt64()
      mu.Unlock()
      wg.Done()
   }
   wg.Add(3)
   go concurrentFn()
```

```
    go concurrentFn()
    go concurrentFn()

    wg.Wait()
    return sum
}
```

The choice between atomic and lock depends on readability, efficiency requirements, and what operation you want to synchronize. For example, if you want to concurrently perform a simple operation on a number like value write or read, addition, substitution, or compare and swap, you can consider the atomic package (*https://oreil.ly/NZnXr*). Atomic is often more efficient than mutexes (lock) since the compiler will translate them into special atomic CPU operations (*https://oreil.ly/8g0yM*) that can change data under a single memory address in a thread-safe way.[23]

If, however, using atomic impacts the readability of our code, the code is not on a critical path, or we have a more complex operation to synchronize, we can use a lock. Go offers `sync.Mutex`, which allows simple locking, and `sync.RWMutex`, which allows locking for reads (`RLock()`) and writes (`Lock()`). If you have many goroutines that do not modify shared memory, lock them with `RLock()` so there is no lock contention between them, since concurrent read of shared memory is safe. Only when a goroutine wants to modify that memory can it acquire a full lock using `Lock()` that will block all readers.

On the other hand, lock and atomic are not the only choices. The Go language has another ace in its hand on this subject. On top of the coroutine concept, Go also utilizes C. A. R. Hoare's Communicating Sequential Processes (CSP) (*https://oreil.ly/5KXA9*) paradigm, which can also be seen as a type-safe generalization of Unix pipes.

> Do not communicate by sharing memory; instead, share memory by communicating.
>
> —"Effective Go" (*https://oreil.ly/G4Lmq*)

This model encourages sharing data by implementing a communication pipeline between goroutines using a channel concept. Sharing the same memory address to pass some data requires extra synchronization. However, suppose one goroutine sends that data to some channel, and another receives it. In that case, the whole flow naturally synchronizes itself, and shared data is never accessed by two goroutines simultaneously, ensuring thread safety.[24] Example channel communication is presented in Example 4-8.

23 Funny enough, even atomic operations on CPU require some kind of locking. The difference is that instead of specialized locking mechanisms like spinlock (*https://oreil.ly/ZKXuN*), atomic instruction can use faster memory bus lock (*https://oreil.ly/9jchk*).

24 Assuming the programmer keeps to that rule. There is a way to send a pointer variable (e.g., `*string`) that points to shared memory, which violates the rule of sharing information through communicating.

Example 4-8. An example of memory-safe multigoroutine communication through the channel

```
func sharingWithChannel() (sum int64) {
   result := make(chan int64) ❶

   concurrentFn := func() {
      // ...
      result <- randInt64() ❷
   }
   go concurrentFn()
   go concurrentFn()
   go concurrentFn()

   for i := 0; i < 3; i++ { ❸
      sum += <-result ❹
   }
   close(result) ❺
   return sum
}
```

❶ Channel can be created in Go with the `ch := make(chan <type>, <buffer size>)` syntax.

❷ We can send values of a given type to our channel.

❸ Notice that in this example, we don't need `sync.WaitGroup` since we abuse the knowledge of how many exact messages we expect to receive. If we did not have that information, we would need a waiting group or another mechanism.

❹ We can read values of a given type from our channel.

❺ Channels should also be closed if we don't plan to send anything through them anymore. This releases resources and unblocks certain receiving and sending flows (more on that later).

The important aspect of channels is that they can be buffered. In such a case, it behaves like a queue. If we create a channel with, e.g., a buffer of three elements, a sending goroutine can send exactly three elements before it gets blocked until someone reads from this channel. If we send three elements and close the channel, the receiving goroutine can still read three elements before noticing the channel was closed. A channel can be in three states. It's important to remember how the goroutine sending or receiving from this channel behaves when switching between these states:

Allocated, open channel

If we create a channel using `make(chan <type>)`, it's allocated and open from the start. Assuming no buffer, such a channel will block an attempt to send a value until another goroutine receives it or when we use the `select` statement with multiple cases. Similarly, the channel receive will block until someone sends to that channel unless we receive in a `select` statement with multiple cases or the channel was closed.

Closed channel
: If we `close(ch)` the allocated channel, a send to that channel will cause panic and receives will return zero values immediately. This is why it is recommended to keep responsibility for the closing channel in the goroutine that sends the data (sender).

Nil channel
: If you define channel type (`var ch chan <type>`) without allocating it using `make(chan <type>)`, our channel is nil. We can also "nil" an allocated channel by assigning nil (`ch = nil`). In this state, sending and receiving will block forever. Practically speaking, it's rarely useful to nil channels.

Go channels is an amazing and elegant paradigm that allows for building very readable, event-based concurrency patterns. However, in terms of CPU efficiency, they might be the least efficient compared to the `atomic` package and mutexes. Don't let that discourage you! For most practical applications (if not overused!), channels can structure our application into robust and efficient concurrent implementation. We will explore some practical patterns of using channels in "Optimizing Latency Using Concurrency" on page 402.

Before we finish this section, it's important to understand how we can tune concurrency efficiency in the Go program. Concurrency logic is implemented by the Go scheduler in the Go runtime package (*https://oreil.ly/q3iCp*), which is also responsible for other things like garbage collection (see "Garbage Collection" on page 185), profiles, or stack framing. The Go scheduler is pretty automatic. There aren't many configuration flags. As it stands at the current moment, there are two practical ways developers can control concurrency in their code:[25]

A number of goroutines
: As developers, we usually control how many goroutines we create in our program. Spawning them for every small workpiece is usually not the best idea, so don't overuse them. It's also worth noting that many abstractions from standard or third-party libraries can spawn goroutines, especially those that require `Close`

25 I omitted two additional mechanisms on purpose. First of all, `runtime.Gosched()` exists, which allows yielding the current goroutine so others can do some work in the meantime. This command is less useful nowadays since the current Go scheduler is preemptive, and manual yielding has become impractical. The second interesting operation, `runtime.LockOSThread()`, sounds useful, but it's not designed for efficiency; rather, it pins the goroutine to the OS thread so we can read certain OS thread states from it.

or cancellation. Notably, common operations like `http.Do`, `context.WithCancel`, and `time.After` create goroutines. If used incorrectly, the goroutines can be easily leaked (leaving orphan goroutines), which typically wastes memory and CPU effort. We will explore ways to debug numbers and snapshots of goroutines in "Goroutine" on page 365.

First Rule of Efficient Code

Always close or release the resources you use. Sometimes simple structures can cause colossal and unbounded waste of memory and goroutines if we forget to close them. We will explore common examples in "Don't Leak Resources" on page 426.

`GOMAXPROCS`

This important environmental variable can be set to control the number of virtual CPUs you want to leverage in your Go program. The same configuration value can be applied via the `runtime.GOMAXPROCS(n)` function. The underlying logic on how the Go scheduler uses this variable is fairly complex,[26] but it generally controls how many parallel OS thread executions Go can expect (internally called a "proc" number). The Go scheduler will then maintain `GOMAXPROCS/proc` number of queues and try to distribute goroutines among them. The default value of `GOMAXPROCS` is always the number of virtual CPU cores your OS exposes, and that is typically what will give you the best performance. Trim the `GOMAXPROCS` value down if you want your Go program to use fewer CPU cores (less parallelism) in exchange for potentially higher latency.

Recommended GOMAXPROCS Configuration

Set `GOMAXPROCS` to the number of virtual cores you want your Go program to utilize at once. Typically, we want to use the whole machine; thus, the default value should work.

For virtualized environments, especially using lightweight virtualization mechanisms like containers, use Uber's `automaxprocs` library (*https://oreil.ly/ysr40*), which will adjust `GOMAXPROCS` based on the Linux CPU limits the container is allowed to use, which is often what we want.

Multitasking is always a tricky concept to introduce into a language. I believe the goroutines with channels in Go are quite an elegant solution to this problem, which

26 I recommend watching Chris Hines's talk from GopherCon 2019 (*https://oreil.ly/LoFiH*) to learn the low-level details around the Go scheduler.

allows many readable programming patterns without sacrificing efficiency. We will explore practical concurrency patterns in "Optimizing Latency Using Concurrency" on page 402, by improving the latency of Example 4-1 presented in this chapter.

Let's now look into when concurrency might be useful in our Go programs.

When to Use Concurrency

As with any efficiency optimization, the same classic rules apply when transforming a single goroutine code to a concurrent one. No exceptions here. We have to focus on the goal, apply the TFBO loop, benchmark early, and look for the biggest bottleneck. As with everything, adding concurrency has trade-offs, and there are cases where we should avoid it. Let's summarize the practical benefits and disadvantages of concurrent code versus sequential:

Advantages

- Concurrency allows us to speed up the work by splitting it into pieces and executing each part concurrently. As long as the synchronization and shared resources are not a significant bottleneck, we should expect an improved latency.
- Because the Go scheduler implements an efficient preemptive mechanism, concurrency improves CPU core utilization for I/O-bound tasks, which should translate into lower latency, even with a `GOMAXPROCS=1` (a single CPU core).
- Especially in virtual environments, we often reserve a certain CPU time for our programs. Concurrency allows us to distribute work across available CPU time in a more even way.
- For some cases, like asynchronous programming and event handling, concurrency represents a problem domain well, resulting in improved readability despite some complexities. Another example is the HTTP server. Treating each HTTP incoming request as a separate goroutine not only allows efficient CPU core utilization but also naturally fits into how code should be read and understood.

Disadvantages

- Concurrency adds significant complexity to the code, especially when we transform existing code into concurrency (instead of building API around channels from day one). This hits readability since it almost always obfuscates execution flow, but even worse, it limits the developer's ability to predict all edge cases and potential bugs. This is one of the main reasons why I recommend postponing adding concurrency as long as possible. And once you have to introduce concurrency, use as few channels as possible for the given problem.
- With concurrency, there is a risk of saturating resources due to unbounded concurrency (uncontrolled amount of goroutines in a single moment) or leaking

goroutines (orphan goroutines). This is something we also need to care about and test against (more on this in "Don't Leak Resources" on page 426).

- Despite Go's very efficient concurrency framework, goroutines and channels are not free of overhead. If used wrongly, it can impact our code efficiency. Focus on providing enough work to each goroutine that will justify its cost. Benchmarks are a must-have.
- When using concurrency, we suddenly add three more nontrivial tuning parameters into our program. We have a `GOMAXPROCS` setting, and depending on how we implement things, we can control the number of goroutines we spawn and how large a buffer of the channel we should have. Finding correct numbers requires hours of benchmarking and is still prone to errors.
- Concurrent code is hard to benchmark because it depends even more on the environment, possible noisy neighbors, multicore settings, OS version, and so on. On the other hand, sequential, single-core code has much more deterministic and portable performance, which is easier to prove and compare against.

As we can see, using concurrency is not the cure for all performance problems. It's just another tool in our hands that we can use to fulfill our efficiency goals.

Adding Concurrency Should Be One of Our Last Deliberate Optimizations to Try

As per our TFBO cycle, if you are still not meeting your RAERs, e.g., in terms of speed, make sure you try more straightforward optimization techniques before adding concurrency. The rule of thumb is to think about concurrency when our CPU profiler (explained in Chapter 9) shows that our program spends CPU time only on things that are crucial to our functionality. Ideally, before we hit our readability limit, is the most efficient way we know.

The mentioned list of disadvantages is one reason, but the second is that our program's characteristics might differ after basic (without concurrency) optimizations. For example, we thought our task was CPU bound, but after improvements, we may find most of the time is now spent waiting on I/O. Or we might realize we did not need the heavy concurrency changes after all.

Summary

The modern CPU hardware is a nontrivial component that allows us to run our software efficiently. With ongoing operating systems, Go language development, and advancements in hardware, only more optimization techniques and complexities will arise to decrease running costs and increase processing power.

In this chapter, I hopefully gave you basics that will help you optimize your usage of CPU resources and, generally, your software execution speed. First, we discussed the Assembly language and how it can be useful during Go development. Then, we explored Go compiler functionalities, optimizations, and ways to debug its execution.

Later, we jumped into the main challenge for CPU execution: memory access latency in modern systems. Finally, we discussed the various low-level optimizations like L-caches, pipelining, CPU branch prediction, and Hyper-Threading.

Last, we explored the practical problems of executing our programs in production systems. Unfortunately, our machine's program is rarely the only process, so efficient execution matters. Finally, we summarized Go's concurrency framework's benefits and disadvantages.

In practice, CPU resource is essential to optimize in modern infrastructure to achieve faster execution and the ability to pay less for our workloads. Unfortunately, CPU resource is only one aspect. For example, our choice optimization might prefer using more memory to reduce CPU usage or vice versa.

As a result, our programs typically use a lot of memory resources (plus I/O traffic through disk or network). While execution is tied to CPU resources like memory and I/O, it might be the first on our list of optimizations depending on what we want (e.g., cheaper execution, faster execution, or both). Let's discuss the memory resource in the next chapter.

CHAPTER 5

How Go Uses Memory Resource

In Chapter 4, we started looking under the hood of the modern computer. We discussed the efficiency aspects of using the CPU resource. Efficient execution of instructions in the CPU is important, but the sole purpose of performing those instructions is to modify the data. Unfortunately, the path of changing data is not always trivial. For example, in Chapter 4 we learned that in the von Neumann architecture (presented in Figure 4-1), we experience the CPU and memory wall problem when accessing data from the main memory (RAM).

The industry invented numerous technologies and optimization layers to overcome challenges like that, including memory safety and ensuring large memory capacities. As a result of those inventions, accessing eight bytes from RAM to the CPU register might be represented as a simple `MOVQ <destination register> <address XYZ>` instruction. However, the actual process done by the CPU to get that information from the physical chip storing those bytes is very complex. We discussed mechanisms like the hierarchical cache system, but there is much more.

In some ways, those mechanisms are abstracted from programmers as much as possible. So, for example, when we define a variable in Go code, we don't need to think about how much memory has to be reserved, where, and in how many L-caches it has to fit. This is great for development speed, but sometimes it might surprise us when we need to process a lot of data. In those cases, we need to revive our mechanical sympathy (*https://oreil.ly/Co2IM*) toward memory resource, optimizing TFBO flow ("Efficiency-Aware Development Flow" on page 102), and good tooling.

This chapter will focus on understanding the RAM resource. We will start by exploring overall memory relevance. Then we will set the context in "Do We Have a Memory Problem?" on page 152. Next, we will explain the patterns and consequences of each element involved in the memory access from bottom to top. The data journey for memory starts in "Physical Memory" on page 153, the hardware memory chips.

Then we will move to operating system (OS) memory management techniques that allow managing limited physical memory space in multiprocess systems: "Virtual Memory" on page 158 and "OS Memory Mapping" on page 168, with a more detailed explanation of the "mmap Syscall" on page 162.

With the lower layers of memory access explained, we can move to the key knowledge for Go programmers looking to optimize memory efficiency—the explanation of "Go Memory Management" on page 172. This includes the necessary elements like memory layout, what "Values, Pointers, and Memory Blocks" on page 176 mean, and the basics of the "Go Allocator" on page 181 with its measurable consequences. Finally, we will explore "Garbage Collection" on page 185.

We will go into many details about memory in this chapter, but the key aim is to build an instinct toward the patterns and behavior of Go programs when it comes to memory usage. For example, what problems can occur while accessing memory? How do we measure memory usage? What does it mean to allocate memory? How can we release it? We will explore answers to those questions in this chapter. But let's start this chapter by clarifying why RAM is relevant to our program execution. What makes it so important?

Memory Relevance

All Linux programs require more resources than just the CPU to perform their programmed functionalities. For example, let's take a web server like NGINX (*https://oreil.ly/7F0cZ*) (written in C) or Caddy (*https://oreil.ly/MpHMZ*) (written in Go). Those programs allow serving static content from disk or proxy HTTP requests, among other functionalities. They use the CPU to execute written code. However, a web server like this also interacts with other resources, for example:

- With RAM to cache basic HTTP responses
- With a disk to load configuration, static content, or write log lines for observability needs
- With a network to serve HTTP requests from remote clients

As a result, the CPU resource is only one part of the equation. This is the same for most programs—they are created to save, read, manage, operate, and transform data from different mediums.

One would argue that the "memory" resource, often called RAM,[1] sits at the core of those interactions. The RAM is the backbone of the computer because every external

1 In this book when I say "memory," I mean RAM and vice versa. Other mediums offer "memorizing" data in computer architecture (e.g., L-caches), but we tend to treat RAM as the "main" memory resource.

piece of data (bytes from disk, network, or another device) has to be buffered in memory to be accessible to the CPU. So, for example, the first thing the OS does to start a new process is load part of the program's machine code and initial data to memory for the CPU to execute it.

Unfortunately, we must be aware of three main caveats when using memory in our programs:

- RAM access is significantly slower than CPU operational speed.
- There is always a finite amount of RAM in our machines (typically from a few GB to hundreds of GB per machine), which forces us to care about space efficiency.[2]
- Unless the persistent type of memory (*https://oreil.ly/uaPiN*) will be commoditized with RAM-like speeds, pricing, and robustness, our main memory is strictly volatile. When the computer power goes down, all information is completely lost.[3]

The ephemeral characteristics of memory and its finite size are why we are forced to add an auxiliary, persistent I/O resource to our computer, i.e., a disk. These days we have relatively fast solid state drive (SSD) disks (yet still around 10x slower than RAM) with a limited lifetime (~five years). On the other hand, we have a slower and cheaper hard disk drive (HDD). While cheaper than RAM, the disk resource is also a scarce resource.

Last but not least, for scalability and reliability reasons, our computers rely on data from remote locations. Industry invented different networks and protocols that allow us to communicate with remote software (e.g., databases) or even remote hardware (via iSCSI or NFS protocols). We typically abstract this type of I/O as a network resource usage. Unfortunately, the network is one of the most challenging resources to work with because of its unpredictable nature, limited bandwidth, and bigger latencies.

While using any of those resources, we use it through the memory resource. As a result, it is essential to understand its mechanics. There are many things a programmer can do to impact the application's memory usage. But unfortunately, without

2 Not only because of physical limitations like not enough chip pins, space, and energy for transistors, but also because managing large memory poses huge overhead as we will learn in "OS Memory Management" on page 156.

3 In some way, RAM volatility can sometimes be treated as a feature, not a bug! Have you ever wondered why restarting a computer or process often fixes your problem? The memory volatility forces programmers to implement robust initialization techniques that rebuild the state from backup mediums, enhancing reliability and mitigating potential program bugs. In extreme cases, crash-only software (*https://oreil.ly/DAbDs*) with the restart is the primary way of failure handling.

proper education, our implementations tend to be prone to inefficiencies and unnecessary waste of computer resources or execution time. This problem is amplified by the vast amount of data our programs have to process these days. This is why we often say that efficient programming is all about the data.

Memory Inefficiency Is Usually the Most Common Problem in Go Programs

Go is a garbage collected language, which allows Go to be an extremely productive language. However, the garbage collector (GC) sacrifices some visibility and control over memory management (more on that in "Garbage Collection" on page 185).

But even when we forget about GC overhead, for cases where we need to process a significant amount of data or are under some resource constraints, we have to take more care with how our program uses memory. Therefore, I recommend reading this chapter with extra care since most first-level optimizations are usually around memory resources.

When should we start the memory optimization process? A few common symptoms might reveal that we might have a memory efficiency issue.

Do We Have a Memory Problem?

It's useful to understand how Go uses the computer's main memory and its efficiency consequences, but we must also follow the pragmatic approach. As with any optimizations, we should refrain from optimizing memory until we know there is a problem. We can define a set of situations that should trigger our interest in Go memory usage and potential optimizations in this area:

- Our physical computer, virtual machine, container, or process crashed because of an out-of-memory (OOM) signal, or our process is about to hit that memory limit.[4]
- Our Go program is executing slower than usual, while the memory usage is higher than average. Spoiler: our system might be under memory pressure causing trashing or swapping, as explained in "OS Memory Mapping" on page 168.

4 We can resolve that problem by simply adding more memory to the system or switching to the server (or virtual machine) with more memory resource. That might be a solid solution if we are willing to pay additionally if it's not a memory leak and if such a resource can be increased (e.g., the cloud has virtual machines with more memory). Yet I suggest investigating your program memory usage, especially if you continuously have to expand the system memory. Then there might be easy wins, thanks to trivially wasted space we could optimize.

- Our Go program is executing slower than usual, with high spikes of CPU utilization. Spoiler: allocation or releasing memory slows our programs if an excessive number of short-lived objects is created.

If you encounter any of those situations, it might be time to debug and optimize the memory usage of your Go program. As I will teach you in "Complexity Analysis" on page 240, if you know what you are looking for, a set of early warning signals can indicate huge memory problems that could be avoided easily. Moreover, building such a proactive instinct can make you a valuable team asset!

But we can't build anything without good foundations. As with the CPU resource, you won't be able to apply optimizations without actually understanding them! We have to understand the reasons behind those optimizations. For example, Example 4-1 allocates 30.5 MB of memory for 1 million integers in the input. But what does it mean? Where was that space reserved? Does it mean we used exactly 30.5 MB of physical memory, or more? Was this memory released at some point? This chapter aims to give you awareness, allowing you to answer all of these questions. We will learn why memory is often the issue and what we can do about it.

Let's start with the basics of memory management from the point of view of hardware (HW), operating system (OS), and the Go runtime. Let's start with essential details about physical memory directly impacting our program execution. On top of that, this knowledge might help you better understand the specifications and documentation of modern physical memory!

Physical Memory

We store information digitally in the form of bits, the basic computer storage unit. A bit can have one of two values, 0 or 1. With enough bits, we can represent any information: integer, floating value, letters, messages, sounds, images, videos, programs, metaverses (*https://oreil.ly/il8Tz*), etc.

The main physical memory that we use when we execute our programs (RAM) is based on dynamic random-access memory (DRAM (*https://oreil.ly/hbo59*)). These chips are soldered into modules, often referred to as RAM "sticks." When connected to the motherboard, these chips allow us to store and read data bits as long as the DRAM is continuously powered.

DRAM contains billions of memory cells (as many cells as the number of bits DRAM can store). Each memory cell comprises one access transistor acting as a switch and one storage capacitor. The transistor guards the access to the capacitor, which is charged to the store 1 or drained to keep the 0 value. This allows each memory cell to store a single bit of information. This architecture is much simpler and cheaper to produce and use than Static RAM (SRAM), which is generally faster and used for smaller types of memory like registers and hierarchical caches in the CPU.

At the time of this writing, the most popular memory used for RAM is the simpler, synchronous (clock) version in the DRAM family—SDRAM (*https://oreil.ly/07efG*). Particularly, the fifth generation of SDRAM called DDR4.

Eight bits form a "byte." That number came from the fact that in the past, the smallest number of bits that could hold a text character was eight.[5] The industry standardized a "byte" as the smallest meaningful unit of information.

As a result, most hardware is byte addressable. This means that, from a software programmer's point of view, there are instructions to access data through individual bytes. If you want to access a single bit, you need to access the whole byte and use bitmasks (*https://oreil.ly/pFoxI*) to get or write the bit you want.

The byte addressability makes developer life easier when working with data from different mediums like memory, disk, network, etc. Unfortunately, that creates a certain illusion that the data is always accessible with byte granularity. Don't let that mislead you. More often than not, the underlying hardware has to transfer a much larger chunk of data to give you the desired byte.

For example, in "Hierachical Cache System" on page 127, we learned that CPU registers are typically 64 bits (8 bytes), and the cache line is even bigger (64 bytes). Yet we have CPU instructions that can copy a single byte from memory to the CPU register. However, an experienced developer will notice that to copy that single byte, in many cases, the CPU will fetch not 1 byte but at least a complete cache line (64 bytes) from physical memory.

From a high-level point of view, physical memory (RAM) can also be seen as byte addressable, as presented in Figure 5-1.

Memory space can be seen as a contiguous set of one-byte slots with a unique address. Each address is a number from zero to the total memory capacity in the system in bytes. For this reason, 32-bit systems that use only 32-bit integers for memory addresses typically could not handle RAM with more capacity than 4 GB—the largest number we can represent with 32 bits is 2^{32}. This limitation was removed with the introduction of the 64-bit operating systems that use 64-bit (8-byte)[6] integers for memory addressing.

5 Nowadays, popular encodings like UTF-8 can dynamically use from one up to four bytes of memory per single character.

6 By just doubling the "pointer" size, we moved the limit to how many elements we can address to extreme sizes. We could even estimate that 64-bit is enough to address all grains of sand from all beaches on Earth (*https://oreil.ly/By1J3*)!

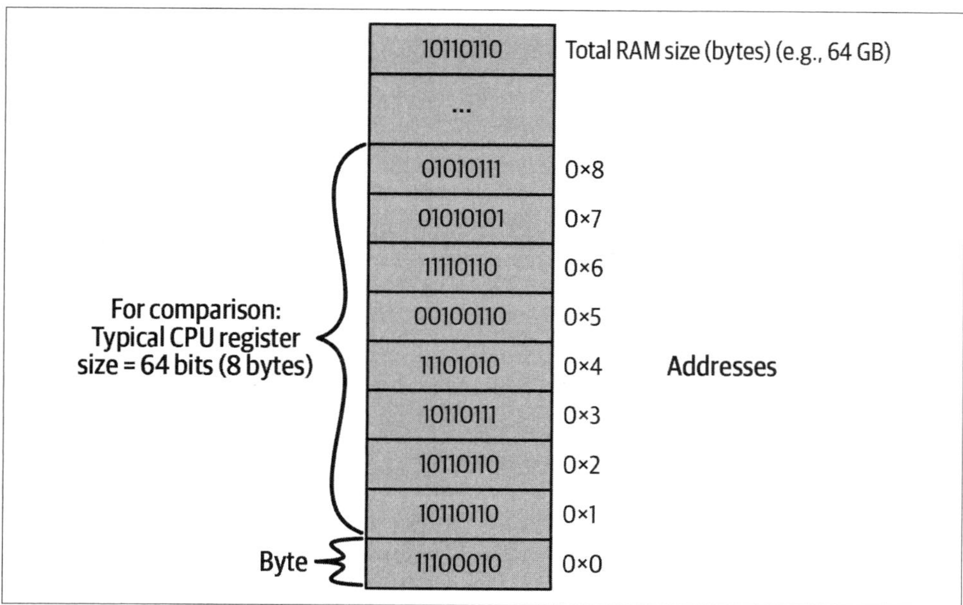

Figure 5-1. Physical memory addresses space

We discussed in "CPU and Memory Wall Problem" on page 126 that memory access is not that fast compared to, for example, CPU speed. But there is more. Addressability, in theory, should allow fast, random access to bytes from the main memory. After all, this is why that main memory is called "random-access memory." Unfortunately, if we look at our napkin math in Appendix A, sequential memory access can be 10 times (or more) faster than random access!

But there is more—we don't expect any improvements in this area in the future. Within the last few decades, we only improved the speed (bandwidth) of the sequential read. We did not improve random access latency at all! The lack of improvement on the latency side is not a mistake. It is a strategic choice—the internal designs of the modern RAM modules have to work against various requirements and limitations, for example:

Capacity
: There is a strong demand for bigger capacities of RAM, e.g., to compute more data or run more realistic games.

Bandwidth and latency
: We want to wait less time to access memory while writing or reading large chunks of data since memory access is the major slowdown for CPU operations.

Voltage
: There is a demand for a lower voltage requirement for each memory chip, which would allow for running more of them while maintaining low power consumption and manageable thermal characteristics (more time on battery for our laptops and smartphones!).

Cost
: RAM is a fundamental piece of the computer required in large quantities; thus, production and usage costs must be kept low.

Slower random access has many implications for the layers of many managers we will learn about in this chapter. For example, this is why the CPU with L-caches fetches and caches bigger chunks of memory up front, even if only one byte is needed for computation.

Let's summarize a few things worth remembering about modern generations of hardware for RAM like DDR4 SDRAM:

- Random access of the memory is relatively slow, and generally, there aren't many good ideas to improve that soon. If anything, lower power consumption, larger capacity, and bandwidth only increase that delay.
- Industry is improving overall memory bandwidth by allowing us to transfer bigger chunks of adjacent (sequential) memory. This means that efforts to align Go data structures and knowing how they are stored in memory matter—ensuring we can access them faster.

Whether sequentially or randomly, our programs never directly access physical memory—the OS manages the RAM space. This is great for developers, as we don't need to understand low-level memory access details. But there are more important reasons why there has to be an OS between our programs and hardware. So let's discuss why and what it means for our Go programs.

OS Memory Management

What are the operating system's goals for memory management? Hiding complexities of physical memory access is only one thing. The other, more important, goal is to allow using the same physical memory simultaneously and securely across thousands of processes and their OS threads.[7] The problem of multiprocess execution on common memory space is nontrivial for multiple reasons:

7 I introduced the *process* and *thread* terms in "Operating System Scheduler" on page 134.

Dedicated memory space for each process
: Programs are compiled assuming nearly full and continuous access to the RAM. As a result, the OS must track which slots from the physical memory from our address space (shown in Figure 5-1) belong to which process. Then we need to find a way to coordinate those "reservations" to the processes so only allocated addresses are accessed.

Avoiding external fragmentation
: Having thousands of processes with dynamic memory usage poses a great risk of waste in memory due to inefficient packing. We call this problem the external fragmentation of memory (*https://oreil.ly/lBfRq*).

Memory isolation
: We have to ensure that no process touches the physical memory address reserved for other processes running on the same machine (e.g., operating system processes!). This is because any accidental write or read from outside of process memory (out-of-bounds memory access) can crash other processes, malform data on persistent mediums (e.g., disk), or crash the whole machine (e.g., if you corrupt the memory used by the OS).

Memory safety
: Operating systems are usually multiuser systems, which means processes can have different permissions to different resources (e.g., files on disk or other process memory space). This is why the mentioned out-of-bounds memory accesses have serious security risks.[8] Imagine a malicious process with no permissions reading credentials from other process memory, or causing a Denial-of-Service (DoS) attack.[9] This is especially important for virtualized environments, where a single memory unit can be shared across different operating systems and even more users.

Efficient memory usage
: Programs never use all the memory they asked for at the same time. For example, instruction code and statically allocated data (e.g., constant variables) can be as large as dozens of megabytes. But for single-threaded applications, a maximum of a few kilobytes of data is used in a given second. Instructions for error handling are rarely used. Arrays are often oversized for worst-case scenarios.

---

8 Many Common Vulnerabilities and Exposures (CVE) issues exist due to various bugs that allow out-of-bounds memory access (*https://oreil.ly/iSbqk*).

9 It might be less intuitive, but the malicious process can perform a DoS if access to another process memory is not restricted. For example, by setting counters to incorrect values or breaking loop invariants, the victim program might error out or exhaust machine resources.

To solve all those challenges, modern OS manages memory using three fundamental mechanisms we will learn about in this section: paged virtual memory, memory mapping, and hardware address translation. Let's start by explaining virtual memory.

Virtual Memory

The key idea behind virtual memory (*https://oreil.ly/RBiCV*) is that every process is given its own logical, simplified view of the RAM. As a result, programming language designers and developers can effectively manage process memory space as if they had an entire memory space for themselves. Even more, with virtual memory, the process can use a full range of addresses from 0 to 2^{64} - 1 for its data, even if the physical memory has, for example, the capacity to accommodate only 2^{35} addresses (32 GB of memory). This frees the process from coordinating the memory among other processes, bin packing challenges, and other important tasks (e.g., physical memory defragmentation, security, limits, and swap). Instead, all of these complex and error-prone memory management tasks can be delegated to the kernel (a core part of the Linux operating system).

There are a few ways of implementing virtual memory, but the most popular technique is called *paging*.[10] The OS divides physical and virtual memory into fixed-size chunks of memory. The virtual memory chunks are called *pages* (*https://oreil.ly/JTWoU*), whereas physical memory chunks are called *frames*. Both pages and frames can be then individually managed. The default page size is usually 4 KB,[11] but it can be changed to larger page sizes with respect to specific CPU capabilities.[12] It is also possible to use 4 KB pages for normal workloads and dedicated (sometimes transparent to processes!) huge pages (*https://oreil.ly/7KuGx*) from 2 MB to 1 GB.

---

10 In the past, segmentation (*https://oreil.ly/8BFmb*) was used to implement virtual memory. This has proven to have less versatility, especially the inability to move this space around for defragmentation (better packing of memory). Still, even with paging, segmentation is applied to virtual memory by the process itself (with underlying paging). Plus, the kernel sometimes still uses nonpaged segmentation for its part of critical kernel memory.

11 You can check the current page size on the Linux system using the `getconf PAGESIZE` command.

12 For example, typically, Intel CPUs are capable of hardware-supported 4 KB, 2 MB, or 1 GB pages (*https://oreil.ly/mxlry*).

The Importance of Page Size

The 4 KB number was chosen in the 1980s, and many say that it's time to bump this number up, given modern hardware and cheaper RAM (in terms of dollars per byte).

Yet the choice of page size is a game of trade-offs. Larger pages inevitably waste more memory space,[13] which is often referred to as the internal memory fragmentation (*https://oreil.ly/PnOuT*). On the other hand, keeping a 4 KB page size or making it smaller makes memory access slower and memory management more expensive, eventually blocking the ability to use larger RAM modules in our computers.

The OS can dynamically map pages in virtual memory to specific physical memory frames (or other mediums like chunks of disk space), mostly transparently to the processes. The mapping, state, permissions, and additional metadata of the page are stored in the page entry in the many hierarchical page tables maintained by the OS.[14]

To achieve an easy-to-use and dynamic virtual memory, we need to have a versatile address translation mechanism. The problem is that only the OS knows about the current memory space mapping between virtual and physical space (or lack of it). Our running program's process only knows about virtual memory addresses, so all CPU instructions in machine code use virtual addresses. Our programs will be even slower if we try to consult the OS for every memory access to translate each address, so the industry figured out dedicated hardware support for translating memory pages.

From the 1980s, almost every CPU architecture started to include the Memory Management Unit (MMU) used for every memory access. MMU translates each memory address referenced by CPU instructions to a physical address based on the OS page table entries. To avoid accessing RAM to search for the relevant page tables, engineers added the Translation Lookaside Buffer (TLB). TLB is a small cache that can cache a few thousand page table entries (typically 4 KB of entries). The overall flow looks like Figure 5-2.

13 Even naive and conservative calculations indicate around 24% of total memory is wasted for 2 MB pages (*https://oreil.ly/iklRd*).

14 We won't discuss the implementation of page tables since it's pretty complex and not something Go developers have to worry about. Yet this topic is quite interesting as the trivial implementation of paging would have a massive overhead in memory usage (what's the point of memory management that would take the majority of memory space it manages?). You can learn more here (*https://oreil.ly/jU9Is*).

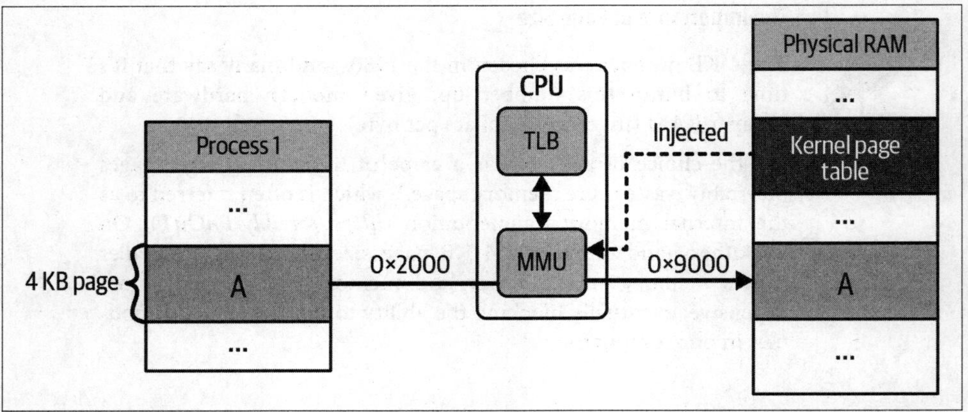

Figure 5-2. Address translation mechanism done by MMU and TLB in CPU. OS has to inject the relevant page tables so MMU knows what virtual addresses correspond to physical addresses.

TLB is very fast, but it has limited capacity. If MMU cannot find the accessed virtual address in the TLB, we have a TLB miss. This means that either the CPU (hardware TLB management) or OS (software-managed TLB) has to walk through page tables in RAM, which causes significant latency (around one hundred CPU clock cycles)!

It is essential to mention that not every "allocated" virtual memory page will have a reserved physical memory page behind it. In fact, most of the virtual memory is not backed up by RAM at all. As a result, we can almost always see large amounts of virtual memory used by the process (called `VSS` or `VSZ` in various Linux tools like `ps`). Still, the actual physical memory (often called `RSS` or `RES` from "resident memory") reserved for this process might be tiny. There are often cases where a single process allocates more virtual memory than is available to the whole machine! See an example situation like this on my machine in Figure 5-3.

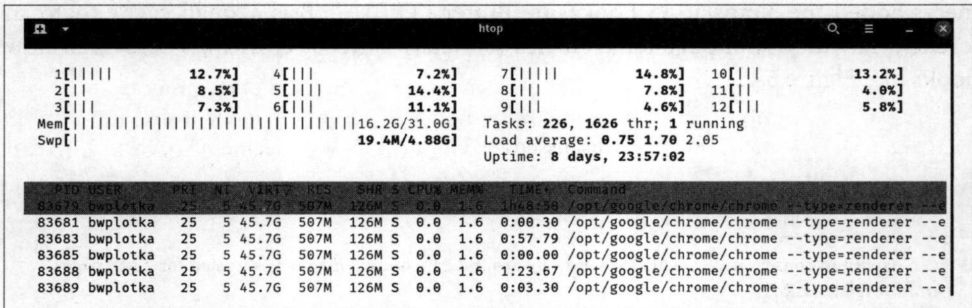

Figure 5-3. First few lines of `htop` output, showing the current usage of a few Chrome browser processes, sorted by virtual memory size

As we can see in Figure 5-3, my machine has 32 GB of physical memory, with 16.2 GB currently used. Yet we see Chrome processes using 45.7 GB of virtual memory each! However, if you look at the `RES` column, it has only 507 MB resident, with 126 MB of it shared with other processes. So how this is possible? How can the process think that it has 45.7 GB of RAM available, given the machine has only 32 GB and the system actually allocated just a few hundred MBs in RAM?

We can call such a situation a memory overcommitment (*https://oreil.ly/wbZGf*), and it exists because of the very same reasons airlines often overbook seats for their flights (*https://oreil.ly/El9iy*). On average, many travelers cancel their trips at the last minute or do not show up for their flight. As a result, to maximize the plane's used capacity, it is more profitable for airlines to sell more tickets than seats in the airplane and handle the rare "out of seats" situations "gracefully" (e.g., by moving the unlucky customer to another flight). This means that the true "allocation" of seats happens when travelers actually "access" them during the flight onboarding process.

The OS performs the same overcommitment strategy by default[15] for processes trying to allocate physical memory. The physical memory is only allocated when our program accesses it, not when it "creates" a big object, for example, `make([]byte, 1024)` (you will see a practical example of this in "Go Allocator" on page 181).

Overcommitment is implemented with the pages and memory mapping techniques. Typically, memory mapping refers to a low-level memory management capability offered with the `mmap` (*https://oreil.ly/m5n7A*) system call on Linux (and the similar `MapViewOfFile` function in Windows).

Developers Can Utilize mmap Explicitly in Programs for Specific Use Cases

The `mmap` call is used extensively in almost every database software, e.g., in MySQL (*https://oreil.ly/o8a5o*) and PostgreSQL (*https://oreil.ly/scByc*) as well as those written in Go, like Prometheus (*https://oreil.ly/2Sa3P*), Thanos (*https://oreil.ly/tFBUf*), and M3db (*https://oreil.ly/Jg3wb*) projects. The `mmap` (among other memory allocation techniques) is also what Go runtime and other programming languages use under the hood to allocate memory from OS, e.g., for the heap (discussed in "Go Memory Management" on page 172).

15 There is also an option to disable an overcommitment mechanism (*https://oreil.ly/h82uS*) on Linux. When disabled, the virtual memory size (VSS) is not allowed to be bigger than the physical memory used by the process (RSS). You might want to do this so the process will have generally faster memory accesses, but the waste of memory is enormous. As a result, I have never seen such an option used in practice.

Using explicit mmap for most Go applications is not recommended. Instead, we should stick to the Go runtime's standard allocation mechanisms, which we will learn in "Go Memory Management" on page 172. As our "Efficiency-Aware Development Flow" on page 102 said, only if we see indications through benchmarking that this is not enough, might we consider moving to more advanced methods like mmap. This is why mmap is not even on my Chapter 11 list!

However, there is a reason why I explain mmap at the start of our journey with the memory resource. Even if we don't use it explicitly, the OS uses the same memory mapping mechanism to manage all allocated pages in our system. The data structures we use in our Go programs are indirectly saved to certain virtual memory pages, which are then mmap-like managed by the OS or Go runtime. As a result, understanding the explicit mmap syscall will conveniently explain the on-demand paging and mapping techniques Linux OS uses to manage virtual memory.

Let's focus on the Linux mmap syscall next.

mmap Syscall

To learn about OS memory mapping patterns, let's discuss the mmap (*https://oreil.ly/m5n7A*) syscall. Example 5-1 shows a simplified abstraction, using mmap OS syscall, that allows allocating a byte slice in our process virtual memory without Go memory management coordination.

Example 5-1. The adapted snippet of Linux-specific Prometheus mmap abstraction (https://oreil.ly/KJ4dD) that allows creating and maintaining read-only memory-mapped byte arrays

```
import (
    "os"

    "github.com/efficientgo/core/errors"
    "github.com/efficientgo/core/merrors"
    "golang.org/x/sys/unix"
)

type MemoryMap struct {
    f *os.File // nil if anonymous.
    b []byte
}

func OpenFileBacked(path string, size int) (mf *MemoryMap, _ error) { ❶
    f, err := os.Open(path)
    if err != nil {
        return nil, err
    }
```

```
    b, err := unix.Mmap(int(f.Fd()), 0, size, unix.PROT_READ, unix.MAP_SHARED) ❷
    if err != nil {
        return nil, merrors.New(f.Close(), err).Err() ❸
    }

    return &MemoryMap{f: f, b: b}, nil
}

func (f *MemoryMap) Close() error {
    errs := merrors.New()
    errs.Add(unix.Munmap(f.b)) ❹
    errs.Add(f.f.Close())
    return errs.Err()
}

func (f *MemoryMappedFile) Bytes() []byte { return f.b }
```

❶ `OpenFileBacked` creates explicit memory mapped backed up by the file from the provided path.

❷ `unix.Mmap` is a Unix-specific Go helper that uses the `mmap` syscall to create a direct mapping between bytes from the file on disk (between 0 and the `size` address) and virtual memory allocated by the returned `[]byte` array in the `b` variable. We also pass the read-only flag (`PROT_READ`) and shared flag (`MAP_SHARED`).[16] We can also skip the passing file descriptor, and pass 0 as the first argument and `MAP_ANON` as the last argument to create anonymous mapping (more on that later).[17]

❸ We use the `merrors` (*https://oreil.ly/lnrJM*) package to ensure the we capture both errors if `Close` also returns an error.

❹ `unix.Munmap` is one of the few ways to remove mapping and de-allocate `mmap`-ed bytes from virtual memory.

The returned byte slice from the open-ed `MemoryMap.Bytes` structure can be read as a regular byte slice acquired in typical ways, e.g., `make([]byte, size)`. However, since we marked this memory-mapped location as read-only (`unix.PROT_READ`), writing to such a slice will cause the OS to terminate the Go process with the `SIGSEGV` reason.[18]

16 `MAP_SHARED` means that any other process can reuse the same physical memory page if it accesses the same file. This is harmless if the mapped file does not change over time, but it has more complex nuances for mapping modifiable content.

17 A full list of options can be found in the `mmap` documentation (*https://oreil.ly/m5n7A*).

18 `SIGSEV` means a segmentation fault. This tells us that the process wants to access an invalid memory address.

Furthermore, a segmentation fault will also happen if we read from this slice after doing `Close` (`Unmap`) on it.

At first glance, the `mmap`-ed byte array looks like a regular byte slice with extra steps and constraints. So what's unique about it? It's best to explain that using an example! Imagine that we want to buffer a 600 MB file in the `[]byte` slice so we can quickly access a couple of bytes on demand from random offsets of that file. The 600 MB might sound excessive, but such a requirement is commonly seen in databases or caches where reading from a disk on demand might be too slow.

The naive solution without an explicit `mmap` could look like Example 5-2. Every few instructions, we will look at what the OS memory statistics told us about the allocated pages on physical RAM.

Example 5-2. Buffering 600 MB from a file to access three bytes from three different locations

```
f, err := os.Open("test686mbfile.out") ❶
if err != nil {
   return err
}

b := make([]byte, 600*1024*1024)
if _, err := f.Read(b); err != nil { ❷
   return err
}

fmt.Println("Reading the 5000th byte", b[5000]) ❸
fmt.Println("Reading the 100 000th", b[100000]) ❸
fmt.Println("Reading the 104 000th byte", b[104000]) ❸

if err := f.Close(); err != nil {
   return err
}
```

❶ We open the 600+ MB file. At this point, if you ran the `ls -l /proc/$PID/fd` (where `$PID` is the process ID of this executed program) command on a Linux machine, you would see file descriptors telling you that this process has used these files. One of the descriptors is a symbolic link to our `test686mbfile.out` file we just opened. The process will hold that file descriptor until the file is closed.

❷ We read 600 MB into a pre-allocated `[]byte` slice. After the `f.Read` method execution, the RSS of the process shows 621 MB.[19] This means that we need over 600 MB of free physical RAM to run this program. The virtual memory size (VSZ) increased too, hitting 1.3 GB.

❸ No matter what bytes we access from our buffer, our program will not allocate any more bytes on RSS for our buffer (however, it might need extra bytes for the `Println` logic).

Generally, Example 5-2 proves that without an explicit `mmap`, we would need to reserve at least 600 MB of memory (~150,000 pages) on physical RAM from the very beginning. We also keep all of them reserved for our process until it is collected by the garbage collection process.

What would the same functionality look like with the explicit `mmap`? Let's do something similar in Example 5-3 using the Example 5-1 abstraction.

Example 5-3. Memory mapping 600 MB from file to access three bytes from three different locations, using Example 5-1

```
f, err := mmap.OpenFileBacked("test686mbfile.out," 600*1024*1024) ❶
if err != nil {
   return err
}
b := f.Bytes() ❷

fmt.Println("Reading the 5000th byte", b[5000]) ❸
fmt.Println("Reading the 100 000th byte", b[100000]) ❹
fmt.Println("Reading the 104 000th byte", b[104000]) ❺

if err := f.Close(); err != nil { ❻
   return err
}
```

❶ We open our test file and memory map 600 MB of its content into the `[]byte` slice. At this point, similar to Example 5-2, we see a related file descriptor for our `test686mbfile.out` file in the *fd* directory. More importantly, however, if you executed the `ls -l /proc/$PID>/map_files` (again, `$PID` is the process ID) command, you would also have another symbolic link to the `test686mbfile.out` file we just referenced. This represents a file-backed memory map.

19 On Linux, you can find this information by doing `ps -ax --format=pid,rss,vsz | grep $PID`, where `$PID` is process ID.

❷ After this statement, we have the byte buffer `b` with the file content. However, if we check the memory statistics for this process, the OS did not allocate any page in physical memory for our slice elements.[20] So the total RSS is as small as 1.6 MB, despite having 600 MB of content accessible in `b`! The VSZ, on the other hand, is around 1.3 GB, which indicates the OS is telling the Go program that it can access this space.

❸ After accessing a single byte from our slice, we see an increase in RSS, around 48–70 KB worth of RAM pages for this mapping. This means that the OS only allocated a few (10 or so) pages on RAM when our code wanted to access a single, concrete byte from `b`.

❹ Accessing a different byte far away from already allocated pages triggers the allocation of extra pages. RSS reading would show 100–128 KB.

❺ If we access a single byte 4,000 bytes away from the previous read, OS does not allocate any additional pages. This might be for a few reasons.[21] For instance, when our program read the file's contents at offset 100,000, the OS already allocated a 4 KB page with the byte we accessed here. Thus RSS reading would still show 100–128 KB.

❻ If we remove the memory mapping, all our related pages will eventually be unmapped from RAM. This means our process total RSS number should be smaller.[22]

20 How do I know? We can have exact statistics for each memory mapping process we use on Linux thanks to the `/proc/<PID>/smaps` file.

21 There are many reasons why accessing nearby bytes might not need allocating more pages on RAM in the memory-mapped situation. For example, the cache hierarchy (discussed in "Hierachical Cache System" on page 127), the OS, and compiler deciding to pull more at once, or such a page being already a shared or private page because of previous accesses.

22 Note that physical frames for this file can still be allocated on physical memory by the OS (just not accounted for our process). This is called `page cache` and can be useful if any process tries to memorize the same file. Page cache is stored as best effort in the memory that would otherwise not be used. It can be released when the system is under high memory pressure or manually by the administrator, e.g., with `sysctl -w vm.drop_caches=1`.

An Underrated Way to Learn More About Your Process and OS Resource Behavior

Linux provides amazing statistics and debugging information for the current process or thread state. Everything is accessible as special files inside */proc/<PID>*. The ability to debug each detailed statistic (e.g., every little memory mapping status) and configuration was eye-opening for me. Learn more about what you can do by reading the proc (*https://oreil.ly/jxBig*) (process pseudofilesystem) documentation.

I recommend getting familiar with the Linux pseudofilesystem or the tools using it if you plan to work more on low-level Linux software.

One of the main behaviors highlighted when we used explicit `mmap` in Example 5-3 is called on-demand paging. When the process asks the OS for any virtual memory using `mmap`, the OS will not allocate any page on RAM, no matter how large. Instead, the OS will only give the process the virtual address range. Further along, when the CPU performs the first instruction that accesses memory from that virtual address range (e.g., our `fmt.Println("Reading the 5000th byte," b[5000])` in Example 5-3), the MMU will generate a page fault. Page fault is a hardware interrupt that is handled by the OS kernel. The OS can then respond in various ways:

Allocate more RAM frames
: If we have free frames (physical memory pages) in RAM, the OS can mark some of them as used and map them to the process that triggered the page fault. This is the only moment when the OS actually "allocates" RAM (and increases the `RSS` metric).

De-allocate unused RAM frames and reuse them
: If no free frame exists (high memory usage on the machine), the OS can remove a couple of frames that belong to file-backed mappings for any process as long as the frames are not currently accessed. As a result, many pages can be unmapped from physical frames before OS has to resort to more brutal methods. Still, this will potentially cause other processes to generate another page fault. If this situation happens very often, our whole OS with all processes will be seriously slowed down (memory trashing situation).

Triggering out-of-memory (OOM) situation
: If the situation worsens and all unused file-backed memory-mapped pages are freed, and we still have no free pages, the OS is essentially out of memory. Handling that situation can be configured in the OS, but generally, there are three options:

- The OS can start unmapping pages from physical memory for memory mappings backed by anonymous files. To avoid data loss, a swap disk partition can be configured (the `swapon --show` command will show you the existence and usage of swap partitions in your Linux system). This disk space is then used to back up virtual memory pages from the anonymous file memory map. As you can imagine, this can cause a similar (if not worse) memory trashing situation and overall system slowdown.[23]
- A second option for the OS is to simply reboot the system, generally known as the system-level OOM crash (*https://oreil.ly/BboW0*).
- The last option is to recover from the OOM situation by immediately terminating a few lower-priority processes (e.g., from the user space). This is typically done by the OS sending the `SIGKILL` signal (*https://oreil.ly/SLWOv*). The detection of what processes to kill varies,[24] but if we want more determinisms, the system administrator can configure specific memory limits per process or group of processes using, for example, `cgroups` (*https://oreil.ly/E72wh*)[25] or `ulimit` (*https://oreil.ly/fF12F*).

On top of the on-demand paging strategy, it's worth mentioning that the OS never releases any frame pages from RAM at the moment of process termination or when it explicitly releases some virtual memory. Only virtual mapping is updated at that point. Instead, physical memory is mainly reclaimed lazily (on demand) with the help of a page frame reclaiming algorithm (PFRA) (*https://oreil.ly/ruKUM*) that we won't discuss in this book.

Generally, the `mmap` syscall might seem complex to use and understand. Yet, it explains what it means when our program allocates some RAM by asking the OS. Let's now compose what we learned into the big picture of how the OS manages the RAM and talk about the consequences we developers might observe when dealing with a memory resource.

OS Memory Mapping

The explicit memory mapping presented in Example 5-3 is just one example of the possible OS memory mapping techniques. Besides, rare file-backed mapping and advanced off-heap solutions, there is almost no need to explicitly use such `mmap` syscalls in our Go programs. However, to manage virtual memory efficiently, the OS is transparently

23 Swapping is usually turned off by default on most machines.

24 "Teaching the OOM killer" (*https://oreil.ly/AFDh0*) explains some problems in choosing what process to kill first. The lesson here is that the global OOM killer is often hard to predict (*https://oreil.ly/4rPzk*).

25 Exact implementation of memory controller can be found here (*https://oreil.ly/Ken3G*).

using the same technique of page memory mapping for nearly all the RAM! The example memory mappings situation is presented in Figure 5-4, which pulls into one graphic a few common page mapping situations we could have in our machine.

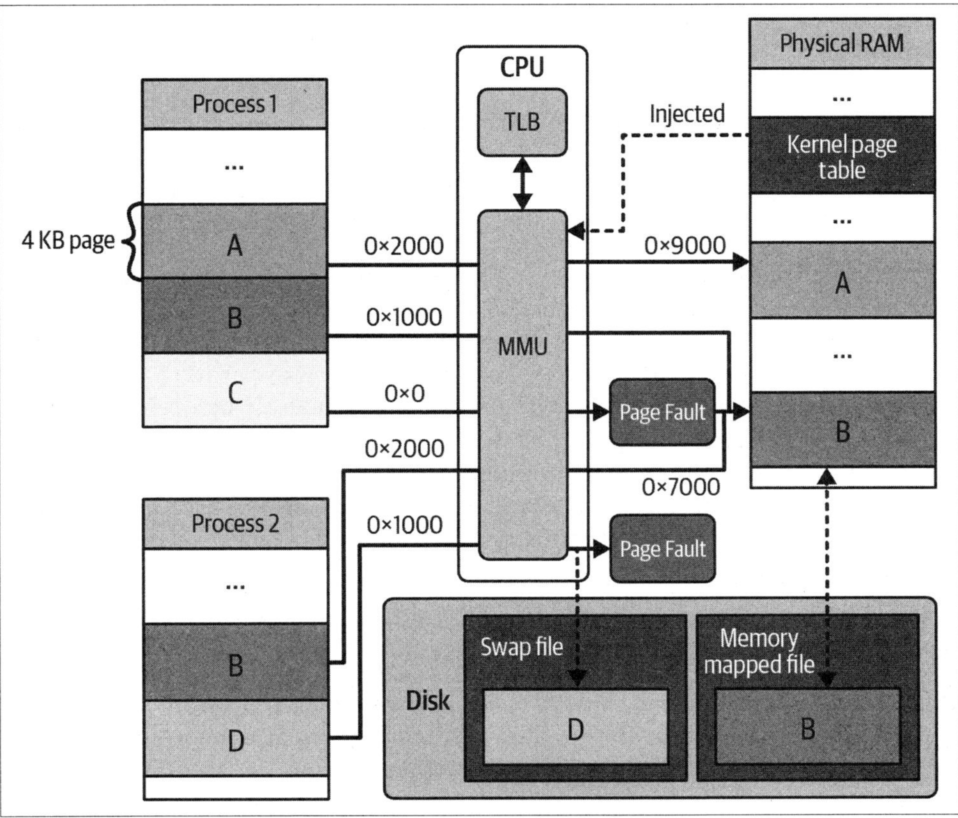

Figure 5-4. Example MMU translation of a few memory pages from the virtual memory of two processes

The situation in Figure 5-4 might look complicated, but we have already discussed some of those cases. Let's enumerate them from the perspective of Process 1 or 2:

Page `A`
: Represents the simplest case of *anonymous file mapping* that has already mapped the frame on RAM. So, for example, if Process 1 writes or reads a byte from an address between `0x2000` and `0x2FFF` in its virtual space, the MMU will translate the address to RAM physical address `0x9000`, plus the required offset. As a result, the CPU will be able to fetch or write it as a cache line to its L-caches and desired register.

Page `B`
: Represents a *file-based memory page* mapped to a physical frame like we created in Example 5-3. This frame is also shared with another process since there is no need to keep two copies of the same data as both mappings map to the same file on a disk. This is only allowed if the mapping is not set as `MAP_PRIVATE`.

Page `C`
: This is an anonymous file mapping that wasn't yet accessed. For example, if Process 1 writes a byte to an address between `0x0` and `0xFFF`, a page fault hardware interrupt is generated by the CPU, and the OS will need to find a free frame.

Page `D`
: This is an anonymous page like `C`, but some data was already written on it. Yet the OS seems to have `swap` enabled and unmaps it from RAM because this page was not used for a long time by Process 2, or the system is under memory pressure. The OS backed the data to swap files in the swap partition to avoid data loss. Process 2 accessing any byte from a virtual address between `0x1000` and `0x1FFF` would result in a page fault, which will tell the OS to find a free frame on RAM and read page `D` content from the swap file. Only then can data be available to Process 2. Note that such swap logic for anonymous pages is disabled by default on most operating systems.

You should now have a clearer view of OS memory management basics and virtual memory patterns. So let's now go through a list of important consequences those pose on Go (and any other programming language):

Practically speaking, observing the size of virtual memory is never useful.
: On-demand paging is why we always see larger virtual memory usage (represented by virtual set size, or VSS) than resident memory usage (RSS) for a process (e.g., the browser memory usage in Figure 5-3). While the process thinks that all pages it sees on the virtual address space are in RAM, most of them might be currently unmapped and stored on disk (mapped file or swap partition). In most cases, you can ignore (*https://oreil.ly/u9l5k*) the VSS metric when assessing the amount of memory your Go program uses.

It is impossible to tell precisely how much memory a process (or system) has used in a given time.
: What metric can we use if the VSS metric does not help assess process memory usage? For Go developers interested in the memory efficiency of their programs, knowing the current and past memory usage is essential information. It tells how efficient our code is and if our optimizations work as expected.

: Unfortunately, because of the on-demand paging and memory mapping behavior we learned in this section, this is currently very hard—we can only roughly estimate. We will discuss the best available metrics in "Memory Usage" on page

234, but don't be surprised if the RSS metric shows a few kilobytes or even megabytes more or less than you expected.

OS memory usage expands to all available RAM.
: Due to lazy release and page caches, even if our Go process released all memory, sometimes the RSS will still look very high if there's generally low memory pressure on the system. This means that there's enough physical RAM to satisfy the rest of the processes, so the OS doesn't bother to release our pages. This is often why the RSS metric is not very reliable, as discussed in "Memory Usage" on page 234.

Tail latency of our Go program memory access is much slower than just physical DRAM access latency.
: There is a high price to pay for using OS with virtual memory. In the worst cases, already slow memory access caused by DRAM design (mentioned in "Physical Memory" on page 153) is even slower. If we stack up things that can happen, like TLB miss, page fault, looking for a free page, or on-demand memory loading from disk, we have extreme latency, which can waste thousands of CPU cycles. The OS does as much as possible to ensure those bad cases rarely happen, so the amortized (average) access latency is as low as possible.

  As Go developers, we have some control to reduce the risk of those extra latencies happening more often. For example, we can use less memory in our programs or prefer sequential memory access (more on that later).

High usage of RAM might cause slow program execution.
: When our system executes many processes that want to access large quantities of pages close to RAM capacity, memory access latencies and OS cleanup routines can take most of the CPU cycles. Furthermore, as we discussed, things like memory trashing, constant memory swaps, and page reclaim mechanisms will slow the whole system. As a result, if your program latency is high, it is not necessarily doing too much work on the CPU or executing slow operations (e.g., I/O), it might just use a lot of the memory!

Hopefully, you understand the impact of OS memory management on how we should think about the memory resource. As in "Physical Memory" on page 153, I only explained the basics of memory management. This is because the kernel algorithms evolve, and different OSes manage memory differently. The information I provided should give you a rough understanding of the standard techniques and their consequences. Such a foundation should also give you a kick-start toward learning more from materials like *Understanding the Linux Kernel* (*https://oreil.ly/Wr1nY*) by Daniel P. Bovet and Marco Cesati (O'Reilly) or LWN.net (*https://lwn.net*).

With that knowledge, let's discuss how Go has chosen to leverage the memory functionalities the OS and hardware offer. It should help us find the right optimizations to try in our TFBO flow if we have to focus on the memory efficiency of our Go program.

Go Memory Management

The programming language task here is to ensure that developers who write programs can create variables, abstractions, and operations that use memory safely, efficiently, and (ideally) without fuss! So let's dig into how the Go language enables that.

Go uses a relatively standard internal process memory management pattern that other languages (e.g., C/C++) share, with some unique elements. As we learned in "Operating System Scheduler" on page 134, when a new process starts, the operating system creates various metadata about the process, including a new dedicated virtual address space. The OS also creates initial memory mappings for a few starting segments based on information stored in the program binary. Once the process starts, it uses `mmap` or `brk/sbrk` (*https://oreil.ly/31emh*)[26] to dynamically allocate more pages on virtual memory when needed. An example organization of the virtual memory in Go is presented in Figure 5-5.

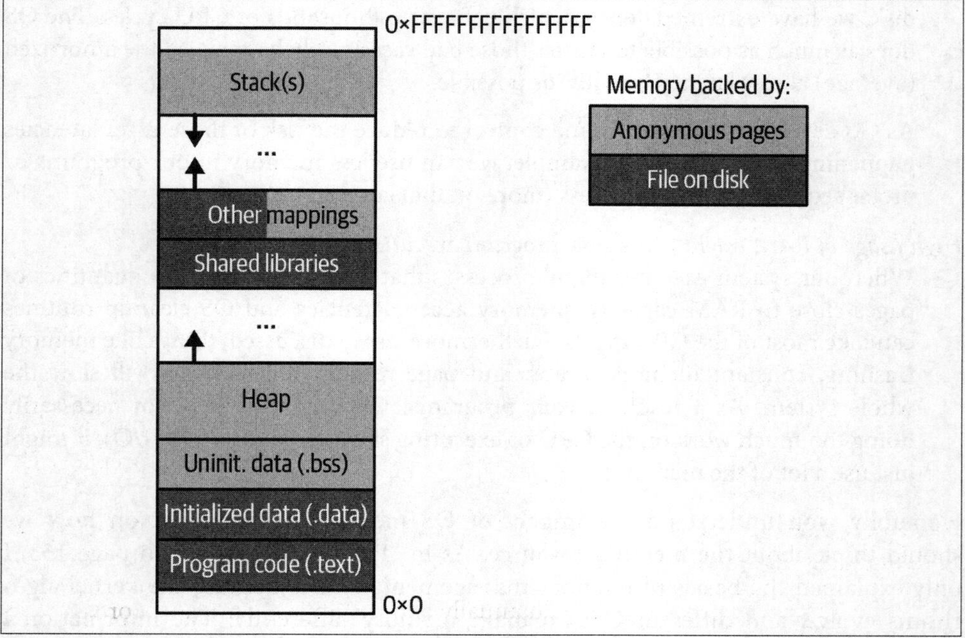

Figure 5-5. Memory layout of an executed Go program in virtual address space

26 Remember, whatever type or amount of virtual memory the OS is giving to the process, it uses the memory mapping technique. `sbrk` allows simpler resizing of the virtual memory section typically covered by the heap. However, it behaves like any other `mmap` using anonymous pages.

We can enumerate a couple of common sections:

`.text`, `.data`, *and shared libraries*
: Program code and all global data like global variables are automatically memory mapped by the OS when the process starts (whether it takes 1 MB or 100 GB of virtual memory). This data is read-only, backed up by the binary file. Additionally, only a small contiguous part of the program is executed at a time by the CPU so that the OS can keep a minimal amount of pages with code and data in the physical memory. Those pages are also heavily shared (more processes are started using the same binary, plus some dynamically linked shared libraries).

Block starting symbol (`.bss`)
: When OS starts a process, it also allocates anonymous pages for uninitialized data (`.bss`). The amount of space used by `.bss` is known in advance—for example, the `http` package defines the `DefaultTransport` (*https://oreil.ly/7m0Wv*) global variable. While we don't know the value of this variable, we know it will be a pointer, so we need to prepare eight bytes of memory for it. This type of memory allocation is called static allocation. This space is allocated once, backed by anonymous pages, and is never freed (from virtual memory at least; if swapping is enabled, it can be unmapped from RAM).

Heap
: The first (and probably the most important) dynamic segment in Figure 5-5 is the memory reserved for dynamic allocations, typically called the *heap* (do not confuse it with the data structure (*https://oreil.ly/740nv*) with the same name). Dynamic allocations are required for program data (e.g., variables) that have to be available outside a single function scope. As a result, such allocations are unknown in advance and must be stored in memory for an unpredictable time. When the process starts, the OS prepares the initial number of anonymous pages for the heap. After that, the OS gives the process some control over that space. It can then increase or decrease its size using the `sbrk` syscall or by preparing or removing extra virtual memory using the `mmap` and `unmmap` syscalls. It's up to the process to organize and manage the heap in the best possible way, and different languages do that differently:

  - C forces the programmer to manually allocate and free memory for variables (using `malloc` and `free` functions).
  - C++ adds smart pointers like `std::unique_ptr` (*https://oreil.ly/QS9zj*) and `std::shared_ptr` (*https://oreil.ly/QbQqQ*), which offer simple counting mechanisms to track the object lifecycle (reference counting).[27]

27 Of course no one blocks anyone from implementing external garbage collection on top of those mechanisms in C and C++.

- Rust has a powerful memory ownership mechanism (*https://oreil.ly/MajFo*), but it makes programming much more difficult for nonmemory critical code areas.[28]
- Finally, languages like Python, C#, Java, and others implement advanced heap allocators and garbage collector mechanisms. Garbage collectors periodically check if any memory is unused and can be released.

  In this sense, Go is closer to Java with memory management than C. Go implicitly (transparently to the programmer) allocates memory that requires dynamic allocation on the heap. For that purpose, Go has its unique components (implemented in Go and Assembly); see "Go Allocator" on page 181 and "Garbage Collection" on page 185.

Most of the Time, It's Enough to Optimize the Heap Usage

Heap is the memory that usually stores the largest amounts of data in physical memory pages. It is so significant that it's enough to look at the heap size to assess the Go process memory usage in most cases. On top of that, the overhead of heap management with runtime garbage collection is significant too. Both make the heap our first choice to analyze when optimizing memory use.

Manual process mappings

Both Go runtime and the developer writing Go code can manually allocate additional memory-mapped regions (e.g., using our Example 5-1 abstraction). Of course, it's up to the process what kind of memory mapping to use (private or shared, read or write, anonymous or file backed), but all of them have a dedicated space in the process's virtual memory, presented in Figure 5-5.

Stack

The last section of the Go memory layout is reserved for function stacks. The stack is a simple yet fast structure allowing accessing values in last in, first out (LIFO) order. Programming languages use them to store all the elements (e.g., variables) that can use automatic allocation. As opposed to dynamic allocations fulfilled by the heap, automatic allocations work well for local data like local variables, function input, or return arguments. Allocations of those elements can be "automatic" because the compiler can deduce their lifespan before the program starts.

---

28 It's hard that the ownership model in Rust requires the programmer to be hyperaware of every memory allocation and what part owns it. Despite that, I am a huge fan of the Rust ownership model if we could scope this memory management only to a certain part of our code. I believe it would be beneficial to bring some ownership pattern to Go, where a small amount of code could use that, whereas the rest would use GC. Wish list for someday? :)

Some programming languages might have a single stack or a stack per thread. Go is a bit unique here. As we learned in "Go Runtime Scheduler" on page 138, the Go execution flow is designed around goroutines. Thus Go maintains a single dynamically sized stack per Go routine. This might even mean hundreds of thousands of stacks (*https://oreil.ly/zrqhj*). Whenever the goroutine invokes another function, we can push its local variables and arguments to stack in a stack frame. We can pop those elements (de-allocate the stack frame) from the stack when we leave the function. If stack structures require more space than what's reserved in virtual memory, Go will ask the OS for more memory attributed to the stack segment, e.g., via the `mmap` syscall.

Stacks are incredibly fast as there is no extra overhead to figure out when memory used by certain elements must be removed (no usage tracking). Thus ideally, we write our algorithms so that they allocate primarily on the stack instead of the heap. Unfortunately, this is impossible in many cases due to stack limitations (we can't allocate too-large objects) or when the variable has to live longer than the function's scope. Therefore, the compiler decides which data can be allocated automatically (on the stack) and which must be allocated dynamically (on the heap). This process is called escape analysis, which you saw in Example 4-3.

All the mechanisms discussed (except manual mappings) are helping Go developers. We don't need to care where and how we should allocate memory for our variables. That is a huge win—for example, when we want to make some HTTP calls, we simply create an HTTP client using a standard library, e.g., with the `client := http.Client{}` code statement. As a result of Go's memory design, we can immediately start using `client`, focusing on our code's functionality, readability, and reliability. In particular:

- We don't need to ensure that the OS has a free virtual memory page to hold the `client` variable. Likewise, we don't need to find a valid segment and virtual address for it. Both will be done automatically by the compiler (if the variable can be stored on the stack) or runtime allocator (dynamic allocation on the heap).
- We don't need to remember to release memory kept by the `client` variable when we stop using it. Instead, suppose the `client` would go beyond code reach (nothing references it). In that case, the data in Go will be released—immediately when stored on the stack or in the next garbage collection execution cycle if stored on the heap (more on that in "Garbage Collection" on page 185).

  Such automation is much less error-prone to potential memory leaks ("I forgot to release memory for `client`") or dangling pointers ("I released memory for `client`, but actually some code still uses it").

Generally, we don't need to care what segment is used for our objects for everyday use of the Go language.

> How do I know whether a variable is allocated on the heap or the stack? From a correctness standpoint, you don't need to know. Each variable in Go exists as long as there are references to it. The storage location chosen by the implementation is irrelevant to the semantics of the language.
>
> The storage location does have an effect on writing efficient programs.
>
> —The Go Team, "Go: Frequently Asked Questions (FAQ)" (*https://oreil.ly/UUGgI*)

However, since allocations are so effortless, there is a risk of not noticing the memory waste.

Transparent Allocations Mean There Is a Risk of Overdoing Them

Allocations are implicit in Go, making coding much easier, but there are trade-offs. One is around memory efficiency: if we don't see explicit memory allocations and releases, it's easier to miss apparent high memory usage in our code.

It's similar to going shopping with cash versus a credit card. You will likely overspend with a credit card than with cash since you don't see that money flowing. With a credit card, money spent is almost transparent to us—it is the same with allocations in Go.

To sum up, Go is a very productive language because, when programming, we don't need to worry about where and how the data held by our variables and abstractions is stored. Yet sometimes when our measurements indicate efficiency problems, it's useful to have a basic awareness of the parts of our program that might allocate some memory, how this occurs, and how the memory is released. So let's uncover that.

Values, Pointers, and Memory Blocks

Let's get this straight before we start—you don't need to know what type of statements trigger memory allocation, where (on a stack or heap), and how much memory was allocated. But, as you will learn in Chapters 7 and 9, many robust tools can tell us all that accurately and quickly. In most cases, we can find what code line and roughly how much was allocated within seconds. Thus, there is generally a common theme: we should not guess that information (since humans tend to guess wrong) because there are tools for that.

This is generally true, but there is no harm in building some basic allocation awareness. On the contrary, it might make us more effective while using those tools to analyze memory usage. The aim is to build a healthy instinct for what pieces of code can potentially allocate the suspicious amount of memory and where we need to be careful.

Many books try to teach this by listing examples of common statements that allocate. This is great, but it's a bit like giving someone a fish instead of a fishing rod (*https://oreil.ly/utQIG*). So again, it's helpful, but only for "common" statements. Ideally, I want you to understand the underlying rules for why something allocates.

Let's dive into how we reference objects in Go to start noticing that allocation more quickly. Our code can perform certain operations on objects stored in some memory. Therefore, we must link those objects to operations, and we typically do that via variables. We describe those variables using Go's type system to make it even easier for the compiler and developers.

However, Go is value oriented (*https://oreil.ly/lgy2S*) rather than reference oriented (like many managed runtime (*https://oreil.ly/ben85*) languages). This means that Go variables never reference objects. Instead, the variables always store the whole *value* of the object. There is no exception to this rule!

To understand this better, the memory representation of three variables is shown in Figure 5-6.

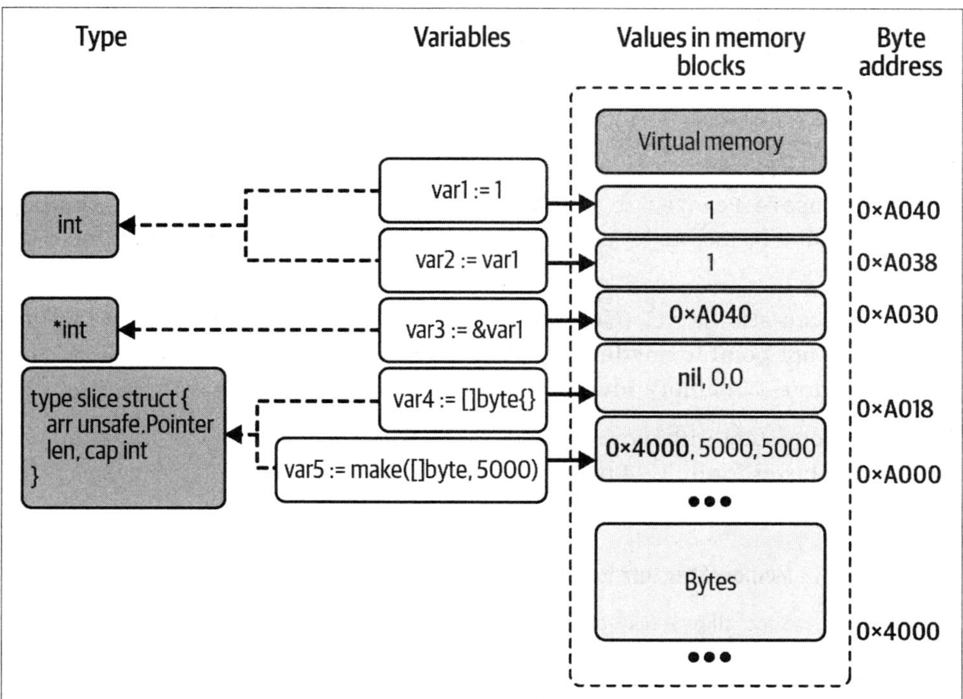

Figure 5-6. Representation of three variables allocated on the process's virtual memory

Think About Variables as Boxes Holding Values

Whenever the compiler sees a definition of the `var` variable or function arguments (including parameters) in the invocation scope, it allocates a contiguous "memory block" for a box. The box is big enough to contain the whole value of the given type. For example, `var var1 int` and `var var2 int` will need a box for eight bytes.[29]

Thanks to our available space in "boxes," we can copy some values. In Figure 5-6, we can copy an integer `1` to `var1`. Now, Go does not have reference variables, so even if we assign the `var1` value to another box named `var2`, this is yet another box with unique space. We can confirm that by printing `&var1` and `&var2`. It should print `0xA040` and `0xA038`, respectively. As a result, a simple assignment is always a copy, which adds latency proportional to the value's size.

> Unlike C++, each variable defined in a Go program occupies a unique memory location. It is not possible to create a Go program where two variables share the same storage location in memory. It is possible to create two variables whose contents point to the same storage location, but that is not the same thing.
>
> —Dave Cheney, "There Is No Pass-By-Reference in Go" (*https://oreil.ly/iPu5w*)

The `var3` box is a pointer to the integer type. A "pointer" variable is a box that stores the value representing the memory address. The type of memory address is just `uintptr` or `unsafe.Pointer`, so simply a 64-bit unsigned integer that allows pointing to another value in memory. As a result, any pointer variable needs a box for eight bytes.

The pointer can also be `nil` (Go's NULL value), a special value indicating that the pointer does not point to anything. In Figure 5-6, we can see that the `var3` box contains a value too—a memory address of the `var1` box.

This is also consistent with more complex types. For example, both `var var4` and `var var5` require boxes for only 24 bytes. This is because the `slice` struct value has three integers.

Memory Structure for Go Slice

Slice allows easy dynamic behavior of the underlying array of a given type. A slice data structure requires a memory block that can hold `length`, `capacity`, and `pointer` to the desired array.[30]

---

29 You can reveal the box size with the `unsafe.Sizeof` (*https://oreil.ly/QtpSf*) function.

30 See the handy `reflect.SliceHeader` (*https://oreil.ly/9unR4*) struct that represents a slice.

Generally, the slice is just a more complex struct. You can think about a struct as a cabinet—it is full of drawers (struct fields) that are simply boxes that share a memory block with other drawers in the same cabinet. So, for example, the `slice` type has three drawers. One of them is of pointer type.

There are two special behaviors of `slice` and a few other special types:

- You can use the `make` (*https://oreil.ly/Mlx6Q*) built-in function that only works for `map`, `chan`, and `slice` types. It returns the type's value[31] and allocates underlying structures, like an array for slices, a buffer for channels, and a hashmap for maps.
- We can put `nil` into boxes of types, like `func`, `map`, `chan`, or `slice`, although they are not strictly pointers, e.g., `[]byte(nil)`.

One drawer of the `var4` and `var5` cabinets is a type of pointer that holds the memory address. Thanks to `make([]byte, 5000)` in `var5`, it points to another memory block containing a 5,000-element byte array.

Structure Padding

The slice structure with three 64-bit fields requires a 24-byte long memory block. But the memory block size for a structure type is not always the sum of the size of its fields!

Smart compilers like in Go might attempt to align type sizes to the typical cache lines or the OS or internal Go allocator page sizes. For this reason, Go compilers sometimes add padding between fields.[32]

To reinforce that knowledge, let's ask a common question when designing a new function or method: should my arguments be pointers of values? Of course, the first thing we should answer is obviously, if we want the caller to see the modifications of that value. But there is an efficiency aspect as well. Let's discuss the difference in Example 5-4, assuming we don't need to see modifications of those arguments from outside.

---

31 Technically speaking, the type `map` variable is a pointer to the hashmap. However, to avoid always typing `*map`, the Go team decided to hide that detail (*https://oreil.ly/mfwDa*).

32 We won't cover struct padding (*https://oreil.ly/1gx5O*) in this edition. There is also an amazing utility that helps you to notice the waste introduced by struct misalignment (*https://oreil.ly/WtYFZ*).

Example 5-4. Different arguments highlight the differences using values, pointers, and special types like `slice`

```
func myFunction(
    arg1 int, arg2 *int, ❶
    arg3 biggie, arg4 *biggie, ❷
    arg5 []byte, arg6 *[]byte, ❸
    arg7 chan byte, arg8 map[string]int, arg9 func(), ❹
) {
   // ...
}

type biggie struct { ❷
    huge [1e8]byte
    other *biggie
}
```

❶ Function arguments are like any newly declared variable: boxes. So for `arg1`, it will create an eight-byte box (most likely allocate it on the stack) and copy the passed integer during the `myFunction` invocation. For `arg2`, it will create a similar eight-byte box that will copy the pointer instead.

For such simple types, avoiding the pointer makes more sense if you don't need to modify the value. You use the same amount of memory and the same copying overhead. The only difference is that the value pointed to by `arg2` has to live on the heap, which is more expensive and, in many cases, can be avoided.

❷ The rule is the same for custom `struct` arguments, but the size and copying overhead might matter more. For example, `arg3` is of `biggie struct`, which is of extraordinary size. Because of the static array with 100 million elements, the type requires a ~100 MB memory block.

For bigger types like this, we should consider using a pointer when passing through functions. This is because every `myFunction` invocation will allocate 100 MB on the heap for the `arg3` box (it's too large to be on the stack)! On top of that, it will spend CPU time copying large objects between boxes. So, `arg4` will allocate eight bytes on the stack (and copy only that) and point to memory on the heap with the `biggie` object, which can be reused across function calls.

Note that despite `biggie` being copied in `arg3`, the copy is *shallow*, i.e., `arg3.other` will share a memory with the previous box!

❸ The `slice` type behaves like the `biggie` type. We must remember the underlying `struct` type of the slice (*https://oreil.ly/Tla4w*).

As a result, `arg5` will allocate a 24-byte box and copy three integers. In contrast, `arg6` will allocate an eight-byte box and copy only one integer (pointer). From the efficiency point of view, it does not matter. It only matters if we want to expose modifications of the underlying array (both `arg5` and `arg6` allow that) or if we want to also expose changes to the `pointer`, `len`, and `cap` fields as `arg6` allows.

4 Special types like `chan`, `map`, and `func()` can be treated similarly to pointers. They share memory through the heap, and the only cost is to allocate and copy the pointer value into `arg7`, `arg8`, or `arg9` boxes.

The same decision flow can be applied to decide about pointer versus value types for:

- Return arguments
- The `struct` fields
- Elements of map, slice, or channels
- The method receiver (e.g., `func (receiver) Method()`)

Hopefully, the preceding information will give you an understanding of which Go code statements allocate memory and roughly how much. Generally:

- Every variable declaration (including function arguments, return arguments, and method receiver) allocates the whole type or just a pointer to it.
- `make` allocates special types and their underlying (pointed) structures.
- new(<type>) is the same as &<type>, so it allocates a pointer box and the type on the heap in the separate memory block.

Most program memory allocations are only known in runtime; thus, dynamic allocation (in a heap) is needed. Therefore, when we optimize memory in Go programs, 99% of the time we just focus on the heap. Go comes with two important runtime components: Allocator and GC, responsible for heap management. Those components are nontrivial pieces of software that often introduce certain waste in terms of extra CPU cycles by the program runtime and some memory waste. Given its nondeterministic and nonimmediate memory release nature, it's worth discussing this in detail. Let's do that in the next two sections.

Go Allocator

It's far from easy to manage the heap, as it poses similar challenges as the OS has toward physical memory. For example, the Go program runs multiple goroutines, and each wants a few (dynamically sized!) segments of the heap memory for a different amount of time.

The Go Allocator is a piece of internal runtime Go code maintained by the Go team. As the name suggests, it can dynamically (in runtime) allocate the memory blocks required to operate on objects. In addition, it is optimized to avoid locking and fragmentation, and to mitigate slow syscalls to the OS.

During compilation, the Go compiler performs a complex stack escape analysis to detect if the memory for objects can be automatically allocated (mentioned in Example 4-3). If yes, it adds appropriate CPU instructions that store related memory blocks in the stack segment of the memory layout. However, in most cases the compiler can't avoid putting most of our memory on the heap. In these cases, it generates different CPU instructions invoking the Go Allocator code.

The Go Allocator is responsible for bin packing (*https://oreil.ly/l27Jv*) the memory blocks in the virtual memory space. It also asks for more space from the OS if needed using `mmap` with private, anonymous pages, which are initialized by zero.[33] As we learned in "OS Memory Mapping" on page 168, those pages are also allocated on the physical RAM only when accessed.

Generally, the Go developer can live without learning details about Go Allocator internals. However, it's enough to remember that:

- It is based on a custom Google `C++ malloc` implementation called TCMalloc (*https://oreil.ly/AZ5S7*).
- It is OS virtual memory page aware, but it operates with 8 KB pages.
- It mitigates fragmentation by allocating memory blocks to certain spans that hold one or multiple 8 KB pages. Each span is created for class memory block sizes. For example, in Go 1.18, there are 67 different size classes (*https://oreil.ly/tMlnv*) (size buckets), the largest being 32 KB.
- Memory blocks for objects that do not contain a pointer are marked with the `noscan` type, making it easier to track nested objects in the garbage collection phase.
- Objects with over 32 KB memory block (e.g., 600 MB byte array) are treated specially (allocated directly without span).
- If runtime needs more virtual space from OS for the heap, it allocates a bigger chunk of memory at once (at least 1 MB), which amortizes the latency of the syscall.

All of the preceding points are constantly changing, with the open source community and Go team adding various small optimizations and features.

33 This is one of the reasons why in Go, every new structure has defined zero value or nil at the start, instead of random value.

They say one code snippet is worth a thousand words, so let's visualize and explain some of these allocation characteristics caused by a mix of Go, OS, and hardware using an example. Example 5-5 shows the same functionality as Example 5-3, but instead of explicit `mmap`, we will rely on Go memory management and no underlying file.

Example 5-5. Allocation of a large `[]byte` slice followed by different access patterns

```
b := make([]byte, 600*1024*1024) ❶
b[5000] = 1
b[100000] = 1
b[104000] = 1 ❷
for i := range b { ❸
    b[i] = 1
}
```

❶ The `b` variable is declared as a `[]byte` slice. The following `make` statement is tasked to create a byte array with 600 MB of data (~600 million elements in the array). This memory block is allocated on the heap.[34]

If we would analyze this situation closely, the Go Allocator seemed to create three contiguous anonymous mappings for that slice with different (virtual) memory sizes: 2 MB, 598 MB, and 4 MB. (The total size is usually bigger than the requested 600 MB because of the Go Allocator internal bucketed algorithm.) Let's summarize the interesting statistics:

- The RSS for three memory mappings used by our slice: 548 KB, 0 KB, and 120 KB (much lower than VSS numbers).
- Total RSS of the whole process shows 21 MB. Profiling shows that most of this comes from outside the heap.
- Go reports 600.15 MB of the heap size (despite RSS being significantly lower).

❷ Only after we start accessing the slice elements (either by writing or reading) will the OS start reserving actual physical memory surrounding those elements. Our statistics:

- The RSS for three memory mappings: 556 KB, (still) 0 KB, and 180 KB (only a few KB more than before accessing).
- Total RSS still shows 21 MB.

34 We know that because `go build -gcflags="-m=1" slice.go` outputs the `./slice.go:11:11: make([]byte, size) escapes to heap` line.

- Go reports 600.16 MB of the heap size (actually a few KB more, probably due to background goroutines).

❸ After we loop over all elements to access it, we will see that the OS mapped on demand all pages for our b slice in physical memory. Our statistics prove this:

- The RSS for three memory mappings: 1.5 MB, (fully mapped) 598 MB, and 1.2 MB.
- Total RSS of the whole process shows 621.7 MB (finally, same as heap size).
- Go reports the same 600.16 MB of the heap size.

This example might feel similar to Examples 5-2 and 5-3, but it's a bit different. Notice that in Example 5-5, there is no (explicit) file involved that could store some data if the page is not mapped. We also utilize the Go Allocator to organize and manage different anonymous page mappings most efficiently, whereas in Example 5-3, the Go Allocator is unaware of that memory usage.

Internal Go Runtime Knowledge Versus OS Knowledge

The Go Allocator tracks certain information we can collect through different observability mechanisms discussed in Chapter 6.

Be mindful when using those. In the preceding example, we saw that the heap size tracked by the Go Allocator was significantly larger than the actual amount of memory used on physical RAM (RSS)![35] Similarly, the memory used by explicit `mmap`, as in Example 5-3, is not reflected in any Go runtime metrics. This is why it's good to rely on more than one metric on our TFBO journey, as discussed in "Memory Usage" on page 234.

The behavior of Go heap management backed up by on-demand paging tends to be indeterministic and fuzzy. We cannot control it directly either. For instance, if you tried to reproduce Example 5-5 on your machine, you would most likely observe slightly different mappings, more or less different RSS numbers (with a tolerance of few MBs), and different heap sizes. It all depends on the Go version you build a program with, the kernel version, the RAM capacity and model, and the load on your system. This poses important challenges to the assessment step of our TFBO process, which we will discuss in "Reliability of Experiments" on page 256.

35 This behavior was often leveraged by more advanced memory ballasting, which generally is less needed after Go 1.19 introduced the memory soft limit discussed in "Garbage Collection" on page 185.

Don't Be Bothered by a Small Memory Increase

Don't try to understand where every hundred bytes or kilobytes of your process RSS memory came from. In most cases, it is impossible to tell or control at that low level. Heap management overhead, speculative page allocations by both the OS and the Go Allocator, dynamic OS mapping behavior, and eventual memory collection (we will learn about that in the next section) make things indeterministic on such a "micro" kilobyte level.

Even if you spot some pattern in one environment, it will be different in others unless we talk about bigger numbers like hundreds of megabytes or more!

The lesson here is that we have to adjust our mindsets. There will always be a few unknowns. What matters is to understand bigger unknowns that contribute the most to the potentially too-high memory usage situation. Together with this allocator awareness, you will learn how to do that in Chapters 6 and 9.

So far, we have discussed how to efficiently reserve memory for our memory blocks through the Go Allocator and how to access it. However, we can't just reserve more memory indefinitely if there is no logic for removing the memory blocks our code doesn't need anymore. That's why it's critical to understand the second part of heap management responsible for releasing unused objects from the heap—garbage collection. Let's explore that in the next section.

Garbage Collection

> You pay for memory allocation more than once. The first is obviously when you allocate it. But you also pay every time the garbage collection runs.
>
> —Damian Gryski, "go-perfbook" (*https://oreil.ly/yg1LK*)

The second part of heap management is similar to vacuuming your house. It is related to a process that removes the proverbial garbage—unused objects from the program's heap. Generally speaking, the garbage collector (GC) is an additional background routine that executes "collection" at certain moments. The cadence of collections is critical:

- If the GC runs less often, we risk allocating a significant amount of new RAM space without the ability to reuse the memory pages currently allocated by garbage (unused objects).
- If the GC runs too often, we risk spending most of the program time and CPU on GC work instead of moving our functionality forward. As we will learn later, the GC is relatively fast but can directly or indirectly impact the execution of

other goroutines in the system, especially if we have many objects in a heap (if we allocate a lot).

The interval of the GC runs is not based on time. Instead, two configuration variables (working independently) define the pace: `GOGC` and, from Go 1.19, `GOMEMLIMIT`. To learn more about them, read an official detailed guide about GC tuning (*https://oreil.ly/f2F6H*). For this book, let's explain both very briefly:

The `GOGC` *option represents the "GC percentage."*
: `GOGC` is enabled by default with a 100 value. It means that the next GC collection will be done when the heap size expands to 100% of the size it has at the end of the last GC cycle. GC's pacing algorithm estimates when that goal will be reached based on current heap growth. It can also be set programmatically with the `debug.SetGCPercent` function (*https://oreil.ly/7khRe*).

The `GOMEMLIMIT` *option controls the soft memory limit.*
: The `GOMEMLIMIT` option was introduced in Go 1.19. It is disabled by default (set to `math.MaxInt64`), and offers running GC more often when we are close (or above) the set memory limit. It can be used with `GOGC=off` (disabled) or together with `GOGC`. This option can also be set programmatically with the `debug.SetMemoryLimit` function (*https://oreil.ly/etDUv*).

GOMEMLIMIT Does Not Prevent Your Program from Allocating More than the Set Value!

The GC's soft memory limit configuration is called "soft" for a reason. It tells the GC how much memory overhead space there is for the GC "laziness" to save the CPU.

However, when your program allocates and uses more memory than the desired limit, with the `GOMEMLIMIT` option set, it will only make things worse. This is because the GC will run nearly continuously, taking up 25% of the precious CPU time from other functionalities.

We still have to optimize the memory efficiency of our programs!

Manual trigger.
: Programmers can also trigger another GC collection on demand by invoking `runtime.GC()` (*https://oreil.ly/znoCL*). It is mostly used in testing or benchmarking code, as it can block the entire program. Other pacing configurations like `GOGC` and `GOMEMLIMIT` might run in between.

The Go GC implementation can be described as the concurrent, nongenerational, tricolor mark and sweep collector (*https://oreil.ly/vvOgl*) implementation. Whether invoked by the programmer or by the runtime-based `GOGC` or `GOMEMLIMIT` option, the `runtime.GC()` implementation comprises a few phases. The first one is a mark phase that has to:

1. Perform a "stop the world" (STW) event to inject an essential write barrier (*https://oreil.ly/Sl9PI*) (a lock on writing data) into all goroutines. Even though STW is relatively fast (10–30 microseconds on average), it is pretty impactful—it suspends the execution of all goroutines in our process for that time.
2. Try to use 25% of the CPU capacity given to the process to concurrently mark all objects in the heap that are still in use.
3. Terminate marking by removing the write barrier from the goroutines. This requires another STW event.

After the mark phase, the GC function is generally complete. As interesting as it sounds, the GC doesn't release any memory! Instead, the sweeping phase releases objects that were not marked as in use. It is done lazily: every time a goroutine wants to allocate memory through the Go Allocator, it must perform a sweeping work first, then allocate. This is counted as an `allocation` latency, even though it is technically a garbage collection functionality—worth noting!

Generally speaking, the Go Allocator and GC compose a sophisticated implementation of bucketed object pooling (*https://oreil.ly/r1K18*), where each pool of slots of different sizes are prepared for incoming allocations. When an allocation is not needed anymore, it is eventually released. The memory space for this allocation is not immediately released to the OS since it can be assigned to another incoming allocation soon (this is similar to the pooling pattern using `sync.Pool` we will discuss in "Memory Reuse and Pooling" on page 449). When the number of free buckets is big enough, Go releases memory to the OS. But even then, it does not necessarily mean that runtime deletes mapped regions straight away. For example, on Linux, Go runtime typically "releases" memory through the `madvise` syscall (*https://oreil.ly/pxXum*) with the `MADV_DONTNEED` argument by default.[36] This is because our mapped region might be needed again pretty soon, so it's faster to keep them just in case and ask the OS to take them back only if other processes require this physical memory.

---

36 It's also possible to change Go memory release strategy by changing the `GODEBUG` environment variable (*https://oreil.ly/ynNXr*). For example, we can set `GODEBUG=madvdontneed=0`, so `MADV_FREE` will be used instead to notify the OS about unneeded memory space. The difference between `MADV_DONTNEED` and `MADV_FREE` is precisely around the point mentioned in the Linux Community quote. For `MADV_FREE`, memory release is even faster for Go programs, but the resident set size (RSS) metric of the calling process might not be immediately reduced until the OS reclaims that space. This has proven to cause a massive problem on some systems (e.g., lightly virtualized systems like Kubernetes) that rely on RSS to manage the processes. This happened in 2019 when Go defaulted to `MADV_FREE` for a couple of versions. More on that is explained in my blog post (*https://oreil.ly/UYXJy*).

> Note that, when applied to shared mappings, MADV_DONTNEED might not lead to immediate freeing of the pages in the range. The kernel is free to delay freeing the pages until an appropriate moment. The resident set size (RSS) of the calling process will be immediately reduced, however.
>
> — Linux Community, "madvise(2), Linux Manual Page" (*https://oreil.ly/JDuS7*)

With the theory behind the GC algorithm, it will be easier for us to understand in Example 5-6 what happens if we try to clean the memory used for the large, 600 MB byte slice we created in Example 5-5.

Example 5-6. Memory release (de-allocation) of large slice created in Example 5-5

```
b := make([]byte, 600*1024*1024)
for i := range b { ❶
    b[i] = 1
}

b[5000] = 1 ❷
b = nil ❸
runtime.GC() ❹

// Let's allocate another one, this time 300 MB!
b = make([]byte, 300*1024*1024)
for i := range b { ❺
    b[i] = 2
}
```

❶ As we discussed in Example 5-5, the statistics after allocating a large slice and accessing all elements might look as follows:

- Slice is allocated in three memory mappings with the corresponding virtual memory size (VSS) numbers: 2 MB, 598 MB, and 4 MB.
- The RSS for three memory mappings: 1.5 MB, 598 MB, and 1.2 MB.
- Total RSS of the whole process shows 621.7 MB.
- Go reports 600.16 MB of the heap size.

❷ After the last statement where data from b is accessed, even before b = nil, the Mark phase of GC would consider b as a "garbage" to clean. Yet, the GC has its own pace; thus, immediately after this statement, no memory will be released—memory statistics will be the same.

❸ In typical cases when you no longer use the b value and the function scope ends, or you will replace b content with a pointer to a different object, there is no need for an explicit b = nil statement. The GC will know that the array pointed to by b is garbage. Yet sometimes, especially on long-living functions (e.g., a goroutine that performs background job items delivered by the Go channel), it is useful to set the variable to `nil` to make sure the next GC run will mark it for cleaning earlier.

❹ In our tests, let's invoke the GC manually to see what happens. After this statement, the statistics will look as follows:

- All three memory mappings still exist, with the same VSS values. This proves what we mentioned about the Go Allocator only advising on memory mappings, not removing those straightaway!
- The RSS for three memory mappings: 1.5 MB, 0 (RSS released), and 60 KB.
- Total RSS of the whole process shows 21 MB (back to the initial number).
- Go reports 159 KB of the heap size.

❺ Let's allocate another twice smaller slice. The following memory statistics prove the theory that Go will try to reuse previous memory mappings!

- Same three memory mappings still exist, with the same VSS values.
- The RSS for three memory mappings: 1.5 MB, 300 MB, and 60 KB.
- Total RSS of the whole process shows 321 MB.
- Go reports 300.1 KB of the heap size.

As we mentioned earlier, the beauty of GC is that it simplifies programmer life thanks to carefree allocations, memory safety, and solid efficiency for most applications. Unfortunately, it also makes our life a bit harder when our program violates our efficiency expectations, and the reason is not what you might think. The main problem with the Go Allocator and GC pair is that they hide the root cause of our memory efficiency problems—in almost all cases, our code allocates too much memory!

> Think of a garbage collector like a Roomba: Just because you have one does not mean you tell your children not to drop arbitrary pieces of garbage onto the floor.
>
> —Halvar Flake, Twitter (*https://oreil.ly/ukXDV*)

Let's explore the potential symptoms we might notice in Go when we are not careful with the number and type of the allocations:

CPU overhead

First and foremost, the GC must go through all the objects stored on the heap to tell which ones are in use. This can use a significant portion of the CPU resource, especially if there are many objects in heap.[37]

This is especially visible if the objects stored on the heap are rich in pointer types, which forces the GC to traverse them to check if they don't point to an object that was not yet marked as "in use." Given the limited CPU resources in our computers, the more work we have to do for the GC, the less work we can perform toward the core program functionality, which translates to higher program latency.

> In platforms with garbage collection, memory pressure naturally translates into increased CPU consumption.
>
> —Google Teams, *Site Reliability Engineering* (*https://oreil.ly/PhZaD*)

Additional increase in program latency

CPU time spent on GC is one thing, but there is more. First, the STW event performed twice slows down all goroutines. This is because the GC must stop all goroutines and inject (and then remove) a write barrier. It also prevents some goroutines that have to store some data in memory from doing any further work for the moment of GC marking.

There is also a second, often missed effect. The GC collection runs are destructive to the hierarchical cache system efficiency.

> For your program to be fast, you want everything you're doing to be in the cache. ... There are technical and physical reasons in the silicon why allocating memory, throwing it away and GC cleaning that for you, is going to not only slow your program down, because GC is doing its work, but it slows the rest of your program down, because it kicked everything out of [the CPU] cache.
>
> —Bryan Boreham, "Make Your Go Go Faster!" (*https://oreil.ly/cDw6c*)

Memory overhead

Since Go 1.19, there has been a way to set a soft memory limit for the GC. This still means that we have to often implement on our side checks against unbounded allocations (e.g., rejecting reading too-large HTTP body requests), but at least the GC is more prompt if you need to avoid that overhead.

Still, the collection phase is eventual. This means we might be unable to release some memory blocks before new allocations come in. Changing the `GOGC` option

37 To be strict, Go ensures that a maximum of 25% of the total CPU assigned for the process is used for the GC (*https://oreil.ly/9rtOs*). This is, however, not a silver-bullet solution. By reducing the maximum CPU time used, we simply use the same amount, just over longer periods.

to run GC less often only amplifies the problem but might be a good trade-off if you optimize for the CPU resource and have spare RAM on your machines.

Additionally, in extreme cases, our program might even leak memory if the GC is not fast enough to deal with all new allocations (*https://oreil.ly/4giW6*)!

The GC can sometimes have surprising effects on our program efficiency. Hopefully, after this section, you will be able to notice when you are affected. You will also be able to notice the GC bottlenecks with the observability tools explained in Chapter 9.

The Solution to Most Memory Efficiency Issues

Produce less garbage!

It's easy to overallocate memory in Go. This is why the best way to solve GC bottleneck or other memory efficiency issues is to allocate less. I will introduce "The Three Rs Optimization Method" on page 421, which goes through different optimizations that help with those efficiency problems.

Summary

It was a long chapter, but you made it! Unfortunately, memory resource is one of the hardest to explain and master. Probably that's why there are so many opportunities to reduce the size or number of our Go program's allocations.

You learned the long, multilayer path between our code that needs to allocate bits on memory and bits landing on the DRAM chip. You learned about many memory trade-offs, behaviors, and consequences on the OS level. Finally, you now know how Go uses those mechanisms and why memory allocations in Go are so transparent.

Perhaps you can already figure out the root causes of why Example 4-1 was using 30.5 MB of the heap for every single operation when the input file was 3 MB large. In "Optimizing Memory Usage" on page 395, I will propose the algorithm and code improvements to Example 4-1 that allow it to use memory in numbers that are a fraction of the input file size, while also improving the latency.

It is important to note that this space is evolving. Go compiler, Go garbage collector, and Go Allocator are constantly being improved, changed, and scaled for the needs of Go users. Yet most of the incoming changes will likely be only iterations of what we have now in Go.

Ahead of us are Chapters 6 and 7, which I consider two of the most crucial chapters in the book. I have already mentioned many tools I used to explain the main concepts in past chapters: metrics, benchmarking, and profiling. It's time to learn them in detail!

CHAPTER 6
Efficiency Observability

In "Efficiency-Aware Development Flow" on page 102, you learned to follow the TFBO (test, fix, benchmark, and optimize) flow to validate and achieve the required efficiency results with the least effort. Around the elements of the efficiency phase, observability takes one of the key roles, especially in Chapters 7 and 9. We focus on that phase in Figure 6-1.

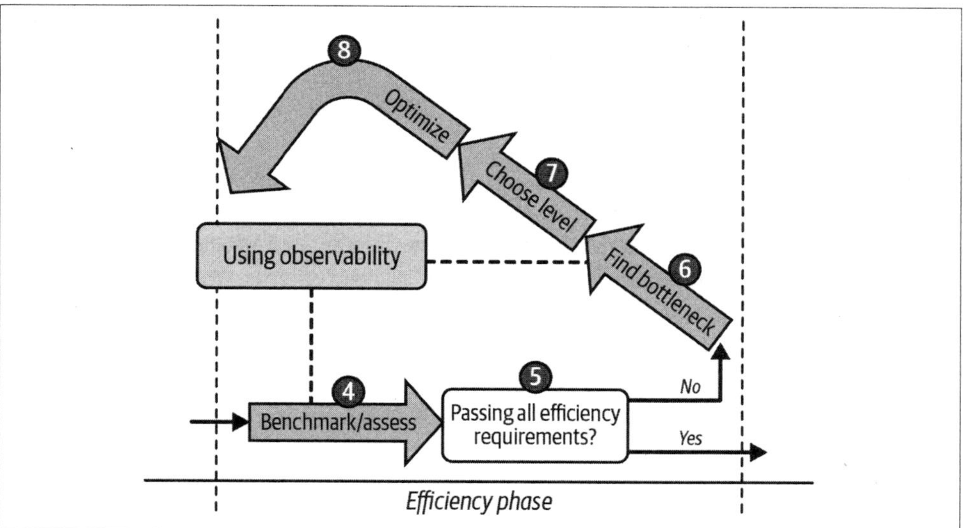

Figure 6-1. An excerpt from Figure 3-5 focusing on the part that requires good observability

In this chapter, I will explain the required observability and monitoring tools for this part of the flow. First, we will learn what observability is and what problems it solves. Then, we will discuss different observability signals, typically divided into logs, tracing, metrics, and, recently, profiles. Next, we will explain the first three signals in "Example: Instrumenting for Latency" on page 199, which takes latency as an example of the efficiency information we might want to measure (profiling is explained in Chapter 9). Last but not least, we will go through the specific semantics and sources of metrics related to our program efficiency in "Efficiency Metrics Semantics" on page 220.

You Can't Improve What You Don't Measure!

This quote, often attributed to Peter Drucker, is a key to improving anything: business revenues, car efficiency, family budget, body fat, or even happiness (*https://oreil.ly/eKiIR*).

Especially when it comes to invisible waste that our inefficient software is producing, we can say that it's impossible to optimize software without assessing and measuring before and after the change. Every decision must be data driven, as our guesses in this virtual space are often wrong.

With no further ado, let's learn how to measure the efficiency of our software in the easiest possible way—with the concept the industry calls observability.

Observability

To control software efficiency, we first need to find a structured and reliable way to measure the latency and resource usage of our Go applications. The key is to count these as accurately as possible and present them at the end as easy to understand numeric values. This is why for consumption measurements, we sometimes (not always!) use a "metric signal," which is a pillar of the essential software (or system) characteristics called observability.

Observability

In the cloud-native infrastructure world, we often talk about the observability of our applications. Unfortunately, observability is a very overloaded word.[1] It can be summarized as follows: an ability to deduce the state of a system inferred from external signals.

The external signals the industry uses nowadays can be generally categorized into four types: metrics, logs, traces, and profiling.[2]

Observability is a huge topic nowadays as it can help us in many situations while developing and operating our software. Observability patterns allow us to debug failures or unexpected behaviors of our programs, find root causes of incidents, monitor healthiness, alert on unforeseen situations, perform billing, measure SLIs (service level indicators) (*https://oreil.ly/hsdXJ*), run analytics, and much more. Naturally, we will focus only on the parts of observability that will help us ensure that our software efficiency matches our requirements (the RAERs mentioned in "Efficiency Requirements Should Be Formalized" on page 83). So what is an observability signal?

> - Metrics are a numeric representation of data measured over intervals of time. Metrics can harness the power of mathematical modeling and prediction to derive knowledge of the behavior of a system over intervals of time in the present and future.
> - An event log is an immutable, timestamped record of discrete events that happened over time. Event logs in general come in three forms but are fundamentally the same: a timestamp and a payload of some context.
> - A trace is a representation of a series of causally related distributed events that encode the end-to-end request flow through a distributed system. Traces are a representation of logs; the data structure of traces looks almost like that of an event log. A single trace can provide visibility into both the path traversed by a request as well as the structure of a request.
>
> —Cindy Sridharan, *Distributed Systems Observability* (*https://oreil.ly/YrSIE*) (O'Reilly, 2018)

1 Some of you might ask why I am sticking to the word *observability* and don't mention monitoring. In my eyes, I have to agree with my friend Björn Rabenstein (*https://oreil.ly/9ado0*) that the difference between monitoring and observability tends to be driven by marketing needs too much. One might say that observability has become meaningless these days. In theory, monitoring means answering known unknown problems (known questions), whereas observability allows learning about unknown unknowns (any question you might have in the future). In my eyes, monitoring is a subset of observability. In this book, we will stay pragmatic. Let's focus on how we can leverage observability practically, not using theoretical concepts.

2 The fourth signal, profiling, just started to be considered by some as an observability signal. This is because only recently did the industry see a value and need for gathering profiling continuously.

Generally, all those signals can be used to observe our Go applications' latency and resource consumption for optimization purposes. For example, we can measure the latency of a specific operation and expose it as a metric. We can send that value encoded into a log line or trace annotations (e.g., "baggage" (*https://oreil.ly/V5sQ6*) items). We can calculate latency by subtracting the timestamps of two log lines—when the operation started and when it finished. We can use trace spans, which track the latency of a span (individual unit of work done) by design.

However, whatever we use to deliver that information to us (via metric-specific tools, logs, traces, or profiles), in the end, it has to have metric semantics. We need to derive information to a numeric value so we can gather it over time; subtract; find max, min, or average; and aggregate over dimensions. We need the information to visualize and analyze. We need it to allow tools to reactively alert us when required, potentially build further automation that will consume it, and compare other metrics. This is why an efficiency discussion will mostly navigate through metric aggregations: the tail latency of our application, maximum memory usage over time, etc.

As we discussed, to optimize anything, you have to start measuring it, so the industry has developed many metrics and instruments to capture the usage of various resources. The process of observing or measuring always starts with the instrumentation.

Instrumentation

Instrumentation is a process of adding or enabling instruments for our code that will expose the observability signals we need.

Instrumentation can have many forms:

Manual instrumentation
: We can add a few statements to our code that import a Go module that generates an observability signal (for example, Prometheus client for metrics (*https://oreil.ly/AoWkJ*), go-kit logger (*https://oreil.ly/adTO3*), or a tracing (*https://oreil.ly/o7uYH*) library) and hook it to the operations we do. Of course, this requires modifying our Go code, but it usually leads to more personalized and rich signals with more context. Usually, it represents open box (*https://oreil.ly/qMjUP*) information because we can collect information tailored to the program functionality.

Autoinstrumentation
: Sometimes instrumentation means installing (and configuring) a tool that can derive useful information by looking at outside effects. For example, a service mesh gathers observability by looking at HTTP requests and responses, or a tool hooks to the operating system and gathers information through

cgroups (*https://oreil.ly/aCe6S*) or eBPF (*https://oreil.ly/QjxV9*).[3] Autoinstrumentation does not require changing and rebuilding code and usually represents closed box information (*https://oreil.ly/UO0gK*).

On top of that, it's helpful to categorize instrumentation based on the granularity of the information:

Capturing raw events
: Instrumentation in this category will try to deliver a separate piece of information for each event in our process. For example, suppose we would like to know how many and what errors are happening in all HTTP requests served by our process. In that case, we could have instrumentation that delivers a separate piece of information about each request (e.g., as a log line). Furthermore, this information usually has some metadata about its context, like the status code, user IP, timestamp, and the process and code statement in which it happened (target metadata).

  Once ingested to some observability backend, such raw data is very rich in context and, in theory, allows any ad hoc analysis. For example, we can scan through all events to find an average number of errors or the percentile distributions (more on that in "Latency" on page 221). We can navigate to every individual error representing a single event to inspect it in detail. Unfortunately, this kind of data is generally the most expensive to use, ingest, and store. We often risk an inaccuracy here since it's likely we'll miss an individual event or two. In extreme cases, it requires complex skills and automation for big data and data mining explorations to find the information you want.

Capturing aggregated information
: We can capture pre-aggregated data instead of raw events. Every piece of information delivered by such instrumentation represents certain information about a group of events. In our HTTP server example, we could count successful and failed requests, and periodically deliver that information. Before forwarding this information, we could go even further and pre-calculate the error ratio inside our code. It's worth mentioning that this kind of information also requires metadata, so we can summarize, aggregate further, compare, and analyze those aggregated pieces of information.

  Pre-aggregated instrumentation forces Go processes or autoinstrumentation tools to do more work, but the results are generally easier to use. On top of this, because of the smaller amount of data, the complexity of the instrumentation, signal delivery, and backend is lower, thereby increasing reliability and

3 As a recent example, we can give this repository (*https://oreil.ly/sPlPe*) that gathers information through eBPF probes and tries to search popular functions or libraries.

decreasing cost significantly. There are trade-offs here as well. We lose some information (commonly called the cardinality). The decision of what information to prebuild is made up front, and is coded into instrumentation. If you suddenly have different questions to be answered (e.g., how many errors an individual user had across your processes) and your instrumentation was not set to pre-aggregate that information, you have to change it, which takes time and resources. Yet if you roughly know what you will be asking for ahead of time, aggregated type of information is an amazing win and a more pragmatic approach.[4]

Last but not least, generally speaking we can design our observability flows into push-and-pull collection models:

Push
: A system where a centralized remote process collects observability signals from your applications (including your Go programs).

Pull
: A system where application processes push the signal to a remote centralized observability system.

Push Versus Pull

Each of the conventions has its pros and cons. You can push your metrics, logs, and traces, but you can also pull all of them from your process. We can also use a mixed approach, different for each observability signal.

Push versus pull method is sometimes a controversial topic. The industry is polarized as to what is generally better, not only in observability but also for any other architectures. We will discuss the pros and cons in "Metrics" on page 211, but the difficult truth is that both ways can scale equally well, just with different solutions, tools, and best practices.

After learning about those three categories, we should be ready to dive further into observability signals. To measure and deliver observability information for efficiency optimizations, we can't avoid learning more about instrumenting the three common observability signals: logging, tracing, and metrics. In the next section, let's do that while keeping a practical goal in mind—measuring latency.

4 In some way, I am trying in this book to establish helpful processes around optimizations and efficiency, which by design yield standard questions we know up front. This aggregated information is usually enough for us here.

Example: Instrumenting for Latency

All three signals you will learn in this section can be used to build observability that will fit in any of the three categorizations we discussed. Each signal can:

- Be manually or autoinstrumented
- Give aggregated information or raw events
- Be pulled (collected, tailed, or scraped) from the process or pushed (uploaded)

Yet every signal—logging, tracing, or metric—might be better or worse fitted in any of those jobs. In this section, we will discuss these predispositions.

The best way to learn how to use observability signals and their trade-offs is to focus on the practical goal. Let's imagine we want to measure the latency of a specific operation in our code. As mentioned in the introduction, we need to start measuring the latency to assess it and decide if our code needs more optimizations during every optimization iteration. As you will learn in this section, we can get latency results using any of those observability signals. The details around how information is presented, how complex instrumentation is, and so on will help you understand what to choose in your journey. Let's dive in!

Logging

Logging might be the clearest signal to understand an instrument. So let's explore the most basic instrumentation that we might categorize as logging to collect latency measurements. Taking basic latency measurements for a single operation in Go code is straightforward, thanks to the standard `time` package (*https://oreil.ly/t9FDr*). Whether you do it by hand or use standard or third-party libraries to obtain latencies, if they are written in Go, they use the pattern presented in Example 6-1 using the `time` package.

Example 6-1. Manual and simplest latency measurement of a single operation in Go

```
import (
    "fmt"
    "time"
)

func ExampleLatencySimplest() {
    for i := 0; i < xTimes; i++ {
        start := time.Now() ❶
        err := doOperation()
        elapsed := time.Since(start) ❷

        fmt.Printf("%v ns\n", elapsed.Nanoseconds()) ❸
```

```
        // ...
    }
}
```

❶ `time.Now()` captures the current wall time (clock time) from our operating system clock in the form `time.Time`. Note the `xTime`, example variable that specifies the desired number of runs.

❷ After our `cooperation` functions finish, we can capture the time between `start` and current time using `time.Since(start)`, which returns the handy `time.Duration`.

❸ We can leverage such an instrument to deliver our metric sample. For example, we can print the duration in nanoseconds to the standard output using the `.Nanoseconds()` method.

Arguably, Example 6-1 represents the simplest form of instrumentation and observability. We take a latency measurement and deliver it by printing the result into standard output. Given that every operation will output a new line, Example 6-1 represents manual instrumentation of raw event information.

Unfortunately, this is a little naive. First of all, as we will learn in "Reliability of Experiments" on page 256, a single measurement of anything can be misleading. We have to capture more of those—ideally hundreds or thousands for statistical purposes. When we have one process, and only one functionality we want to test or benchmark, Example 6-1 will print hundreds of results that we can later analyze. However, to simplify the analysis, we could try to pre-aggregate some results. Instead of logging raw events, we could pre-aggregate using a mathematical average function and output that. Example 6-2 presents a modification of Example 6-1 that aggregates events into an easier-to-consume result.

Example 6-2. Instrumenting Go to log the average latency of an operation in Go

```
func ExampleLatencyAggregated() {
    var count, sum int64
    for i := 0; i < xTimes; i++ {
        start := time.Now()
        err := doOperation()
        elapsed := time.Since(start)

        sum += elapsed.Nanoseconds() ❶
        count++

        // ...
    }
```

```
    fmt.Printf("%v ns/op\n", sum/count) ❷
}
```

❶ Instead of printing raw latency, we can gather a sum and number of operations in the sum.

❷ Those two pieces of information can be used to calculate the accurate average and present that for a group of events instead of the unique latency. For example, one run printed the `188324467 ns/op` string on my machine.

Given that we stop presenting latency for raw events, Example 6-2 represents a manual, aggregated information observability. This method allows us to quickly get the information we need without complex (and time-consuming) tools analyzing our logging outputs.

This example is how the Go benchmarking tool will do the average latency calculations. We can achieve exactly the same logic as in Example 6-2 using the snippet in Example 6-3 in a file with the *_test.go* suffix.

Example 6-3. Simplest Go benchmark that will measure average latency per operation

```
func BenchmarkExampleLatency(b *testing.B) {
    for i := 0; i < b.N; i++ { ❶
        _ = doOperation()
    }
}
```

❶ The `for` loop with the `N` variable is essential in the benchmarking framework. It allows the Go framework to try different `N` values to perform enough test runs to fulfill the configured number of runs or test duration. For example, by default, the Go benchmark runs to fit one second, which is often too short for meaningful output reliability.

Once we run Example 6-3 using `go test` (explained in detail in "Go Benchmarks" on page 277), it will print certain output. One part of the information is a result line with a number of runs and average nanoseconds per operation. One of the runs on my machine gave an output latency of `197999371 ns/op`, which generally matches the result from Example 6-2. We can say that the Go benchmark is an autoinstrumentation with aggregated information using logging signals for things like latency.

On top of collecting latency about the whole operation, we can gain a lot of insight from having different granularity of those measurements. For example, we might wish to capture the latency of a few suboperations inside our single operation. Finally, for more complex deployments, when our Go program is part of a distributed system, as discussed in "Macrobenchmarks" on page 306, we have potentially many processes we have to measure across. For those cases, we have to use more sophisticated logging that will give us more metadata and ways to deliver a logging signal, not only by simply printing to a file, but by other means too.

The amount of information we have to attach to our logging signal results in the pattern called a logger in Go (and other programming languages). A logger is a structure that allows us to manually instrument our Go application with logs in the easiest and most readable way. A logger hides complexities like:

- Formatting of the log lines.
- Deciding if we should log or not based on the logging level (e.g., debug, warning, error, or more).
- Delivering the log line to a configured place, such as the output file. Optionally, more complex, push-based logging delivery is possible to remote backends, which must support back-off retries, authorization, service discovery, etc.
- Adding context-based metadata and timestamps.

The Go standard library is very rich with many useful utilities, including logging. For example, the `log` package (*https://oreil.ly/JEUjT*) contains a simple logger. It can work well for many applications, but it is prone to some usage pitfalls.[5]

Be Mindful While Using the Go Standard Library Logger

There are a few things to remember if you want to use the standard Go logger from the `log` package:

- Don't use the global `log.Default()` logger, so `log.Print` functions, and so on. Sooner or later, it will bite you.
- Never store or consume `*log.Logger` directly in your functions and structures, especially when you write a library.[6] If you do, users will be forced to use a very limited `log` logger instead of their own logging libraries. Use a custom interface instead (e.g., go-kit logger (*https://oreil.ly/tCs2g*)), so users can adapt their loggers to what you use in your code.
- Never use the `Fatal` method outside the main function. It panics, which should not be your default error handling.

5 Given Go compatibility guarantees, even if the community agrees to improve it, we cannot change it until Go 2.0.

6 A nonexecutable module or package intended to be imported by others.

To not accidentally get hit by these pitfalls, in the projects I worked on, we decided to use the third-party popular go-kit (*https://oreil.ly/ziBdb*)[7] logger. An additional advantage of the go-kit logger is that it is easy to maintain some structure. Structure logic is essential to have reliable parsers for automatic log analysis with logging backends like OpenSearch (*https://oreil.ly/RohpZ*) or Loki (*https://oreil.ly/Fw9I3*). To measure latency, let's go through an example of logger usage in Example 6-4. Its output is shown in Example 6-5. We use the `go-kit` module (*https://oreil.ly/vOafG*), but other libraries follow similar patterns.

Example 6-4. Capturing latency though logging using the `go-kit` logger (https://oreil.ly/9uCWi)

```
import (
    "fmt"
    "time"

    "github.com/go-kit/log"
    "github.com/go-kit/log/level"
)

func ExampleLatencyLog() {
    logger := log.With( ❶
        log.NewLogfmtLogger(os.Stderr), "ts", log.DefaultTimestampUTC,
    )

    for i := 0; i < xTimes; i++ {
        now := time.Now()
        err := doOperation()
        elapsed := time.Since(now)

        level.Info(logger).Log( ❷
            "msg", "finished operation",
            "result", err,
            "elapsed", elapsed.String(),
        )

        // ...
    }
}
```

7 There are many Go libraries for logging. `go-kit` has a good enough API that allows us to do all kinds of logging we need in all the Go projects I have helped with so far. This does not mean `go-kit` is without flaws (e.g., it's easy to forget you have to put an even number of arguments for the key-value–like logic). There is also a pending proposal from the Go community on structure logging in standard libraries (`slog` package) (*https://oreil.ly/qnJ6y*). Feel free to use any other libraries, but make sure their API is simple, readable, and useful. Also make sure that the library of your choice is not introducing efficiency problems.

❶ We initialize the logger. Libraries usually allow you to output the log lines to a file (e.g., standard output or error) or directly push it to some collections tool, e.g., to fluentbit (*https://oreil.ly/pUcmX*) or vector (*https://oreil.ly/S0aqR*). Here we choose to output all logs to standard error[8] with a timestamp attached to each log line. We also choose to format logs in the human-accessible way with `NewLogfmtLogger` (still structured so that it can be parsed by software, with space as the delimiter).

❷ In Example 6-1, we simply printed the latency number. Here we add certain metadata to it to use that information more easily across processes and different operations happening in the system. Notice that we maintain a certain structure. We pass an even number of arguments representing key values. This allows our log line to be structured for easier use by automation. Additionally, we choose `level.Info`, meaning this log line will be not printed if we choose levels like errors only.

Example 6-5. Example output logs generated by Example 6-4 (wrapped for readability)

```
level=info ts=2022-05-02T11:30:46.531839841Z msg="finished operation" \
result="error other" elapsed=83.62459ms ❶
level=info ts=2022-05-02T11:30:46.868633635Z msg="finished operation" \
result="error other" elapsed=336.769413ms
level=info ts=2022-05-02T11:30:47.194901418Z msg="finished operation" \
result="error first" elapsed=326.242636ms
level=info ts=2022-05-02T11:30:47.51101522Z msg="finished operation" \
result=null elapsed=316.088166ms
level=info ts=2022-05-02T11:30:47.803680146Z msg="finished operation" \
result="error first" elapsed=292.639849ms
```

❶ Thanks to the log structure, it's both readable to us and automation can clearly distinguish among different fields like `msg`, `elapsed`, `info`, etc. without expensive and error-prone fuzzy parsing.

Logging with a logger might still be the simplest way to deliver our latency information manually to us. We can tail the file (or use `docker log` if our Go process was running in Docker, or `kubectl logs` if we deployed it on Kubernetes) to read those log lines for further analysis. It is also possible to set up an automation that tails those from files or pushes them directly to the collector, adding further information. Collectors can be then configured to push those log lines into free and open source logging backends like OpenSearch (*https://oreil.ly/RohpZ*), Loki (*https://oreil.ly/Fw9I3*),

8 It's a typical pattern allowing processes to print something useful to standard output and keep logs separate in the `stderr` Linux file.

Elasticsearch (*https://oreil.ly/EUlts*), or many of the paid vendors. As a result, you can keep log lines from many processes in a single place, search, visualize, analyze them, or build further automation to handle them as you want.

Is logging a good fit for our efficiency observability? Yes and no. For microbenchmarks explained in "Microbenchmarks" on page 275, logging is our primary tool of measurements because of its simplicity. On the other hand, on a macro level, like "Macrobenchmarks" on page 306, we tend to use logging for a raw event type of observability, which on such a scale gets very complex and expensive to analyze and keep reliable. Still, because logging is so common, we can find efficiency bottlenecks in a bigger system with logging.

Logging tools are also constantly evolving. For example, many tools allow us to derive metrics from log lines, like Grafana Loki's Metric queries inside LogQL (*https://oreil.ly/fdoNm*). In practice, however, simplicity has its cost. One of the problems stems from the fact that sometimes logs are used directly by humans, and sometimes by automation (e.g., deriving metrics or reacting to situations found in logs). As a result, logs are often unstructured. Even with amazing loggers like go-kit in Example 6-4, logs are inconsistently structured, making it very hard and expensive to parse for automation. For example, things like inconsistent units (as in Example 6-5 for latency measurements), which are great for humans, become almost impossible to derive the value as a metric. Solutions like Google mtail (*https://oreil.ly/Q4wAC*) try to approach this with custom parsing language. Still, the complexity and ever-changing logging structure make it hard to use this signal to measure our code's efficiency.

Let's look at the next observability signal—tracing—to learn in which areas it can help us with our efficiency goals.

Tracing

Given the lack of consistent structure in logging, tracing signals emerged to tackle some of the logging problems. In contrast to logging, tracing is a piece of structured information about your system. The structure is built around the transaction, for example, requests-response architecture. This means that things like status codes, the result of the operation, and the latency of operations are natively encoded, thus easier to use by automation and tools. As a trade-off, you need an additional mechanism (e.g., a user interface) to expose this information to humans in a readable way.

On top of that, operations, suboperations, and even cross-process calls (e.g., RPCs) can be linked together, thanks to context propagation mechanisms working well with standard network protocols like HTTP. This feels like a perfect choice for measuring latency for our efficiency needs, right? Let's find out.

As with logging, there are many different manual instrumentation libraries you can choose from. Popular, open source choices for Go are the OpenTracing (*https://oreil.ly/gJeAV*) library (currently deprecated but still viable), OpenTelemetry (*https://oreil.ly/uxKoW*), or clients from the dedicated tracing vendor. Unfortunately, at the moment of writing, the OpenTelemetry library has a too-complex API to explain in this book, plus it's still changing, so I started a small project called tracing-go (*https://oreil.ly/rs6fQ*) that encapsulates the OpenTelemetry client SDK into minimal tracing instrumentation. While tracing-go is my interpretation of the minimal set of tracing functionalities to use, it should teach you the basics of context propagation and span logic. Let's explore an example manual instrumentation using tracing-go to measure dummy `doOperation` function latency (and more!) using tracing in Example 6-6.

Example 6-6. Capturing latencies of the operation and potential suboperations using tracing-go (https://oreil.ly/1027d)

```
import (
    "fmt"
    "time"

    "github.com/bwplotka/tracing-go/tracing"
    "github.com/bwplotka/tracing-go/tracing/exporters/otlp"
)

func ExampleLatencyTrace() {
    tracer, cleanFn, err := tracing.NewTracer(otlp.Exporter("<endpoint>")) ❶
    if err != nil { /* Handle error... */ }
    defer cleanFn()

    for i := 0; i < xTimes; i++ {
        ctx, span := tracer.StartSpan("doOperation") ❷
        err := doOperationWithCtx(ctx)
        span.End(err) ❸

        // ...
    }
}

func doOperationWithCtx(ctx context.Context) error {
    _, span := tracing.StartSpan(ctx, "first operation") ❹
    // ...
    span.End(nil)

    // ...
}
```

❶ As with everything, we have to initialize our library. In our example, usually, it means creating an instance of `Tracer` that is capable of sending the spans that will form traces. We push spans to some collector and eventually to the tracing backend. This is why we have to specify some address to send to. In this example, you could specify a gRPC `host:port` address of the collector (e.g., OpenTelemetry Collector (*https://oreil.ly/z0Pjt*)) endpoint that supports the gRPC OTLP trace protocol (*https://oreil.ly/4IaBd*).

❷ With the tracer, we can create an initial root `span`. The root means the span that spans the whole transaction. A `traceID` is created during creation, identifying all spans in the trace. Span represents individual work done. For example, we can add a different name or even baggage items like logs or events. We also get a `context.Context` instance as part of creation. This Go native context interface can be used to create subspans if our `doOperation` function will do any subwork pieces worth instrumenting.

❸ In the manual instrumentation, we have to tell the tracing provider when the work was done and with what result. In the `tracing-go` library, we can use `end.Stop(<error or nil>)` for that. Once you stop the span, it will record the span's latency from its start, the potential error, and mark itself as ready to be sent asynchronously by `Tracer`. Tracer exporter implementations usually won't send spans straightaway but buffer them for batch pushes. `Tracer` will also check if a trace containing some spans can be sent to the endpoint based on the chosen sampling strategy (more on that later).

❹ Once you have context with the injected span creator, we can add subspans to it. It's useful when you want to debug different parts and sequences involved in doing one piece of work.

One of the most valuable parts of tracing is context propagation. This is what separates distributed tracing from nondistributed signals. I did not reflect this in our examples, but imagine if our operation makes a network call to other microservices. Distributed tracing allows passing various tracing information like `traceID`, or sampling via a propagation API (e.g., certain encoding using HTTP headers). See a related blog post (*https://oreil.ly/Qz6lF*) about context propagation. For that to work in Go, you have to add a special middleware or HTTP client with propagation support, e.g., OpenTelemetry HTTP transport (*https://oreil.ly/Rvq6i*).

Because of the complex structure, raw traces and spans are not readable by humans. This is why many projects and vendors help users by providing solutions to use tracing effectively. Open source solutions like Grafana Tempo with Grafana UI (*https://oreil.ly/CQ1Aq*) and Jaeger (*https://oreil.ly/enkG9*) exist, which offer nice user interfaces and trace collection so you can observe your traces. Let's look at how our spans from Example 6-6 look in the latter project. Figure 6-2 shows a multitrace search view, and Figure 6-3 shows what our individual `doOperation` trace looks like.

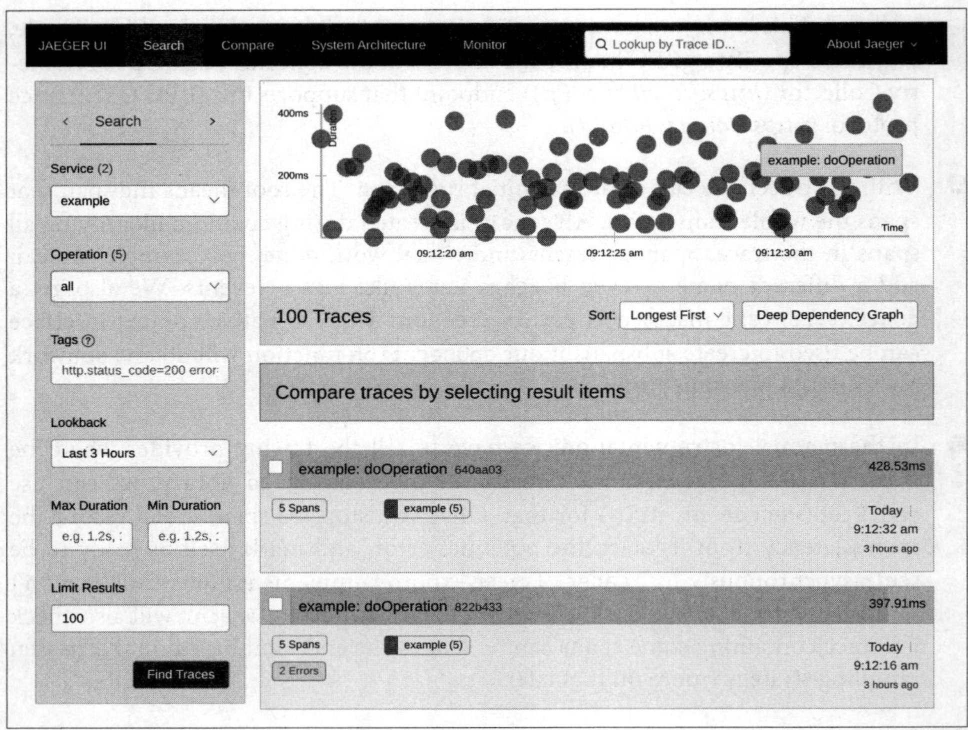

Figure 6-2. View of one hundred operations presented as one hundred traces with their latency results

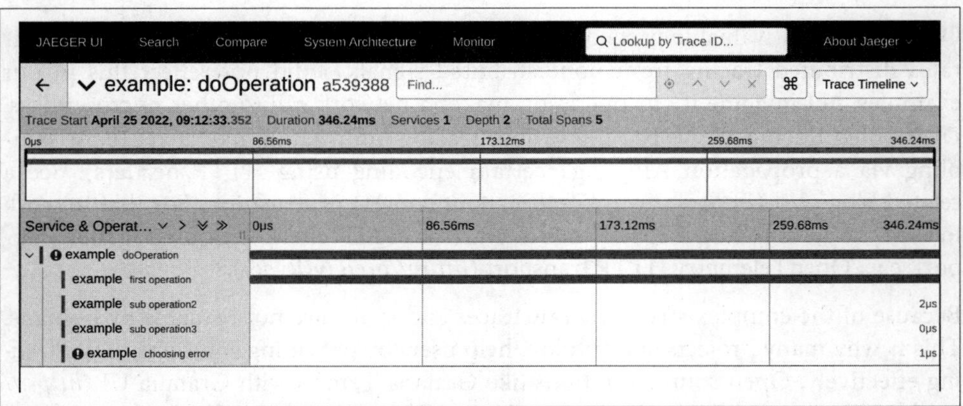

Figure 6-3. Click one trace to inspect all of its spans and associated data

Tools and user interfaces can vary, but generally they follow the same semantics I explain in this section. The view in Figure 6-2 allows us to search through traces based on their timestamp, durations, service involved, etc. The current search matches our one hundred operations, which are then listed on the screen. A convenient, interactive graph of its latencies is placed, so we can navigate to the operation we want. Once clicked, the view in Figure 6-3 is presented. In this view, we can see a distribution of spans for this operation. If the operation spans multiple processes and we used network context propagation, all linked spans will be listed here. For example, from Figure 6-3 we can immediately tell that the first operation was responsible for most of the latency, and the last operation introduced the error.

All the benefits of tracing make it an excellent tool for learning the system interactions, debugging, or finding fundamental efficiency bottlenecks. It can also be used for ad hoc verification of system latency measurements (e.g., in our TFBO flow to assess latency). But unfortunately, there are a few downsides of tracing that you have to be aware of when planning to use it in practice for efficiency or other needs:

Readability and maintainability

The advantage of tracing is that you can put a huge amount of useful context into your code. In extreme cases, you could potentially be able to rewrite the whole program or even system just by looking at all traces and their emitted spans. But there is a catch. All this manual instrumentation requires code lines. More code lines connected to our existing code increases the complexity of our code, which in turn decreases readability. We also need to ensure that our instrumentation stays updated with ever-changing code.

In practice, the tracing industry tends to prefer autoinstrumentation, which in theory can add, maintain, and hide such instrumentation automatically. Proxies like Envoy (especially with service mesh technologies) are great examples of successful (yet simpler) autoinstrumentation tools for tracing that record the interprocess HTTP calls. But unfortunately, more involved auto-instrumentation is not so easy. The main problem is that the automation has to hook on to some generic path like common database or library operations, HTTP requests, or syscalls (e.g., through eBPF probes in Linux). Moreover, it is often hard for those tools to understand what more you would like to capture in your application (e.g., the ID of the client in a specific code variable). On top of that, tools like eBPF are pretty unstable and dependent on the kernel version.

Hiding Instrumentation Under Abstractions

There is a middle ground between manual and fully autonomous instrumentation. We can manually instrument only a few common Go functions and libraries, so all code that uses them will be traced consistently implicitly (automatically!).

For example, we could add a trace for every HTTP or gRPC request to our process. There are already HTTP middlewares (*https://oreil.ly/wZ559*) and gRPC interceptors (*https://oreil.ly/7gXVF*) for that purpose.

Cost and reliability

Traces by design fall into the raw event category of observability. This means that tracing is typically more expensive than pre-aggregated equivalents. The reason is the sheer amount of data we send using tracing. Even if we are very moderate with this instrumentation for a single operation, we ideally have dozens of tracing spans. These days, systems have to sustain many QPS (queries per second). In our example, even for 100 QPS, we would generate over 1,000 spans. Each span must be delivered to some backend to be used effectively, with replication on both the ingestion and storage sides. Then you need a lot of computation power to analyze this data to find, for example, average latency across traces or spans. This can easily surpass your price for running the systems without observability!

The industry is aware of this, and this is why we have tracing sampling, so some decision-making configuration or code decides what data to pass forward and what to ignore. For example, you might want to only collect traces for failed operations or operations that took more than 120 seconds.

Unfortunately, sampling comes with its downsides. For example, it's challenging to perform tail sampling.[9] Last but not least, sampling makes us miss some data (similar to profiling). In our latency example, this might mean that the latency we measure represents only part of all operations that happened. Sometimes it might be enough, but it's easy to get wrong conclusions with sampling (*https://oreil.ly/R4gtX*), which might lead to wrong optimization decisions.

Short duration

We will discuss this in detail in "Latency" on page 221, but tracing won't tell us much when we try to improve very fast functions that last only a few milliseconds or less. Similar to the `time` package, the span itself introduces some

9 Tail sampling is a logic that defers the decision if the trace should be excluded or sampled at the end of the transaction, for example, only after we know its status code. The problem with tail sampling is that your instrumentation might have already assumed that all spans will be sampled.

latency. On top of that, adding span for many small operations can add a huge cost to the overall ingestion, storage, and querying of traces.

This is especially visible in streamed algorithms like chunked encodings, compressions, or iterators. If we perform partial operations, we are still often interested in the latency of the sum of all iterations for certain logic. We can't use tracing for that, as we would need to create tiny spans for every iteration. For those algorithms, "Profiling in Go" on page 331 yields the best observability.

Despite some downsides, tracing becomes very powerful and even replaces the logging signal in many cases. Vendors and projects add more features, for example, Tempo project's metric generator (*https://oreil.ly/SSLye*) that allows recording metrics from traces (e.g., average or tail latency for our efficiency needs). Undoubtedly, tracing would not grow so quickly without the push from the OpenTelemetry (*https://oreil.ly/sPiw9*) community. Amazing things will come from this community if you are into tracing.

The downsides of one framework are often strengths of other frameworks that choose different trade-offs. For example, many tracing problems come from the fact that it naturally represents raw events happening in the system (that might trigger other events). Let's now discuss a signal on the opposite spectrum—designed to capture aggregations changing over time.

Metrics

Metrics is the observability signal that was designed to observe aggregated information. Such aggregation-oriented metric instrumentations might be the most pragmatic way of solving our efficiency goals. Metrics are also what I used the most in my day-to-day job as a developer and SRE to observe and debug production workloads. In addition, metrics are the main signal used for monitoring at Google (*https://oreil.ly/x6rNZ*).

Example 6-7 shows pre-aggregated instrumentation that can be used to measure latency. This example uses Prometheus `client_golang` (*https://oreil.ly/1r2zw*).[10]

Example 6-7. Measuring `doOperation` latency using the histogram metric with Prometheus `client_golang`

```
import (
    "fmt"
    "time"
```

10 I maintain this library together with the Prometheus team. The `client_golang` is also the most used metric client SDK for Go when writing this book, with over 53,000 open source projects (*https://oreil.ly/UW0fG*) using it. It is free and open source.

```
    "github.com/prometheus/client_golang/prometheus"
    "github.com/prometheus/client_golang/prometheus/promauto"
    "github.com/prometheus/client_golang/prometheus/promhttp"
)

func ExampleLatencyMetric() {
    reg := prometheus.NewRegistry() ❶
    latencySeconds := promauto.With(reg).

NewHistogramVec(prometheus.HistogramOpts{ ❷
        Name:    "operation_duration_seconds",
        Help:    "Tracks the latency of operations in seconds.",
        Buckets: []float64{0.001, 0.01, 0.1, 1, 10, 100},
    }, []string{"error_type"}) ❸

    go func() {
        for i := 0; i < xTimes; i++ {
            now := time.Now()
            err := doOperation()
            elapsed := time.Since(now)

            latencySeconds.WithLabelValues(errorType(err)).
                Observe(elapsed.Seconds()) ❹

            // ...
        }
    }()

    err := http.ListenAndServe(
        ":8080",
        promhttp.HandlerFor(reg, promhttp.HandlerOpts{})
    ) ❺
    // ...
}
```

❶ Using the Prometheus library always starts with creating a new metric registry.[11]

❷ The next step is to populate the registry with the metric definitions you want. Prometheus allows a few types of metrics, yet the typical latency measurements for efficiency are best done as histograms. So on top of type, help and histogram buckets are required. We will talk more about buckets and the choice of histograms later.

11 It's tempting to use global `prometheus.DefaultRegistry`. Don't do this. We try to get away from this pattern that can cause many problems and side effects.

❸ As the last parameter, we define the dynamic dimension of this metric. Here I propose to measure latency for different types of errors (or no error). This is useful as, very often, failures have other timing characteristics.

❹ We observe the exact latency with a floating number of seconds. We run all operations in a simplified goroutine, so we can expose metrics while the functionality is performing. The `Observe` method will add such latency into the histogram of buckets. Notice that we observe this latency for certain errors. We also don't take an arbitrary error string—we sanitize it to a type using some custom `errorType` function. This is important because the controlled number of values in the dimension keeps our metric valuable and cheap.

❺ The default way to consume those metrics is by allowing other processes (e.g., Prometheus server (*https://oreil.ly/2Sa3P*)) to pull the current state of the metrics. For example, in this simplified[12] code we serve those metrics from our registry through an HTTP endpoint on the `8080` port.

The Prometheus data model supports four metric types, which are well described in the Prometheus documentation (*https://oreil.ly/mamdO*): counters, gauges, histograms, and summaries. There is a reason why I chose a more complex histogram for observing latency instead of a counter or a gauge metric. I explain why in "Latency" on page 221. For now, it's enough to say that histograms allow us to capture distributions of the latencies, which is typically what we need when observing production systems for efficiency and reliability. Such metrics, defined and instrumented in Example 6-7, will be represented on an HTTP endpoint, as shown in Example 6-8.

Example 6-8. Sample of the metric output from Example 6-7 when consumed from the OpenMetrics compatible HTTP endpoint (https://oreil.ly/aZ6GT)

```
# HELP operation_duration_seconds Tracks the latency of operations in seconds.
# TYPE operation_duration_seconds histogram
operation_duration_seconds_bucket{error_type="",le="0.001"} 0 ❶
operation_duration_seconds_bucket{error_type="",le="0.01"} 0
operation_duration_seconds_bucket{error_type="",le="0.1"} 1
operation_duration_seconds_bucket{error_type="",le="1"} 2
operation_duration_seconds_bucket{error_type="",le="10"} 2
operation_duration_seconds_bucket{error_type="",le="100"} 2
operation_duration_seconds_bucket{error_type="",le="+Inf"} 2
operation_duration_seconds_sum{error_type=""} 0.278675917 ❷
operation_duration_seconds_count{error_type=""} 2
```

12 Always check errors and perform graceful termination on process teardown. See production-grade usage in the Thanos project (*https://oreil.ly/yvvTM*) that leverages the run goroutine helper (*https://oreil.ly/sDIwW*).

❶ Each bucket represents a number (counters) of operations that had latency less than or equal to the value specified in `le`. For example, we can immediately see that we saw two successful operations from the process start. The first was faster than 0.1 seconds; and the second was faster than 1 second, but slower than 0.1 seconds.

❷ Every histogram also captures a number of observed operations and summarized value (sum of observed latencies, in this case).

As mentioned in "Observability" on page 194, every signal can be pulled or pushed. However, the Prometheus ecosystem defaults to the pull method for metrics. Not the naive pull, though. In the Prometheus ecosystem, we don't pull a backlog of events or samples like we would when pulling (tailing) traces of logs from, for example, a file. Instead, applications serve HTTP payload in the OpenMetrics format (like in Example 6-8), which is then periodically collected (scraped) by Prometheus servers or Prometheus compatible systems (e.g., Grafana Agent or OpenTelemetry collector). With the Prometheus data model, we scrape the latest information about the process.

To use Prometheus with our Go program instrumented in Example 6-7, we have to start the Prometheus server and configure the scrape job that targets the Go process server. For example, assuming we have the code in Example 6-7 running, we could use the set of commands shown in Example 6-9 to start metric collection.

Example 6-9. The simplest set of commands to run Prometheus from the terminal to start collecting metrics from Example 6-7

```
cat << EOF > ./prom.yaml
scrape_configs:
- job_name: "local"
  scrape_interval: "15s" ❶
  static_configs:
  - targets: [ "localhost:8080" ] ❷
EOF
prometheus --config.file=./prom.yaml ❸
```

❶ For my demo purposes, I can limit the Prometheus configuration (*https://oreil.ly/4cPSa*) to a single scrape job. One of the first decisions is to specify the scrape interval. Typically, it's around 15–30 seconds for continuous, efficient metric collection.

❷ I also provide a target that points to our tiny instrumented Go program in Example 6-7.

❸ Prometheus is just a single binary written in Go. We install it in many ways (*https://oreil.ly/9CxxD*). In the simplest configuration, we can point it to a created configuration. When started, the UI will be available on the `localhost:9090`.

With the preceding setup, we can start analyzing the data using Prometheus APIs. The simplest way is to use the Prometheus query language (PromQL) documented here (*https://oreil.ly/nY6Yi*) and here (*https://oreil.ly/jH3nd*). With Prometheus server started as in Example 6-9, we can use the Prometheus UI and query the data we collected.

For example, Figure 6-4 shows the result of the simple query fetching the latest latency histogram numbers over time (from the moment of the process start) for our `operation_duration_seconds` metric name that represents successful operations. This generally matches the format we see in Example 6-8.

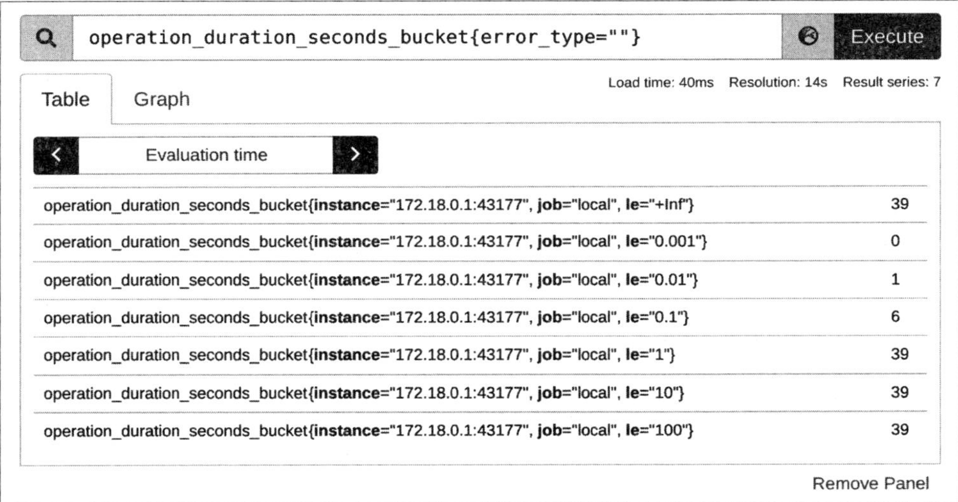

Figure 6-4. PromQL query results for simple query for all `operation_duration_seconds_bucket` metrics graphed in the Prometheus UI

To obtain the average latency of a single operation, we can use certain mathematical operations to divide the rates of `operation_duration_seconds_sum` by `operation_duration_seconds_count`. We use the `rate` function to ensure accurate results across many processes and their restart. `rate` transforms Prometheus counters into a rate per second.[13] Then we can use `/` to divide the rates of those metrics. The result of such an average query is presented in Figure 6-5.

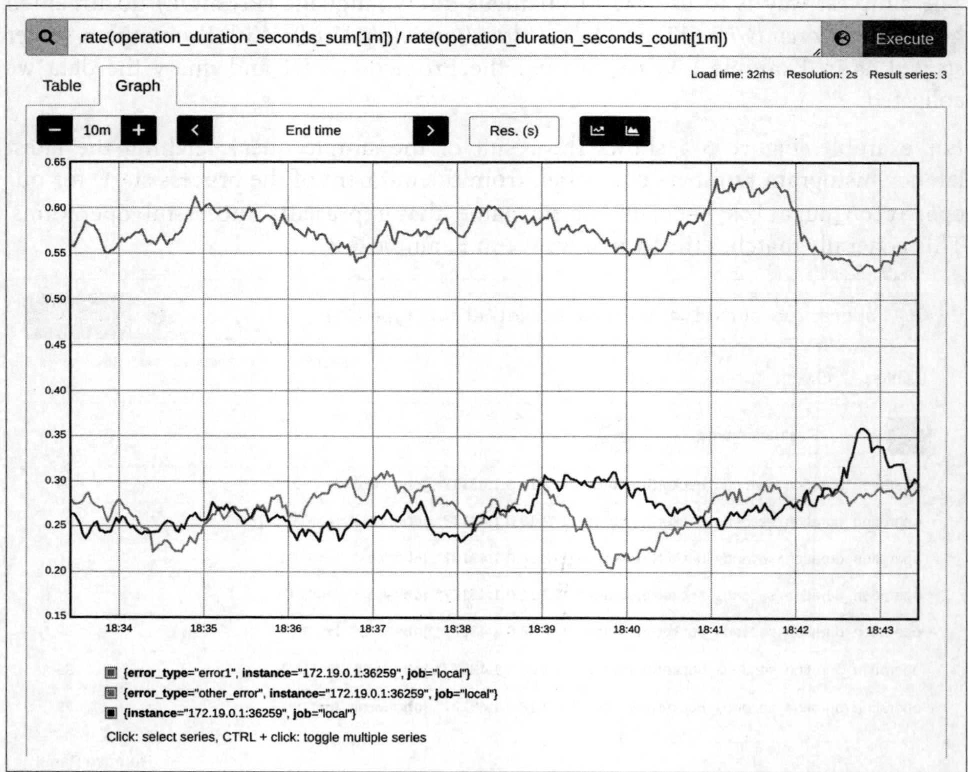

Figure 6-5. PromQL query results representing average latency captured by the Example 6-7 instrumentation graphed in the Prometheus UI

With another query, we can check total operations or, even better, check the rate per minute of those using the `increase` function on our `operation_duration_seconds_count` counter, as presented in Figure 6-6.

13 Note that doing `rate` on the gauges type of metric will yield incorrect results.

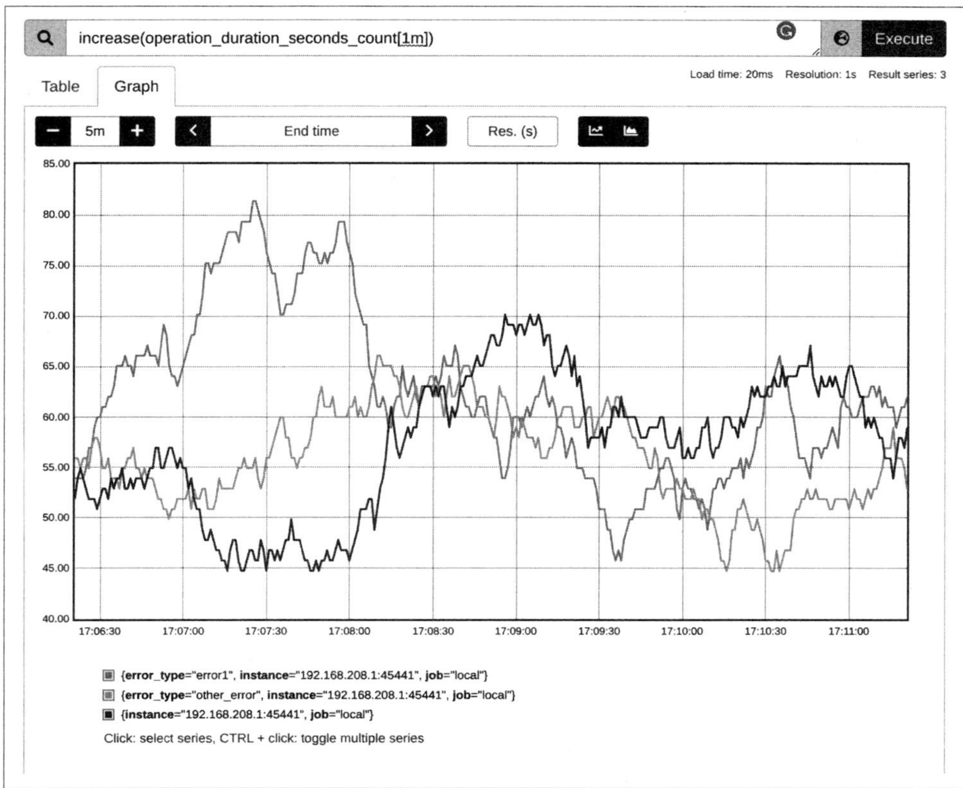

Figure 6-6. PromQL query results representing a rate of operations per minute in our system graphed in the Prometheus UI

There are many other functions, aggregations, and ways of using metric data in the Prometheus ecosystem. We will unpack some of it in later sections.

The amazing part about Prometheus with such a specific scrape technique is that pulling metrics allows our Go client to be ultrathin and efficient. As a result, the Go process does not need to:

- Buffer data samples, spans, or logs in memory or on disk
- Maintain information (and automatically update it!) on where to send potential data
- Implement complex buffering and persisting logic if the metric backend is down temporarily
- Ensure a consistent sample push interval
- Know about any authentication, authorization, or TLS for metric payload

On top of that, the observability experience is better when you pull the data in such a way that:

- Metric users can easily control the scrape interval, targets, metadata, and recordings from a central place. This makes the metric usage simpler, more pragmatic, and generally cheaper.
- It is easier to predict the load of such a system, which makes it easier to scale it and react to the situations that require scaling the collection pipeline.
- Last but not least, pulling metrics allows you to reliably tell your application's health (if we can't scrape metrics from it, it is most likely unhealthy or down). We also typically know what sample is the last one for a metric (staleness).[14]

As with everything, there are some trade-offs. Each pulled, tailed, or scraped signal has its downsides. Typical problems of an observability pull-based system include:

- It is generally harder to pull data from short-lived processes (e.g., CLI and batch jobs).[15]
- Not every system architecture allows ingress traffic.
- It is generally harder to ensure that all the pieces of information will land safely in a remote place (e.g., this pulling is not suitable for auditing).

The Prometheus metrics are designed to mitigate downsides and leverage the strength of the pull model. Most of the metrics we use are counters, which means they only increase. This allows Prometheus to skip a few scrapes from the process but still, in the end, have a perfectly accurate number for each metric within larger time windows, like minutes.

As mentioned before, in the end, metrics (as numeric values) are what we need when it comes to assessing efficiency. It's all about comparing and analyzing numbers. This is why a metric observability signal is a great way to gather required information pragmatically. We will use this signal extensively for "Macrobenchmarks" on page 306 and "Root Cause Analysis, but for Efficiency" on page 330. It's simple, pragmatic, the ecosystem is huge (you can find metric exporters for almost all kinds of software and hardware), it's generally cheap, and it works great with both human users and automation (e.g., alerting).

14 On the contrary, for the push-based system, if you don't see expected data, it's hard to tell if it's because the sender is down or the pipeline to send is down.

15 See our talk from KubeCon EU 2022 (*https://oreil.ly/TtKwH*) about such cases.

Metric observability signals, especially with the Prometheus data model, fit into aggregated information instrumentation. We discussed the benefits, but some limits and downsides are important to understand. All downsides come from the fact that we generally cannot narrow pre-aggregated data down to a state before aggregation, for example, a single event. We might know with metrics how many requests failed, but we don't know the exact stack trace, error message, and so on for a singular error that happened. The most granular information we typically have is a type of error (e.g., status code). This makes the surface of possible questions we can ask a metric system smaller than if we would capture all raw events. Another essential characteristic that might be considered a downside is the cardinality of the metrics and the fact that it has to be kept low.

High Metric Cardinality

Cardinality means the uniqueness of our metric. For example, imagine in Example 6-7 we would inject a unique error string instead of the `error_type` label. Every new label value creates a new, possibly short-lived unique metric. A metric with just a single or a few samples represents more of a raw event, not aggregation over time. Unfortunately, if users try to push event-like information to a system designed for metrics (like Prometheus), it tends to be expensive and slow.

It is very tempting to push more cardinal data to a system designed for metrics. This is because it's only natural to want to learn more from such cheap and reliable signal-like metrics. Avoid that and keep your cardinality low with metric budgets, recording rules, and allow-list relabeling. Switch to event-based systems like logging and tracing if you wish to capture unique information like exact error messages or the latency for a single, specific operation in the system!

Whether gathered from logs, traces, profiles, or metric signals, we already touched on some metrics in previous chapters—for example, CPU core used per second, memory bytes allocated on the heap, or residential memory bytes used per operation. So let's go through some of those in detail and talk about their semantics, how we should interpret them, potential granularity, and example code that illustrates them using signals you have just learned.

There Is No Observability Silver Bullet!

Metrics are powerful. Yet as you learned in this chapter, logging and traces also give enormous opportunities to improve the efficiency observability experience with dedicated tools that allow us to derive metrics from them. In this book, you will see me using all of those tools (together with profiling, which we haven't covered yet) to improve the efficiency of Go programs.

The pragmatic system captures enough of each of those observability signals that fit your use cases. It's unlikely to build metric-only, trace-only, or profiling-only systems!

Efficiency Metrics Semantics

Observability feels like a vast and deep topic that takes years to grasp and set up. The industry constantly evolves, and creating new solutions does not help. However, it will be easier to understand once we start using observability for a specific goal like the efficiency effort. Let's talk about exactly which observability bits are essential to start measuring latency and consumption of the resources we care about, e.g., CPU and memory.

Metrics As Numeric Value Versus Metric Observability Signal

In "Metrics" on page 211, we discussed the metric observability signal. Here we discuss specific metric semantics that are useful to capture for efficiency efforts. To clarify, we can capture those specific metrics in various ways. We can use metric observability signals, but we can also derive them from other signals, like logs, traces, and profiling!

Two things can define every metric:

Semantics
: What's the meaning of that number? What do we measure? With what unit? How do we call it?

Granularity
: How detailed is this information? For example, is it per a unique operation? Is it per a result type of this operation (success versus error)? Per goroutine? Per process?

Metric semantics and granularity both heavily depend on the instrumentation. This section will focus on defining the semantics, granularity, and example instrumentation for the typical metrics we can use to track resource consumption and latency of our software. It is essential to understand the specific measurements we will operate with to work effectively with the benchmark and profiling tools we will learn in "Benchmarking Levels" on page 266 and "Profiling in Go" on page 331. While iterating over those semantics, we will uncover common best practices and pitfalls we have to be aware of. Let's go!

Latency

If we want to improve how fast our program performs certain operations, we need to measure the latency. Latency means the duration of the operation from the start to either success or failure. Thus, the semantics we need feel pretty simple at first glance—we generally want the "amount of time" required to complete our software operation. Our metric will usually have a name containing the words *latency*, *duration*, or *elapsed* with the desired unit. But the devil is in the details, and as you will learn in this section, measuring latency is prone to mistakes.

The preferable unit of the typical latency measurement depends on what kind of operations we measure. If we measure very short operations like compression latency or OS context switch latencies, we must focus on granular nanoseconds. Nanoseconds are also the most granular timing we can count on in typical modern computers. This is why the Go standard library `time.Time` (*https://oreil.ly/QGCme*) and `time.Duration` (*https://oreil.ly/9agLb*) structures measure time in nanoseconds.

Generally speaking, the typical measurements of software operations are almost always in milliseconds, seconds, minutes, or hours. This is why it's often enough to measure latency in seconds, as a floating value, for up to nanoseconds granularity. Using seconds has another advantage: it is a base unit. Using the base unit is often what's natural and consistent across many solutions.[16] Consistency is critical here. You don't want to measure one part of the system in nanoseconds, another in seconds, and another in hours if you can avoid it. It's easy enough to get confused by our data and have a wrong conclusion without trying to guess a correct unit or writing transformations between those.

In the code examples in "Example: Instrumenting for Latency" on page 199, we already mentioned many ways we can instrument latency using various observability signals. Let's extend Example 6-1 in Example 6-10 to show important details that ensure latency is measured as reliably as possible.

16 This is why the Prometheus ecosystem suggests base units (*https://oreil.ly/oJozb*).

Example 6-10. Manual and simplest latency measurement of a single operation that can error out and has to prepare and tear down phases

```
prepare()

for i := 0; i < xTimes; i++ {
    start := time.Now() ❶
    err := doOperation()
    elapsed := time.Since(start) ❷

    // Capture 'elapsed' value using log, trace or metric...

    if err != nil { /* Handle error... */ }
}
tearDown()
```

❶ We capture the `start` time as close as possible to the start of our `doOperation` invocation. This ensures nothing unexpected will get between `start` and operation start that might introduce unrelated latency, which can mislead the conclusion we might take from this metric further on. This, by design, should exclude any potential preparation or setup we have to do for an operation we measure. Let's measure those explicitly as another operation. This is also why you should avoid putting any newline (empty line) between `start` and the invocation of the operation. As a result, the next programmer (or yourself, after some time) won't add anything in between, forgetting about the instrumentation you added.

❷ Similarly, it's important to capture the `finish` time using the `time.Since` helper as soon as we finish, so no unrelated duration is captured. For example, similar to excluding `prepare()` time, we want to exclude any potential close or `tearDown()` duration. Moreover, if you are an advanced Go programmer, your intuition is always to check errors when some functions finish. This is critical, but we should do that for instrumentation purposes after we capture the latency. Otherwise, we might increase the risk that someone will not notice our instrumentation and will add unrelated statements between what we measure and `time.Since`. On top of that, in most cases, you want to make sure you measure the latency of both successful and failed operations to understand the complete picture of what your program is doing.

Shorter Latencies Are Harder to Measure Reliably

The method for measuring operation latency shown in Example 6-10 won't work well for operations that finish under, let's say, 0.1 microseconds (100 nanoseconds). This is because the effort of taking the system clock number, allocating variables, and further computing `time.Now()` and `time.Since` functions can take its time too, which is significant for such short measurements.[17] Furthermore, as we will learn in "Reliability of Experiments" on page 256, every measurement has some variance. The shorter latency, the more impactful this noise can be.[18] This also applies to tracing spans measuring latency.

One solution for measuring very fast functions is used by the Go benchmark as presented by Example 6-3, where we estimate average latency per operation by doing many of them. More on that in "Microbenchmarks" on page 275.

Time Is Infinite; the Software Structures Measuring that Time Are Not!

When measuring latency, we have to be aware of the limitations of time or duration measurements in software. Different types can contain different ranges of numeric values, and not all of them can contain negative numbers. For example:

- `time.Time` can only measure time from January 1, 1885[19] up until 2157.
- The `time.Duration` type can measure time (in nanoseconds) approximately between -290 years before your "starting" point and up to 290 years after your "starting" point.

If you want to measure things outside of those typical values, you need to extend those types or use your own. Last but not least, Go is prone to the leap second problem (*https://oreil.ly/MeZ4b*) and time skews of the operating systems. On some systems, the `time.Duration` (monotonic clock) will also stop if the computer goes to sleep (e.g., laptop or virtual machine suspend), which will lead to wrong measurements, so keep that in mind.

---

17 For example, on my machine `time.Now` and `time.Since` take around 50–55 nanoseconds.

18 This is why it's better to make thousands or even more of the same operation, measure the total latency, and get the average by dividing it by a number of operations. As a result, this is what Go benchmark is doing, as we will learn in "Go Benchmarks" on page 277.

19 Did you know this date was picked simply because of *Back to the Future Part II* (*https://oreil.ly/Oct6X*)?

We discussed some typical latency metric semantics. Now let's move to the granularity question. We can decide to measure the latency of operation A or B in our process. We can measure a group of operations (e.g., transaction) or a single suboperation of it. We can gather this data across many processes or look only at one, depending on what we want to achieve.

To make it even more complex, even if we choose a single operation as our granularity to measure latency, that single operation has many stages. In a single process this can be represented by stack trace, but for multiprocess systems with some network communication, we might need to establish additional boundaries.

Let's take some programs as an example, as the Caddy HTTP web server explained in the previous chapter, with a simple REST (*https://oreil.ly/SHEor*) HTTP call to retrieve an HTML as our example operation. What latencies should we measure if we install such a Go program in a cloud on production to serve our REST HTTP call to the client (e.g., someone's browser)? The example granularities we could measure latency for are presented in Figure 6-7.

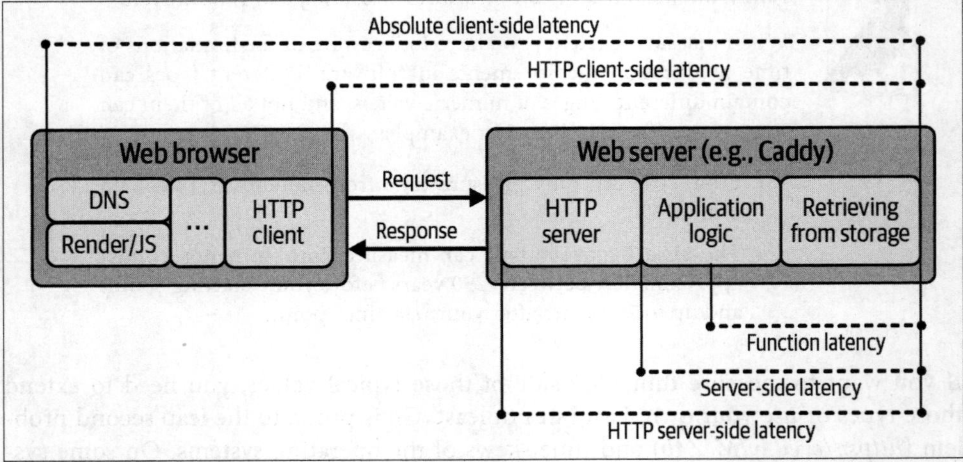

Figure 6-7. Example latency stages we can measure for in our Go web server program communicating with the user's web browser

We can outline five example stages:

Absolute (total) client-side latency
: The latency measured exactly from the moment the user hits Enter in the URL input in the browser, up until the whole response is retrieved, content is loaded, and the browser renders all.

HTTP client-side latency (response time)
: The latency captured from the moment the first bytes of the HTTP request on the client side are being written to a new or reused TCP connection, up until the client receives all bytes of the response. This excludes everything that happens before (e.g., DNS lookup) or after (rendering HTML and JavaScript in the browser) on the client side.

HTTP server-side latency
: The latency is measured from the moment the server receives the first bytes of the HTTP request from the client, up until the server finishes writing all bytes of the HTTP response. This is typically what we are measuring if we use the HTTP middlewares pattern (*https://oreil.ly/JsONO*) in Go.

Server-side latency (service time)
: The latency of server-side computation required to answer the HTTP request, measured without HTTP request parsing and response encoding. Latency is from the moment of having the HTTP request parsed to the moment when we start encoding and sending the HTTP response.

Server-side function latency
: The latency of a single server-side function computation from the moment of invocation, up until the function work is finished and return arguments are in the context of the caller function.

These are just some of the many permutations we can use to measure latency in our Go programs or systems. Which one should we pick for our optimizations? Which matters the most? It turns out that all of them have their use case. The priority of what latency metric granularity we should use and when depends solely on our goals, the accuracy of measurements as explained in "Reliability of Experiments" on page 256, and the element we want to focus on as discussed in "Benchmarking Levels" on page 266. To understand the big picture and find the bottleneck, we have to measure a few of those different granularities at once. As discussed in "Root Cause Analysis, but for Efficiency" on page 330, tools like tracing and profiling can help with that.

Whatever Metric Granularity You Choose, Understand and Document What You Measure!

We waste a lot of time if we take the wrong conclusions from measurements. It is easy to forget or misunderstand what parts of granularity we are measuring. For example, you thought you were measuring server-side latency, but slow client software is introducing latency you felt you didn't include in your metric. As a result, you might be trying to find a bottleneck on the server side, whereas a potential problem might be in a different process.[20] Understand, document, and be explicit with your instrumentation to avoid those mistakes

In "Example: Instrumenting for Latency" on page 199, we discussed how we could gather latencies. We mentioned that generally, we use two main measuring methods for efficiency needs in the Go ecosystem. Those two ways are typically the most reliable and cheapest (useful when performing load tests and benchmarks):

- Basic logging using "Microbenchmarks" on page 275 for isolated functionality, single process measurements
- Metrics such as Example 6-7 for macro measurements that involve larger systems with multiple processes

Especially in the second case, as mentioned previously, we have to measure latency many times for a single operation to get reliable efficiency conclusions. We don't have access to raw latency numbers for each operation with metrics—we have to choose some aggregation. In Example 6-2, we proposed a simple average aggregation mechanism inside instrumentation. With metric instrumentation, this would be trivial to achieve. It's as easy as creating two counters: one for the `sum` of latencies and one for the `count` of operations. We can evaluate collected data with those two metrics into a mean (arithmetic average).

Unfortunately, the average is too naive an aggregation. We can miss lots of important information about the characteristics of our latency. In "Microbenchmarks" on page 275, we can do a lot with the mean for basic statistics (this is what the Go benchmarking tool is using), but in measuring the efficiency of our software in the bigger system with more unknowns, we have to be mindful. For example, imagine we want to improve the latency of one operation that used to take around 10 seconds. We made a potential optimization using our TFBO flow. We want to assess the

20 The noteworthy example from my experience is measuring server-side latency of REST with a large response or HTTP/gRPC with a streamed response. The server-side latency does not depend only on the server but also on how fast the network and client side can consume those bytes (and write back acknowledge packets within TCP control flow (*https://oreil.ly/jcrSF*)).

efficiency on the macro level. During our tests, the system performed 500 operations within 5 seconds (faster!), but 50 operations were extremely slow, with a 40-second latency. Suppose we would stick to the average (8.1 seconds). In that case, we could make the wrong conclusion that our optimization was successful, missing the potential big problem that our optimization caused, leading to 9% of operations being extremely slow.

This is why it's helpful to measure specific metrics (like latency) in percentiles. This is what Example 6-7 instrumentation is for with the metric histogram type for our latency measurements.

> Most metrics are better thought of as distributions rather than averages. For example, for a latency SLI [service level indicator], some requests will be serviced quickly, while others will invariably take longer—sometimes much longer. A simple average can obscure these tail latencies, as well as changes in them. (...) Using percentiles for indicators allows you to consider the shape of the distribution and its differing attributes: a high-order percentile, such as the 99th or 99.9th, shows you a plausible worst-case value, while using the 50th percentile (also known as the median) emphasizes the typical case.
>
> —C. Jones et al., *Site Reliability Engineering*, "Service Level Objectives" (*https://oreil.ly/rMBW3*) (O'Reilly, 2016)

The histogram metric I mentioned in Example 6-8 is great for latency measurements, as it counts how many operations fit into a certain latency range. In Example 6-7, I have chosen[21] exponential buckets `0.001, 0.01, 0.1, 1, 10, 100`. The largest bucket should represent the longest operation duration you expect in your system (e.g., a timeout).[22]

In "Metrics" on page 211, we discussed how we can use metrics using `PromQL`. For the histogram type of metrics and our latency semantics, the best way to understand this is to use the `histogram_quantile` function. See the example output in Figure 6-8 for the median, and Figure 6-9 for the 90th percentile.

21 Right now, the choice of buckets in a histogram if you want to use Prometheus is manual. However, the Prometheus community is working on sparse histograms (*https://oreil.ly/qFdC1*) with a dynamic number of buckets that adjust automatically.

22 More on using histograms can be read here (*https://oreil.ly/VrWGe*).

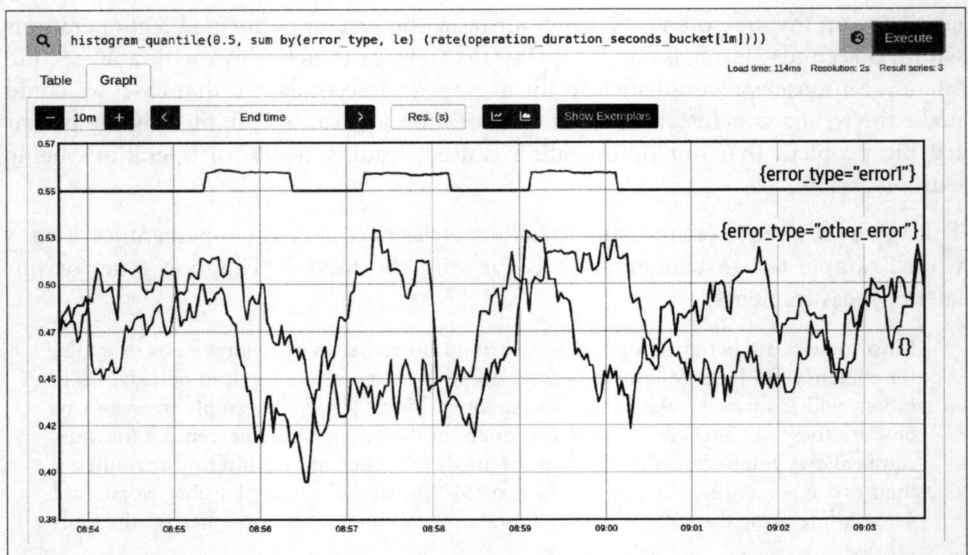

Figure 6-8. Fiftieth percentile (median) of latency across an operation per error type from our Example 6-7 instrumentation

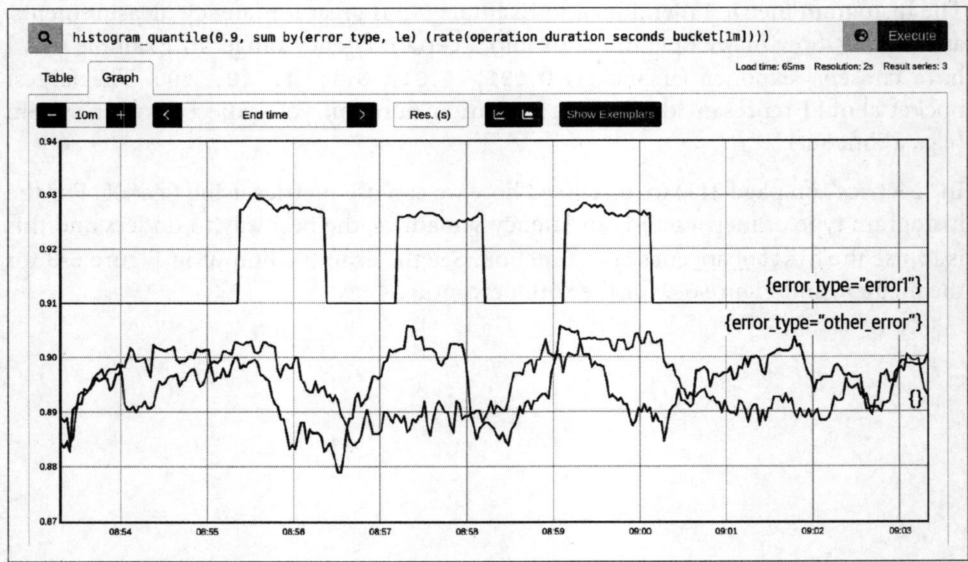

Figure 6-9. Ninetieth percentile of latency across the operation per error type from our Example 6-7 instrumentation

Both results can lead to interesting conclusions for the program I measured. We can observe a few things:

- Half of the operations were generally faster than 590 milliseconds, while 90% were faster than 1 second. So if our RAER ("Resource-Aware Efficiency Requirements" on page 86) states that 90% of operations should be less than 1 second, it could mean we don't need to optimize further.
- Operations that failed with `error_type=error1` were considerably slower (most likely some bottleneck exists in that code path).
- Around 17:50 UTC, we can see a slight increase in latencies for all operations. This might mean some side effect or change in the environment that caused my laptop's operating system to give less CPU to my test.[23]

Such measured and defined latency can help us determine if our latency is good enough for our requirements and if any optimization we do helps or not. It can also help us to find parts that cause slowness using different benchmarking and bottleneck-finding strategies. We will explore those in Chapter 7.

With the typical latency metric definition and example instrumentation, let's move to the next resource we might want to measure in our efficiency journey: CPU usage.

CPU Usage

In Chapter 4, you learned how CPU is used when we execute our Go programs. I also explained that we look at CPU usage to reduce CPU-driven latency[24] and cost, and to enable running more processes on the same machine.

A variety of metrics allow us to measure different parts of our program's CPU usage. For example, with Linux tools like the `proc` filesystem (*https://oreil.ly/MJVHl*) and `perf` (*https://oreil.ly/QPMD9*), we can measure our Go program's miss and hit rates, CPU branch prediction hit rates (*https://oreil.ly/VdENl*), and other low-level statistics. However, for basic CPU efficiency, we usually focus on the CPU cycles, instructions, or time used:

CPU cycles
: The total number of CPU clock cycles used to execute the program thread instructions on each CPU core.

23 It makes sense. I was utilizing my web browser heavily during the test, which confirms the knowledge we will discuss in "Reliability of Experiments" on page 256.

24 As a reminder, we can improve the latency of our program's functionality in many ways other than just by optimizing its CPU usage. We can improve that latency using concurrent execution that often increases total CPU time.

CPU instructions

The total number of CPU instructions of our program's threads executed in each CPU core. On some CPUs from the RISC architecture (*https://oreil.ly/ofvB7*) (e.g., ARM processors), this might be equal to the number of cycles, as one instruction always takes one cycle (amortized cost). However, on the CISC architecture (e.g., AMD and Intel x64 processors), different instructions might use additional cycles. Thus, counting how many instructions our CPU had to do to complete some program's functionality might be more stable.

Both cycles and instructions are great for comparing different algorithms with each other. It is because they are less noisy as:

- They don't depend on the frequency the CPU core had during the program run
- Latency of memory fetches, including different caches, misses, and RAM latency

CPU time

The time (in seconds or nanoseconds) our program thread spends executing on each CPU core. As you will learn in "Off-CPU Time" on page 369, this time is different (longer or shorter) from the latency of our program, as CPU time does not include I/O waiting time and OS scheduling time. Furthermore, our program's OS threads might execute simultaneously on multiple CPU cores. Sometimes we also use CPU time divided by the CPU capacity, often referred to as CPU usage. For example, 1.5 CPU usage in seconds means our program requires (on average) one CPU core for 1 second and a second core for 0.5 seconds.

On Linux, the CPU time is often split into User and System time:

- User time represents the time the program spends executing on the CPU in the user space.
- System time is the CPU time spent executing certain functions in the kernel space on behalf of the user, e.g., syscalls like `read` (*https://oreil.ly/xEQuM*).

Usually, on higher levels such as containers, we don't have the luxury of having all three metrics. We mostly have to rely on CPU time. Fortunately, the CPU time is typically a good enough metric to track down the work needed from our CPUs to execute our workload. On Linux, the simplest way to retrieve the current CPU time counted from the start of the process is to go to */proc/<PID>/stat* (where `PID` means the process ID). We also have similar statistics on the thread level in */proc/<PID>/*

tasks/<TID>/stat (where `TID` means the thread ID). This is exactly what utilities like `ps` or `htop` use.[25]

The `ps` and `htop` tools might be indeed the simplest tools to measure the CPU time in the current moment. However, we usually need to assess the CPU time required for the full functionality we are optimizing. Unfortunately, "Go Benchmarks" on page 277 is not providing CPU time (only latency and allocations) per operation. You could perhaps obtain that number from the `stat` file, e.g., programmatically using the `procfs` Go library (*https://oreil.ly/ZcCDn*), but there are two main ways I would suggest instead:

- CPU profiling, explained in "CPU" on page 367.
- Prometheus metric instrumentation. Let's quickly look at that method next.

In Example 6-7, I showed a Prometheus instrumentation that registers custom latency metrics. It's also very easy to add the CPU time metric, but the Prometheus client library (*https://oreil.ly/1r2zw*) has already built helpers for that. The recommended way is presented in Example 6-11.

Example 6-11. Registering `proc stat` instrumentation about your process for Prometheus use

```
import (
    "net/http"

    "github.com/prometheus/client_golang/prometheus"
    "github.com/prometheus/client_golang/prometheus/collectors"
    "github.com/prometheus/client_golang/prometheus/promhttp"
)

func ExampleCPUTimeMetric() {
    reg := prometheus.NewRegistry()
    reg.MustRegister(
        collectors.NewProcessCollector(collectors.ProcessCollectorOpts{}),
    ) ❶

    go func() {
        for i := 0; i < xTimes; i++ {
             err := doOperation()
             // ...
        }
    }()
```

25 Also a useful `procfs` Go library (*https://oreil.ly/ZcCDn*) that allows retrieving `stats` file data number programmatically.

```
    err := http.ListenAndServe(
        ":8080",
        promhttp.HandlerFor(reg, promhttp.HandlerOpts{}),
    )
    // ...
}
```

❶ The only thing you have to do to have the CPU time metric with Prometheus is to register the `collectors.NewProcessCollector` that uses the `/proc stat` file mentioned previously.

The `collectors.ProcessCollector` provides multiple metrics, like `process_open_fds`, `process_max_fds`, `process_start_time_seconds`, and so on. But the one we are interested in is `process_cpu_seconds_total`, which is a counter of CPU time used from the beginning of our program. What's special about using Prometheus for this task is that it collects the values of this metric periodically from our Go program. This means we can query Prometheus for the process CPU time for a certain time window and map that to real time. We can do that with the `rate` (*https://oreil.ly/8BaUw*) function duration that gives us the per second rate of that CPU time in a given time window. For example, `rate(process_cpu_seconds_ total{}[5m])` will give us the average CPU per second time that our program had during the last five minutes.

You will find an example CPU time analysis based on this kind of metric in "Understanding Results and Observations" on page 316. However, for now, I would love to show you one interesting and common case, where `process_cpu_seconds_total` helps narrow down a major efficiency problem. Imagine your machine has only two CPU cores (or we limit our program to use two CPU cores), you run the functionality you want to assess, and you see the CPU time rate of your Go program looking like Figure 6-10.

Thanks to this view, we can tell that the `labeler` process is experiencing a state of CPU saturation. This means that our Go process requires more CPU time than was available. Two signals tell us about the CPU saturation:

- The typical "healthy" CPU usage is spikier (e.g., as presented in Figure 8-4 later in the book). This is because it's unlikely that typical applications use the same amount of CPU all the time. However, in Figure 6-10, we see the same CPU usage for five minutes.
- Because of this, we never want our CPU time to be so close to the CPU limit (two in our case). In Figure 6-10, we can clearly see a small choppiness around the CPU limit, which indicates full CPU saturation.

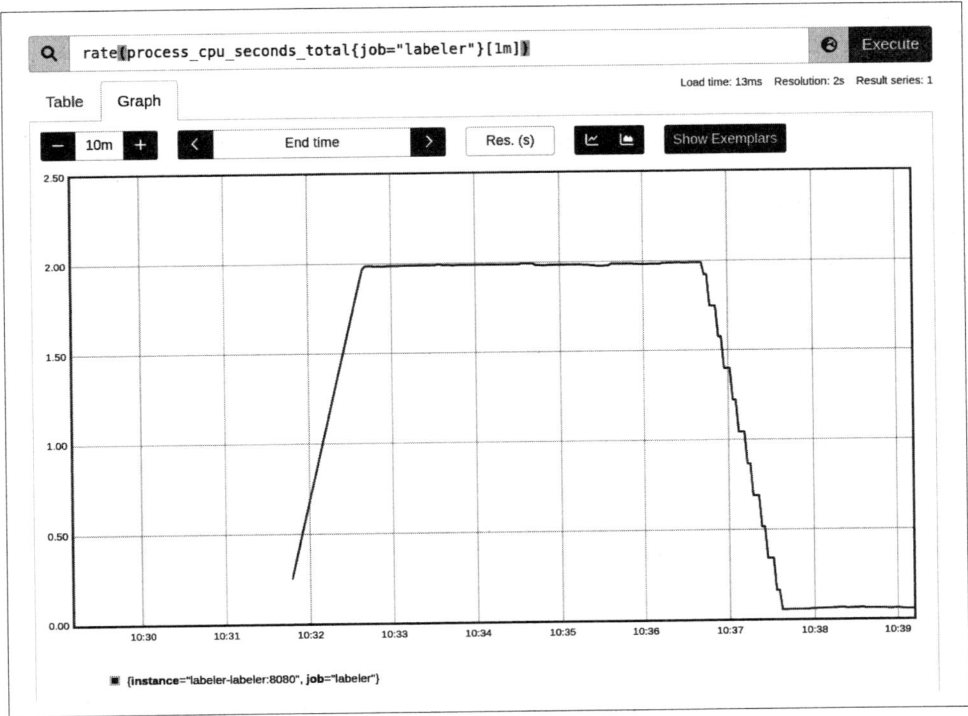

Figure 6-10. The Prometheus graph view of the CPU time for the `labeler` Go program (we will use it in an example in "Macrobenchmarks" on page 306) after a test

Knowing when we are at saturation of our CPU is critical. First of all, it might give the wrong impression that the current CPU time is the maximum that the process needs. Moreover, this situation also significantly slows down our program's execution time (increases latency) or even stalls it completely. This is why the Prometheus-based CPU time metric, as you learned here, has proven to be critical for me in learning about such saturation cases. It is also one of the first things you must find out when analyzing your program's efficiency. When saturation happens, we have to give more CPU cores to the process, optimize the CPU usage, or decrease the concurrency (e.g., limit the number of HTTP requests it can do concurrently).

On the other hand, CPU time allows us to find out about opposite cases where the process might be blocked. For example, if you expect CPU-bound functionality to run with 5 goroutines, and you see the CPU time of 0.5 (50% of one CPU core), it might mean the goroutines are blocked (more on that in "Off-CPU Time" on page 369) or whole machine and OS are busy.

Let's now look at memory usage metrics.

Memory Usage

As we learned in Chapter 5, there are complex layers of different mechanics on how our Go program uses memory. This is why the actual physical memory (RAM) usage is one of the most tricky to measure and attribute to our program. On most systems with an OS memory management mechanism like virtual memory, paging, and shared pages, every memory usage metric will be only an estimation. While imperfect, this is what we have to work with, so let's take a short look at what works best for the Go program.

There are two main sources of memory usage information for our Go process: the Go runtime heap memory statistics and the information that OS holds about memory pages. Let's start with the in-process runtime stats.

runtime heap statistics

As we learned in "Go Memory Management" on page 172, the heap segment of the Go program virtual memory can be an adequate proxy for memory usage. This is because most bytes are allocated on the heap for typical Go applications. Moreover, such memory is also never evicted from the RAM (unless the swap is enabled). As a result, we can effectively assess our functionality's memory usage by looking at the heap size.

We are often most interested in assessing the memory space or the number of memory blocks needed to perform a certain operation. To try to estimate this, we usually use two semantics:

- The total allocations of bytes or objects on the heap allow us to look at memory allocations without often nondeterministic GC impact.
- The number of currently in-use bytes or objects on the heap.

The preceding statistics are very accurate and quick to access because Go runtime is responsible for heap management, so it tracks all the information we need. Before Go 1.16, the recommended way to access those statistics programmatically was using the `runtime.ReadMemStats` function (*https://oreil.ly/AwX75*). It still works for compatibility reasons, but unfortunately, it requires STW (stop the world) events to gather all memory statistics. As a result of Go 1.16, we should all use the `runtime/metrics` (*https://oreil.ly/WYiOd*) package that provides many cheap-to-collect insights about GC, memory allocations, and so on. The example usage of this package to get memory usage metrics is presented in Example 6-12.

Example 6-12. The simplest code prints total heap allocated bytes and currently used ones

```
import(
    "fmt"
    "runtime"
    "runtime/metrics"
)

var memMetrics = []metrics.Sample{
    {Name: "/gc/heap/allocs:bytes"}, ❶
    {Name: "/memory/classes/heap/objects:bytes"},
}

func printMemRuntimeMetric() {
    runtime.GC() ❷
    metrics.Read(memMetrics) ❸

    fmt.Println("Total bytes allocated:", memMetrics[0].Value.Uint64()) ❹
    fmt.Println("In-use bytes:", memMetrics[1].Value.Uint64())
}
```

❶ To read samples from `runtime/metrics`, we must first define them by referencing the desired metric name. The full list of metrics might be different (mostly added ones) across different Go versions, and you can see the list with descriptions at *pkg.go.dev* (*https://oreil.ly/HWGUJ*). For example, we can obtain the number of objects in a heap.

❷ Memory statistics are recorded right after a GC run, so we can trigger GC to have the latest information about the heap.

❸ `metrics.Read` populates the value of our samples. You can reuse the same sample slice if you only care about the latest values.

❹ Both metrics are of `uint64` type, so we use the `Uint64()` method to retrieve the value.

Programmatically accessing this information is useful for local debugging purposes, but it's not sustainable on every optimization attempt. That's why in the community, we typically see other ways to access that data:

- Go benchmarking, explained in "Go Benchmarks" on page 277
- Heap profiling, explained in "Heap" on page 360
- Prometheus metric instrumentation

To register `runtime/metric` as Prometheus metrics, we can add a single line to Example 6-11: `reg.MustRegister(collectors.NewGoCollector())`. The Go collector is a structure that, by default, exposes various memory statistics (*https://oreil.ly/Ib8D2*). For historical reasons, those map to the `MemStats` Go structure, so the equivalents to the metrics defined in Example 6-12 would be `go_memstats_heap_alloc_bytes_total` for a counter, and `go_memstats_heap_alloc_bytes` for a current usage gauge. We will show an analysis of Go heap metrics in "Go e2e Framework" on page 310.

Unfortunately, heap statistics are only an estimation. It is likely that the smaller the heap on our Go program, the better the memory efficiency. However, suppose you add some deliberate mechanisms like large off-heap memory allocations using explicit `mmap` syscall or thousands of goroutines with large stacks. In that case, that can cause an OOM on your machine, yet it's not reflected in the heap statistics. Similarly, in "Go Allocator" on page 181, I explained rare cases where only part of the heap space is allocated on physical memory.

Still, despite the downsides, heap allocations remain the most effective way to measure memory usage in modern Go programs.

OS memory pages statistics

We can check the numbers the Linux OS tracks per thread to learn more realistic yet more complex memory usage statistics. Similar to "CPU Usage" on page 229, `/proc/`*<PID>*`/statm` provides the memory usage statistics, measured in pages. Even more accurate numbers can be retrieved from per memory mapping statistics that we can see in `/proc/`*<PID>*`/smaps` ("OS Memory Mapping" on page 168).

Each page in this mapping can have a different state. A page might or might not be allocated on physical memory. Some pages might be shared across processes. Some pages might be allocated in physical memory and accounted for as memory used, yet marked by the program as "free" (see the `MADV_FREE` release method mentioned in "Garbage Collection" on page 185). Some pages might not even be accounted for in the `smaps` file, because for example, it's part of filesystem Linux cache buffers (*https://oreil.ly/uchws*). For these reasons, we should be very skeptical about the absolute values observed in the following metrics. In many cases, OS is lazy in releasing memory; e.g., part of the memory used by the program is cached in the best way that will be released immediately as long as somebody else is needing that.

There are a few typical memory usage metrics we can obtain from the OS about our process:

VSS

Virtual set size represents the number of pages (or bytes, depending on instrumentation) allocated for the program. Not very useful metrics, as most virtual pages are never allocated on RAM.

RSS

Residential set size represents the number of pages (or bytes) resident in RAM. Note that different metrics might account for that differently; e.g., the cgroups RSS metric (*https://oreil.ly/NL5Ab*) does not include file-mapped memory, which is tracked separately.

PSS

Proportional set size represents memory with shared memory pages divided equally among all users.

WSS

Working set size estimates the number of pages (or bytes) currently used to perform work by our program. It was initially introduced by Brendan Gregg (*https://oreil.ly/rWy8D*) as the hot, frequently used memory—the minimum memory requirement by the program.

The idea is that a program might have allocated 500 GB of memory, but within a couple of minutes, it might use only 50 MB for some localized computation. The rest of the memory could be, in theory, safely offloaded to disk.

There are many implementations of WSS, but the most common I see is the cadvisor interpretation (*https://oreil.ly/mXjA3*) using the cgroup memory controller (*https://oreil.ly/ovSlH*). It calculates the WSS as the RSS (including file mapping), plus some part of the cache pages (cache used for disk reads or writes), minus the `inactive_file` entry—so file mapping that were not touched for some time. It does not include inactive anonymous pages because the typical OS configuration can't offload anonymous pages to disk (swap is disabled).

In practice, RSS or WSS is used to determine the memory usage of our Go program. Which one highly depends on the other workloads on the same machine and follows the flow of the RAM usage expanding to all available space, as mentioned in "Do We Have a Memory Problem?" on page 152. The usefulness of each depends on the current Go version and instrumentation that gives you those metrics. In my experience, with the latest Go version and cgroup metrics, the RSS metric tends to give more reliable results.[26] Unfortunately, accurate or not, WSS is used in systems like Kubernetes

26 One reason is the issue (*https://oreil.ly/LKmSA*) in cadvisor that includes some still-reclaimable memory in the WSS.

to trigger evictions (e.g., OOM) (*https://oreil.ly/lnDkI*), thus we should use it to assess memory efficiency that might lead to OOMs.

Given my focus on infrastructure Go programs, I heavily lean on a metric exporter called cadvisor (*https://oreil.ly/RJzKd*) that converts cgroup metrics to Prometheus metrics. I will explain using it in detail in "Go e2e Framework" on page 310. It allows analyzing metrics like `container_memory_rss + container_memory_mapped_file` and `container_memory_working_set_bytes`, which are commonly used in the community.

Summary

Modern observability offers a set of techniques essential for our efficiency assessments and improvements. However, some argue that this kind of observability designed primarily for DevOps, SREs, and cloud-native solutions can't work for developer use cases (in the past known as Application Performance Monitoring [APM]).

I would argue that the same tools can be used for both developers (for those efficiency and debugging journeys) and system admins, operators, DevOps, and SREs to ensure the programs delivered by others are running effectively.

In this chapter, we discussed the three first observability signals: metrics, logs, and tracing. Then, we went through example instrumentations for those in Go. Finally, I explained common semantics for the latency, CPU time, and memory usage measurements we will use in later chapters.

Now it's time to learn how to use that efficiency observability to make data-driven decisions in practice. First, we will focus on how to simulate our program to assess the efficiency on different levels.

CHAPTER 7

Data-Driven Efficiency Assessment

You learned how to observe our Go program using different observability signals in the previous chapter. We discussed how to transform those signals to numeric values, or metrics, to effectively observe and assess the latency and resource consumption of the program.

Unfortunately, knowing how to measure the current or maximum consumption or latency for running a program does not guarantee the correct assessment of the overall program efficiency for our application. What we are missing here is the experiment part, which might be the most challenging part of optimization generally: how to trigger situations that are worth measuring with the observability tools mentioned in Chapter 6!

The Definition of Measuring

I find the verb "to measure" very imprecise. I have seen this word overused to describe two things: the process of performing an experiment and gathering numeric data from it.

In this book, every time you read about the "measuring" process, I follow the definition used in metrology (the science of measurement) (*https://oreil.ly/5PRMp*). I precisely mean the process of using the instruments to quantify what is happening now (e.g., the latency of the event, or how much memory it required) or what happened in a given time window. Everything that leads to this event that we measure (simulated by us in a benchmark or occurring naturally) is a separate topic, discussed in this chapter.

In this chapter, I will introduce you to the art of experimentation and measurement for efficiency purposes. I will mainly focus on data-driven assessment, more commonly known as benchmarking. This chapter will help you understand the best practices before we jump to writing benchmarking code in Chapter 8. These practices will also be invaluable in Chapter 9, which focuses on profiling.

I start with complexity analysis as a less empirical way of assessing the efficiency of our solutions. Then, I will explain benchmarking in "The Art of Benchmarking" on page 250. We will compare it to functional testing and clarify the common stereotype that claims "benchmarks always lie."

Later in "Reliability of Experiments" on page 256, we will move to the reliability aspect of our experiments for both benchmarking and profiling purposes. I will provide the ground rules to avoid wasting time (or money) by gathering bad data and making wrong conclusions.

Finally, in "Benchmarking Levels" on page 266, I will introduce you to the full landscape of benchmark strategies. In the previous chapters, I already used benchmarks to provide data that explained the behavior of CPU or memory resources. For example, in "Consistent Tooling" on page 45, I mentioned that the Go tooling provides a standard benchmarking framework. But the benchmarking skill I want to teach you in this chapter goes beyond that, and it is just one tool of many discussed in "Microbenchmarks" on page 275. There are many different ways of assessing the efficiency of our Go code. Knowing when to use what is key.

Let's start by introducing the benchmarking tests and what the critical aspects of those are.

Complexity Analysis

We don't always have the luxury of having empirical data that guides us through the efficiency of a certain solution. Your idea of a better system or algorithm might not be implemented yet and would require a lot of effort to do so before we could benchmark it. Additionally, I mentioned the need for complexity estimation in "Example of Defining RAER" on page 90.

This might feel contradictory to what we learned in "Optimization Challenges" on page 79 ("programmers are notoriously bad at estimating exact resource consumption"), but sometimes engineers rely on theoretical analysis to assess the program. One example is when we assess optimizations on the algorithm level (from "Optimization Design Levels" on page 98). Developers and scientists often use complexity analysis to compare and decide what algorithm might fit better to solve certain problems with certain constraints. More specifically, they use asymptotic notations (commonly known as "Big O" complexities). Most likely, you have heard about them, as they are commonly asked about during any software engineering interview.

However, to fully understand asymptotic notations, you must know what "estimated" efficiency complexity means and what it looks like!

"Estimated" Efficiency Complexity

I mentioned in "Resource-Aware Efficiency Requirements" on page 86 that we can represent the CPU time or consumption of any resources as a mathematical function related to specific input parameters. Typically, we talk about *runtime* complexity, which tells us about the CPU time required to perform a certain operation using a particular piece of code and environment. However, we also have *space* complexity, which can describe the required memory, disk space, or other space requirements for that operation.

For example, let's take our `Sum` function from Example 4-1. I can prove that such code has estimated space complexity (representing heap allocations) of the following function, where N is a number of integers in the input file:

$$space(N) \ = (848 + 3.6 * N) + (24 + 24 * N) + (2.8 * N)\ bytes = 872 + 30.4 * N\ bytes$$

Knowing detailed complexity is great, but typically it's impossible or hard to find the true complexity function because there are too many variables. We can, however, try to estimate those, especially for more deterministic resources like memory allocation, by simplifying the variables. For example, the preceding equation is only an estimation with a simplified function that takes only one parameter—the number of integers. Of course, this code also depends on the size of integers, but I assumed the integer is ~3.6 bytes long (statistic from my test input).

"Estimated" Complexity

As I try to teach you in this book—be precise with the wording.

I was so wrong for all those years, thinking that complexity always means Big O asymptotic complexity. Turns out the complexity exists too (*https://oreil.ly/LG5qb*) and can be very useful in some cases. At least we should be aware it exists!

Unfortunately, it's easy to confuse it with asymptotic complexity, so I would propose calling the one that cares about constants—the "estimated" complexity.

How did I find this complexity equation? It wasn't trivial. I had to analyze the source code, do some stack escape analysis, run multiple benchmarks, and use profiling (so all the things you will learn in this and the next two chapters) to discover those complexities.

This Is Just an Example!

Don't worry. To assess or optimize your code, you don't need to perform such detailed complexity analysis, especially in such detail. I did this to show it's possible and what it gives, but there are more pragmatic ways to assess efficiency quickly and find out the next optimizations. You will see example flows in Chapter 10.

Funny enough, at the end of the TFBO flow, when you optimized one part of your program a lot, you might have a detailed awareness of the problem space so that you could find such complexity quickly. However, doing this for every version of your code would be wasteful.

It might be useful to explain the process of gathering the complexity and mapping it to the source code, as shown in Example 7-1.

Example 7-1. Complexity analysis of Example 4-1

```
func Sum(fileName string) (ret int64, _ error) {
   b, err := os.ReadFile(fileName) ❶
   if err != nil {
      return 0, err
   }

   for _, line := range bytes.Split(b, []byte("\n")) { ❷
      num, err := strconv.ParseInt(string(line), 10, 64) ❸
      if err != nil {
         return 0, err
      }

      ret += num
   }

   return ret, nil
}
```

❶ We can attach the 848 + 3.6 * *N* part of the complexity equation to the operation of reading the file content into memory. The test input I used is very stable—the integers have a different number of digits, but on average they have 2.6 digits. Adding a new line (`\n`) character means every line has approximately 3.6 bytes. Since `ReadFile` returns a byte array with the content of the input file, we can say that our program requires exactly 3.6 * *N* bytes for the byte array pointed to by the `b` slice. The constant amount of 848 bytes comes from various objects allocated on the heap in the `os.ReadFile` function—for example, the slice value for `b` (24 bytes), which escaped the stack. To discover that constant, it was enough to benchmark with an empty file and profile it.

❷ As you will learn in Chapter 10, the `bytes.Split` is quite expensive when it comes to both allocations and runtime latency. However, we can attribute most of the allocations to this part, so to the 24 + 24 * *N* complexity part. It's the "majority" because it's the largest constant (24) multiplied by the input size. The reason is the allocation needed to return the `[][]byte` (*https://oreil.ly/Be0OF*) data structure. While we don't copy the underlying byte arrays (we share it with the buffer from `os.ReadFile`), the *N* allocated empty `[]byte` slices require 24 * *N* of the heap in total, plus the 24 for the `[][]byte` slice header. This is a huge allocation if *N* is on the order of billions (22 GB for a billion integers).

❸ Finally, as we learned in "Values, Pointers, and Memory Blocks" on page 176 and as we will uncover in "Optimizing runtime.slicebytetostring" on page 389, we allocate on this line a lot too. It's not visible at first, but the memory required for `string(line)` (which is always a copy) is escaping to heap.[1] This attributes to the 2.8 * *N* part of the complexity because we do this conversion N times for 2.6 digits on average. The source of the remaining 0.2 * *N* is unknown.[2]

I hope that with this analysis, you see what complexity means. Perhaps you already see how useful it is to know. Maybe you already see many optimization opportunities, which we will try in Chapter 10!

Asymptotic Complexity with Big O Notation

The asymptotic complexity ignores the overheads of the implementation, particularly hardware or environment. Instead, it focuses on asymptotic mathematical analysis (*https://oreil.ly/MR0Jz*): how fast runtime or space demands grow in relation to the input size. This allows algorithm classifications based on their scalability, which usually matters for the researchers who search for algorithms solving complex problems (which usually require enormous inputs). For example, in Figure 7-1, we see a small overview of typical functions and an opinionated assessment of what's typically bad and what's good complexity for the algorithm. Note that "bad" complexity here doesn't mean there are algorithms that do better—there are some problems that can't be done in a faster way.

1 This is fixed for this particular `ParseInt` function in Go 1.20 thanks to an amazing improvement (*https://oreil.ly/KLIVM*), but you might be surprised by it in any other function!

2 It only shows up when we do lots of string copies in our programs. Perhaps it comes from some internal byte pools?

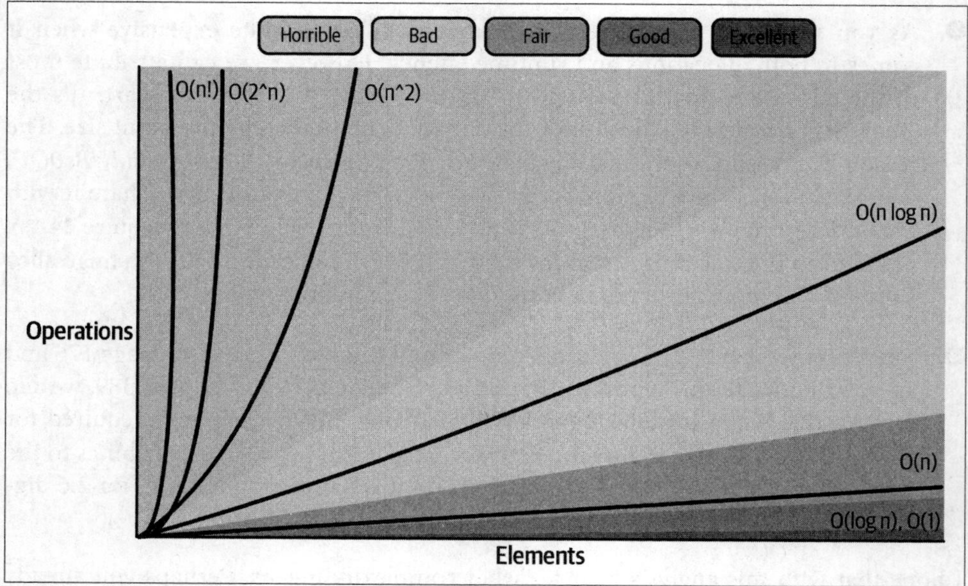

Figure 7-1. Big O complexity chart from https://www.bigocheatsheet.com. Shading indicates the opinionated rates of efficiency for usual problems.

We usually use Big O notation to represent asymptotic complexity. To my knowledge, it was Donald Knuth who attempted to clearly define three notations (O, Ω, Θ)[3] in his article from 1976 (*https://oreil.ly/yeFpW*).

> Verbally, O(f(n)) can be read as "order at most f(n)"; Ω(f(n)) as "order at least f(n)"; Θ(f(n)) as "order exactly f(n)".
>
> —Donald Knuth, "Big Omicron and Big Omega and Big Theta" (*https://oreil.ly/yeFpW*)

The phrase "in order of f(*N*)" means that we are not interested in the exact complexity numbers but rather the approximation:

The upper bound (O)
: Big Oh means the function can't be asymptotically worse than `f(n)`. It is also sometimes used to reflect the worst-case scenario if other input characteristics matter (e.g., in a sorting problem, we usually talk about a number of elements, but sometimes it matters if the input is already sorted).

3 Those "O-notations" are respectively called Big O or Oh, Omega, and Theta. He also defines "o-notations" (o, ω), which means strict upper or lower bound (*https://oreil.ly/S44PO*), so "this function grows slower than `f(N)`, but not exactly `f(N)`." In practice, we don't use o-notations very often.

The tight bound (Θ)
: Big Theta represents the exact asymptotic function or, sometimes, the average, typical case.

The lower bound (Ω)
: Big Omega means the function can't be asymptotically better than `f(n)`. It also sometimes represents the best case.

For example, the quicksort (*https://oreil.ly/a2jhF*) sorting algorithm has the best and average runtime complexity (depending on how input is sorted and where we choose the pivot point) of the $N * \log N$, so $\Omega(N * \log N)$ and $\Theta(N * \log N)$, even though the worst case is $O(N^2)$.

The Industry Is Not Always Using Big O Notation Properly

Generally, during interviews, discussions, and tutorials, you would see people using Big Oh (O) where Big Theta (Θ) should be used to describe a typical case. For example, we often say quicksort is $O(N * \log N)$, which is not true, but in many instances we would accept that answer. Perhaps people try to make this space more accessible by simplifying this topic. I will try to be more precise here, but you can always swap Θ with O (but not in the opposite direction).

For our algorithm in Example 4-1, the asymptotic space complexity is linear:

$$space(N) = 872 + 30.4 * N\ bytes = \Theta(1) + \Theta(N)\ bytes = \Theta(N)\ bytes$$

In asymptotic analysis, constants like 1, 872, and 30.2 do not matter, even though in practice, it might matter if our code allocates 1 MB ($\Theta(N)$) or 30.4 MB.

Note that we don't need precise complexity to figure out the asymptotic one. That's the point: precise complexity depends on too many variables, especially when it comes to runtime complexity. Generally, we can learn to find the theoretical asymptotic complexity based on algorithm pseudocode or description. It takes some practice, but imagine we don't have Example 7-1 implemented; instead, we design an algorithm. For example, the naive algorithm for the sum of all integers in the file can be described as follows:

1. We read the file's content into memory, which has $\Theta(N)$ of asymptotic space complexity, where N is the number of integers or lines. As we read N lines, this also has $\Theta(N)$ runtime complexity.
2. We split the content into subslices. If we do it in place, this means $\Theta(N)$. Otherwise, in theory, it is $\Theta(1)$. This is an interesting one, as we saw in precise

complexity that despite doing this in place, the overhead is 24 * *N*, which suggests $\Theta(N)$. In both cases, the runtime complexity is $\Theta(N)$, as we have to go through all lines.

3. For every subslice (space complexity $\Theta(1)$ and runtime $\Theta(N)$):
   a. We parse the integer. Technically this needs no extra space on the heap, assuming the integers can be kept on the stack. The runtime of this should also be $\Theta(1)$ if we relate to the number of lines and the number of digits is limited.
   b. We add the parsed value into a temporary variable containing a partial sum: $\Theta(1)$ runtime and $\Theta(1)$ space.

With such analysis, we can tell that the space complexity is $\Theta(N) + \Theta(1) + \Theta(N) * \Theta(1)$, so $\Theta(N)$. I also mentioned runtime complexity in step 2, which combines into $\Theta(N) + \Theta(N) + \Theta(N) * \Theta(1)$, so also linear $\Theta(N)$.

Generally, such a `Sum` algorithm is fairly easy to assess asymptotically, but this is not trivial in many cases. It takes some practice and experience. I would love it if some automatic tools detected such complexity. There were interesting attempts (*https://oreil.ly/0h9ff*) in the past, but in practice, they are too expensive.[4] Perhaps there is a way to implement some algorithm that assesses pseudocode for its complexity, but it's our job now!

Practical Applications

Frankly speaking, I was always skeptical about the "complexity" topic. Perhaps I missed the lectures about it at my university,[5] but I was always disappointed when somebody asked me to determine the complexity of some algorithm. I was convinced that it is only used to trick candidates during technical interviews and has almost no use in practical software development.

The first problem was imprecision—when people asked me to determine complexity, they meant asymptotic complexity in Big O notation. Furthermore, what's the point of Big O if, during paid work, I could usually search an element in the array with the linear algorithm instead of a hashmap, and still the code would be fast enough in most cases? Moreover, more experienced developers were rejecting my merge requests because my fancy linked list with better insertion complexity could be just a simpler array with `append`s. Finally, I was learning about all those fast algorithms

4 I would categorize them as "brute force"—they do many benchmarks with different inputs and try to approximate the growth function.

5 I wouldn't be surprised—I had a full-time job in IT from the second year of my computer science studies.

with incredible asymptotic complexity that are not used in practice because of hidden constant costs or other caveats.[6]

I think most of my frustration came from misunderstandings and misuses stemming from the industry's stereotypes and simplifications. I am especially surprised that not a few engineers (*https://oreil.ly/1yxqH*) are willing to perform such "estimated" complexity. Perhaps we often feel demotivated or overwhelmed by how hard it is to estimate beyond asymptotic complexity. For me, reading old programming books was eye-opening—some of them use both complexities in most of their optimization examples!

> The main `for` loop of the program is executed `N-1` times, and contains an inner loop that is itself executed `N` times; the total time required by the program will therefore be dominated by a term proportional to `N^2`. The Pascal running time of Fragment A1 was observed to by approximately 47.0N^2 microseconds.
>
> —Jon Louis Bentley, *Writing Efficient Programs*

When you try to assess or optimize algorithm and code that requires better efficiency, being aware of its estimated complexity and asymptotic complexity has a real value. Let's go through some use cases.

If you know precise complexity, you don't need to measure to know expected resource requirements

In practice, we rarely have precise complexity from the start, but imagine someone giving us such complexity. This gives an enormous win for tasks like capacity planning, where you need to find out the cost of running your system under various loads (e.g., different inputs).

For example, how much memory does the naive implementation of `Sum` use in Example 7-1? It turns out that without any benchmark, I could use the space complexity of 872 + 30.4 * *N* bytes to tell that for various input sizes, for example:

- For 1 million integers, my code would need 30,400,872 bytes, so 30.4 MB if we use the 1,000 multiplier, not the 1,024 (*https://oreil.ly/SYcm8*).[7]
- For 2 million integers, it would need 60.8 MB.

6 For example, quicksort has worse complexity than other algorithms, yet on average it is the fastest. Or the matrix multiplication algorithm like Coppersmith-Winograd (*https://oreil.ly/q9jhn*) has a big constant coefficient hidden by the Big O notation, which makes it only worth doing for matrices that are too big for our modern computers.

7 Be careful: different tools use different conversions; e.g., `pprof` uses the 1,024 multiplier, and the `benchstat` uses the 1,000 multiplier.

This can be confirmed if we would perform a quick microbenchmark (don't worry, I will explain how to perform benchmarks here and in Chapter 8). Results are presented in Example 7-2.

Example 7-2. Benchmark allocation result for Example 4-1 with one million elements and two million elements input, respectively

```
name (alloc/op)    Sum1M           Sum2M
Sum                30.4MB ± 0%     60.8MB ± 0%

name (alloc/op)    Sum1M           Sum2M
Sum                800k ± 0%       1600k ± 0%
```

Based on just those two results, our space complexity is fairly accurate.[8]

It's unlikely you can always find the full, accurate, real complexity. However, usually it's enough to have a very high-level estimation of this complexity, e.g., 30 * N bytes would be detailed enough space complexity for our `Sum` function in Example 7-1.

It tells us if there is any easy optimization to our code

Sometimes we don't need detailed empirical data to know we have efficiency problems.[9] This is great because such techniques can tell us how easy it is to optimize our program further. Such a quick efficiency assessment is something I would love you to know before we move into heavy benchmarking.

For example, when I wrote the naive implementation of the `Sum` in Example 4-1, I expected to write an algorithm with $\Theta(N)$ space (asymptotic) complexity. However, I expected it to have around 3.5 * N of the real complexity because I read the whole file content to memory. Only when I ran benchmarks that gave me output like Example 7-2 did I realize how poor my naive implementation was, with almost 10 times more memory usage than expected (30.5 MB). This expected estimation of the real complexity versus the resulting one is typically a good indication that there might be some trivial optimization if we have to improve the efficiency.

Secondly, if my algorithm space Big O complexity is linear, it is already a bad sign for such simple functionality. My algorithm will use an extreme amount of memory for

8 I was very surprised that we can construct such accurate space complexity and have such accurate memory benchmarking and profiling up to every byte on the heap. Kudos to the Go community and `pprof` community for that hard work!

9 This does not mean we should immediately fix those! Instead, always optimize if you know the problem will affect your goals, e.g., user satisfaction or RAER requirements.

huge inputs. Depending on requirements, that might be fine or it might mean real issues if we want to scale this application.[10] If not a problem right now, the maximum expected input size should be acknowledged and documented as it might be a surprise to somebody who will be using this function in the future!

Finally, suppose the measurements are totally off the expected complexity of the algorithm. In that case, it might signal a memory leak (*https://oreil.ly/ZNB5s*), which is often easy to fix if you have the right tools (as we will discuss in "Don't Leak Resources" on page 426).

Three Clear Indications We Are Wasting Memory Space

- The difference between the theoretical space complexity (asymptotic and estimated) and the reality measured with a benchmark can immediately tell you if something is not as expected.
- Significant space complexity depending on the user (or caller) input is a bad sign that might mean future scalability problems.
- If, with time, the total memory used by the program constantly grows and never goes down, it most likely indicates a memory leak.

It helps us assess ideas for a better algorithm as an optimization

Another amazing use case for complexities is quickly assessing algorithmic optimizations without implementing them. For our `Sum` example, we don't need extreme algorithmic skills to know that we don't need to buffer the whole file in memory. If we want to save memory, we should be able to have a small buffer for parsing purposes. Let's describe an improved algorithm:

1. We open the input file without reading anything.
2. We create a 4 KB buffer, so we need at least 4 KB of memory, which is still a constant amount ($\Theta(1)$).
3. We read the file in 4 KB chunks. For every chunk:
   a. We parse the number.
   b. We add it to a temporary partial sum.

10 Sometimes, there are relatively easy ways to change our code to stream and use external memory (*https://oreil.ly/p6YDD*) algorithms that ensure stable memory usage.

Such an improved algorithm, in theory, should give us the space complexity of ~4 KB, so $O(1)$. As a result, our Example 4-1 could use 7,800 times less space for 1 million integers! So we can tell without implementation that such optimization on an algorithmic level would be very beneficial, and you will see it in action in "Optimizing Memory Usage" on page 395.

Doing such complexity analysis can quickly assess your ideas for improvement without needing the full TFBO loop!

Worse Is Sometimes Better!

If we decide to implement the algorithm with better asymptotic or theoretical complexity, don't forget to assess it at the code level using benchmarks! When designing an algorithm, we often optimize for asymptotic complexity, but when we write code, we optimize the constants of that asymptotic complexity.

Without good measurements, you might implement a good algorithm in terms of Big O complexity, but with the inefficient code, make efficiency optimizations instead of improvement!

It tells us where the bottleneck is and what part of the algorithm is critical

Finally, a quick look at the detailed space complexity, especially when mapped to the source code as in Example 7-1, is a great way to determine the efficiency bottleneck. We can see that the constant 24 is the biggest one, and it comes from the `bytes.Split` function that we will optimize first in Chapter 10. In practice, however, profiling can yield data-driven results much faster, so we will focus on this method in Chapter 9.

To sum up, the wider knowledge about the complexity and ability to mix basic measurements with theoretical asymptotic taught me that complexities could be useful. It can be an excellent tool for more theoretical efficiency assessment if used correctly. However, as you can see, the real value is when we mix empirical measurements with theory. With this in mind, let's learn more about benchmarking!

The Art of Benchmarking

Assessing efficiency is essential in the TFBO flow, represented by step 4 in Figure 3-5. Such evaluation of our code, algorithm, or system is generally a complex problem, achievable in many ways. For example, we discussed assessing efficiency on the algorithm level through research, static analysis, and Big O notations for runtime complexity.

We can assess a lot by performing a theoretical analysis and estimating code efficiency. Still, in many cases, the most reliable way is to get our hands dirty, run some code, and see things in action. As we learned in "Optimization Challenges" on page 79, we are bad at estimating the resource consumption of our code, so empirical assessments allow us to reduce the number of guesses in our evaluations.[11] Ideally, we assume nothing and verify the efficiency using special testing processes that test efficiency instead of correctness. We call those tests *benchmarks*.

Benchmarking Versus Stress and Load Tests

There are many alternative names for benchmarking, such as stress tests, performance tests, and load tests. However, since they generally mean the same, for consistency, I will use benchmarking in this book.

Generally, benchmarking is an effective efficiency assessment method for our software or systems. In abstract, the process of benchmarking is composed of four core parts, which we describe logically as a simple function:

$$Benchmark = N * (Experiment + Measurements) + Comparison$$

At the core of any benchmarking, we have the experimentations and measurements cycle:

Experiment
: The act of simulating a specific functionality of our software to learn about its efficiency behavior. We can scope that experiment to a single Go function or Go structure or even complex, distributed systems. For example, if your team develops the web server, it might mean starting a web server and performing a single HTTP request with realistic data that the user would use.

Measurement
: In Chapter 6, we discussed getting accurate measurements for latency and the consumption of various resources. It's vital to reliably observe our software during the entire experiment to make meaningful conclusions when it ends. For our web server example, this might mean measuring the latency of the operations on various levels (e.g., client and server latencies), as well as the memory consumption of our web server.

11 Unfortunately, we still have to guess a little bit—more on that in "Reliability of Experiments" on page 256. Nothing will get us 100% assurance. Yet benchmarking is probably the best we have as developers for ensuring the software we develop is efficient enough.

Now the unique part of our benchmarking process is that the experiment and measurements cycle has to be performed *N* times with the comparison phase at the end:

The number of test iterations (N)
: *N* is the number of test iterations we must perform to build enough confidence in the results. The exact number of runs depends on many factors, which we will discuss in "Reliability of Experiments" on page 256. Generally, the more iterations we do, the better. In many cases, we have to balance between higher confidence and cost or wait time of a too large number of iterations.

Comparison
: Finally, in the benchmarking definition, we have the comparison aspect (*https://oreil.ly/kzNR3*), which allows us to learn what's improving the efficiency of our software, what's hindering it, and how far we are from the expectations (RAER).

In many ways, you might notice that benchmarking is similar to the testing we do to verify correctness (referred to later as functional testing). As a result, many testing practices apply to benchmarking. Let's look at that next.

Comparison to Functional Testing

Comparison to something we are familiar with is one of the best ways to learn. So, let's compare benchmarking to functional testing. Is there anything we can reuse in terms of methodology or practices? You will learn in this chapter that we can share many things between functional tests and benchmarking. For example, there are a few similar aspects:

- Best practices for forming test cases (e.g., edge cases (*https://oreil.ly/Sw9qB*)), table-driven testing (*https://oreil.ly/Q3bXD*), and regression testing
- Splitting tests into unit, integration, e2e (*https://oreil.ly/tvaMk*), and testing in production (more on that in "Benchmarking Levels" on page 266)
- Automation for continuous testing

Unfortunately, we have to also be aware of significant differences. With benchmarks:

We have to have different test cases and test data (https://oreil.ly/me3cM).
: It might be tempting, but we cannot reuse the same test data (input parameters, potential fake, test data in a database, etc.) as we used for our unit or integrations tests meant for correctness tests. This is because the goals are different. In correctness tests, we tend to focus on different edge cases (*https://oreil.ly/Sw9qB*) from a functional perspective (e.g., failure modes). Whereas in efficiency tests, the edge cases are usually focused on triggering different efficiency issues (e.g., big requests versus many small requests). We will discuss these in "Reproducing Production" on page 258.

> For most systems, though, the programmer should monitor the program on input data that is typical of the data the program will encounter in production. Note that usual test data often does not meet this requirement: while test data is chosen to exercise all parts of the code, profiling [and benchmarking] data should be chosen for its "typicality."
>
> —Jon Louis Bentley, *Writing Efficient Programs*

Embrace the performance nondeterminism
: Modern software and hardware consist of layers of complex optimizations. This can cause nondeterministic conditions to change while performing our benchmarks, which might mean that the results will also be nondeterministic. We will expand on this in "Reliability of Experiments" on page 256, but this is why we usually repeat test iteration cycles hundreds if not thousands of times (our *N* component) to increase confidence in our observations. The main goal here is to figure out how repeatable our benchmark is. If the variance is too high, we know we cannot trust the results and must mitigate the variance. This is why we rely on statistics in our benchmarks, which helps a lot, but also makes it easy to mislead others and ourselves.

> Repeatability: Ensuring that the same operations are benchmarked on all configurations and that metrics are repeatable over many test runs. Rule of thumb is a variation of up to 5% is generally acceptable.
>
> —Bob Cramblitt, "Lies, Damned Lies, and Benchmarks: What Makes a Good Performance Metric" (*https://oreil.ly/ghvJ7*)

It is more expensive to write and run
: As you can imagine, the number of iterations we have to perform increases the running cost and complexity of performing the benchmark, both the compute cost and developer time spent on creating those and waiting. But that is not the only additional cost compared to correctness tests. To trigger efficiency problems, especially for large systems load tests, we have to exhaust different systems capacities, which means buying a lot of computing power just for the sake of tests.

This is why we have to focus on a pragmatic optimization process where we only care about efficiency where necessary. There are also ways to be smart and avoid full-scale macrobenchmarks by using tactical microbenchmarks of isolated functions, as discussed in "Benchmarking Levels" on page 266.

Expectations are less specific
: Correctness tests always end up with some assertions. For example, in Go tests, we check if the result of the functions has the expected value. If not, we use `t.Error` or `t.Fail` to indicate the test should fail (or one-liners like `testutil.Ok` (*https://oreil.ly/ncVhq*) or `testutil.Equals` (*https://oreil.ly/uH1F5*)).

It would be amazing if we could do the same when benchmarking—asserting if the latency and resource consumption are not exceeding the RAER. Unfortunately, we cannot just do `if maxMemoryConsumption < 200 * 1024 * 1024` at the end of a microbenchmark. The typical high variance of the results, challenges in isolating the latency and resource consumption to just one functionality we test, and other problems mentioned in "Reliability of Experiments" on page 256 make it hard to automate the assertion process. Typically, there has to be human or very complex anomaly detection or assertion software to understand whether the results are acceptable. Hopefully, we will see more tools that make it easier in the future.

To make things harder, we might have a RAER for bigger APIs and functionalities. But if the RAER says the latency of the whole HTTP request should be lower than the 20s, what does that mean for the single Go function involved in this request (out of thousands)? How much latency should we expect in microbenchmarks used by this function? There is no good answer.

We Focus More on Relative Results than Absolute Numbers!

In benchmarks, we usually don't assert absolute values. Instead, we focus on comparing results to some baseline (e.g., the previous benchmark before our code change). This way, we know if we improved or negatively affected the efficiency of a single component without looking at the big picture. This is usually enough on the unit microbenchmarks level.

With the basic concept of benchmarking explained, let's address the elephant in the room in the next section—the stereotype that associates benchmarks with lies. Unfortunately, there are solid reasons for this relation (*https://oreil.ly/yotxL*). Let's unpack this and see how we can tell if we can trust the benchmarks that we or others do.

Benchmarks Lie

There is an extension to a famous phrase (*https://oreil.ly/xULP5*) that states that we can order the following words from the best to worst: "lies, damn lies, and benchmarks."

> This interest in performance has not gone unnoticed by the computer vendors. Just about every vendor promotes their product as being faster or having better "bang for the buck." All of this performance marketing begs the question: "How can these competitors all be the fastest?" The truth is that computer performance is a complex phenomenon, and who is fastest all depends upon the particular simplifications being employed to present a particular simplistic conclusion.
>
> —Alexander Carlton, "Lies, Damn Lies, and Benchmarks" (*https://oreil.ly/WClsq*)

Cheating in benchmarks is indeed widespread. The efficiency results through benchmarks have significant importance in a competitive market. Users have too many choices to make, so simplifying the comparison to a simple question, "which is the fastest solution?" or "which one is the most scalable?" is common among decision-makers. As a result, benchmarking became a gamification system that is cheated on (*https://oreil.ly/4NAVh*). The fact that efficiency assessment is very complex to get right and expensive to reproduce makes it easy to get away with a misleading conclusion. There are many examples of companies, vendors, and individuals lying in benchmarks.[12] However, it is essential to highlight that not all cases are done intentionally or with malicious intent. For better or worse, in most cases, the author did not purposely report misleading results. It's only natural to get tricked by statistical fallacies (*https://oreil.ly/jPxnA*) and paradoxes that are counterintuitive to the human brain.

Benchmarks Don't Lie; We Just Misinterpret the Results!

There are many ways we can make wrong conclusions from benchmarks. If done accidentally, it can have severe consequences—usually a big waste of time and money. If done intentionally...well, lies have short legs. :)

We can be misled by benchmarks due to human mistakes, benchmarks performed under conditions irrelevant to us and our problem, or simply statistical error. The benchmark results themselves don't lie; we might have just measured the wrong thing!

The solution is to be a mindful consumer or developer of those benchmarks, plus learn the basics of data science. We will discuss common mistakes and solutions in "Reliability of Experiments" on page 256.

To overcome some biases that are naturally happening in the benchmarks, industries often come up with some standards and certifications. For example, to ensure fair fuel economy efficiency assessments, all light-duty vehicles in the US are required to have their economy results tested by the US Environmental Protection Agency (EPA) (*https://oreil.ly/gKOc2*). Similarly, in Europe, in response to the 40% gap between the fuel economy carmakers' tests and reality, the EU adopted the Worldwide Harmonized Light-Duty Vehicle Test Cycle and Procedure (*https://oreil.ly/LPUXj*). For hardware and software, many independent organizations design consistent benchmarks

12 For example, car makers cheating on emission benchmarks (*https://oreil.ly/WNF1z*) and phone vendors cheating on hardware benchmarks (*https://oreil.ly/sf80C*) (which sometimes results with a ban from the popular Geekbench (*https://oreil.ly/8M4ey*) listing). In the software world, we have a constant battle between various vendors through unfair benchmarks (*https://oreil.ly/RmytC*). Whoever creates them is often one of the fastest on the results list.

for specific requirements. SPEC (*https://oreil.ly/tkV6O*) and Percona HammerDB (*https://oreil.ly/ngRKu*) are two examples out of many.

To overcome both lies and honest mistakes, we must focus on understanding what factors make benchmarks unreliable and what we can do to improve that quality. It's foundational knowledge explaining many benchmark practices we will discuss in Chapter 8. Let's do that in the next section.

Reliability of Experiments

The TFBO cycle takes time. No matter on what level we assess and optimize efficiency, in all cases, it is necessary to spend a nontrivial amount of time on implementing benchmarks, executing them, interpreting results, finding bottlenecks, and trying new optimizations. It is frustrating if all or part of our efforts are wasted due to unreliable assessments.

As mentioned when explaining benchmarking lies, there are many reasons why benchmarks are prone to misleading us. There are a set of common challenges it's useful to be aware of.

The Same Applies to Bottleneck Analysis!

In this chapter, we might be discussing benchmarks, so experiments mainly allow us to measure our efficiency (latency or resource consumption), but similar reliability concerns can be applied to other experiments or measurements around efficiency. For example, profiling our Go programs to find bottlenecks, discussed in Chapter 9.

We can outline three common challenges to the reliability of benchmarks: human errors, the relevance of our experiments to the production environment, and the nondeterministic efficiency of modern computers. Let's go through these in the next sections.

Human Errors

Optimizations and benchmarking routines, as it stands today, involve a lot of manual work from developers. We need to run experiments with different algorithms and code, while caring about reproducing production and performance nondeterminism. Due to the manual nature, this is prone to human error.

It's easy to get lost in what optimizations we already tried, what code you added for debugging purposes, and what is meant to be saved. It is also easy to get confused about what version of code the benchmarking results belong to and what assumptions you already proved wrong.

Many problems with our benchmarks tend to be caused by our sloppiness and lack of organization. Unfortunately, I am guilty of many of those mistakes too! For example, when I thought I was benchmarking optimization X, I discarded it after seeing no significant difference in benchmarking results. Only some hours later did I notice I tested the wrong code, and optimization X was helpful!

Fortunately, there are some ways to reduce those risks:

Keep it simple.
: Try to iterate with code changes related to efficiency in the smallest iterations possible. If you try to optimize multiple elements of your code simultaneously, it most likely will obfuscate your benchmark results. You might miss that one of those optimizations limits the efficiency of the aspect you are interested in.

  Similarly, try to isolate complex parts into smaller separate parts you can optimize and assess separately (divide and conquer).

Know what version of software you are benchmarking.
: It might be trivial, but it's worth repeating—use software versioning (*https://oreil.ly/P0eoP*)! If you try different optimizations, commit them in separate commits and distribute them across separate branches so you can get back to previous versions if needed. Don't lose your optimization effort by forgetting to commit your work at the end of the day.[13]

  This also means you have to be strict about what version of code you just benchmarked. Even a small reorder of seemingly unrelated statements might impact your code's efficiency, so always benchmark your programs in atomic iterations. This also includes all dependencies your code needs, for example, those outlined in your *go.mod* file.

Know what version of benchmark you are using.
: Furthermore, remember to version the code of the benchmark test itself! Avoid comparing results between different benchmark implementations, even if the change was minor (adding an extra check).

  Scripting scripts to execute those benchmarks with the same configuration and versioning those is also a great way not to get lost. In Chapter 8, I mention some best practices around declarative ways to share benchmark options for your future self and others on your team.

13 Some good IDEs also have additional local history (*https://oreil.ly/Ytdi0*) if you forgot to commit your changes in your `git` repository.

Keep your work well organized and structured.

Make notes, design your own consistent workflow, and be explicit in what version of code you experimented with. Track the dependency versions, and track all benchmarking results explicitly in a consistent way. Finally, be clear in communicating your findings with others.

Your code should also be clean during different code attempts. Keep all best practices like DRY (*https://oreil.ly/S887r*), don't keep commented out code, isolate state between tests, etc.

Be skeptical about "too good to be true" benchmarking results.

If you can't explain why your code is suddenly quicker or uses fewer resources, you most certainly did something wrong while benchmarking. It is tempting to celebrate, accept it, and move on without double-checking.

Check common issues like if your benchmark test cases trigger errors instead of successful runs (mentioned in "Test Your Benchmark for Correctness!" on page 290), or perhaps the compiler optimized your microbenchmark away (discussed in "Compiler Optimizations Versus Benchmark" on page 301).

A little bit of laziness in our work is healthy.[14] However, laziness at the wrong moment might significantly increase the number of unknowns and risks to the already difficult subject of program efficiency optimizations.

Now let's look at the second key element of reliable benchmarks, relevance.

Reproducing Production

It might be obvious, but we don't optimize software so it can run faster or consume fewer resources on our development machine.[15] We optimize to ensure the software has efficient enough execution for the target destinations that matter for our business, so-called *production*.

Production might mean a production server environment you deploy if you build a backend application, or a customer device like a PC, laptop, or smartphone if you build an end-user application. Therefore, we can significantly improve the quality of our efficiency assessment for all benchmarks by enhancing their relevance. We can do that by trying our best to simulate (reproduce) situations and environmental conditions of production. Particularly:

---

14 Laziness is actually good (*https://oreil.ly/u8IDm*) for engineers! But it has to be pragmatic, productive, and reasonable laziness toward the efficiency of our work, not purely based on our emotions in the given moment.

15 Unless we write software for fellow developers that runs on similar hardware.

Production conditions
: The characteristics of a production environment. For example, how much RAM and what kind of CPU the production machines will have dedicated for our program. What OS version does it have? What versions and kinds of dependencies will our program use?

Production workload
: The data our program will work with and the behavior of the user traffic it has to handle.

Perhaps the first thing we should do is to gather requirements around the software target destination, ideally in written form in our RAER. Without it, we can't correctly assess the efficiency of our software. Similarly, if you see benchmarks done by a vendor or independent entity, you should check if the benchmark conditions match your production and requirements. Typically, they don't, and to fully trust it, we should try to reproduce such a benchmark on our side.

Assuming we roughly know what the target production for our software looks like, we might start designing our benchmark flow, test data, and cases. The bad news is that it's impossible to fully reproduce every aspect of production in our development or testing environment. There will always be differences and unknowns. There are many reasons why production will be different:

- Even if we run the same kind and version of the OS as production, it is impossible to reproduce the dynamic state of the OS, which impacts efficiency. In fact, we cannot fully reproduce this state between two runs on the same local machine! This challenge is often called nondeterministic performance, and we will discuss it in "Performance Nondeterminism" on page 260.
- It's often too expensive to reproduce all kinds of production workloads that can happen (e.g., forking all production traffic and putting it through testing clusters).
- When developing an end-user application, there are too many permutations of different hardware, dependency software versions, and situations. For example, imagine you create an Android app—tons of smartphone models could potentially run your software, even if we would limit ourselves to smartphones made in the last two years.

The good news is that we don't need to reproduce all aspects of production. Instead, it's often enough to represent key characteristics of the products that might limit our workloads. We might know about it from the start of development—but with time, experiments, and macrobenchmarks (see "Macrobenchmarks" on page 306), or even production—you will learn what matters.

For example, imagine you develop Go code responsible for uploading local files to a remote server, and the users notice unacceptable latency when uploading a large file. Based on that, our benchmark to reproduce this should:

- Focus on test cases that involve big files. Don't try to optimize a large number of small files, all different error cases, and potential encryption layers if that doesn't represent what production users are using the most. Instead, be pragmatic and focus with benchmarks on what your goal is now.
- Be mindful that your local benchmarks are not reproducing potential network latencies and behavior you will see in production. A bug in your code might cause resource leaks only in case of a slow network, which might be hard to reproduce on your machine. For these optimizations, it's worth moving with benchmarks to different levels, as explained in "Benchmarking Levels" on page 266.

Simulating the "characteristics" of production does not necessarily mean the same dataset and workload that will exist on production! For our earlier example, you don't need to create 200 GB test files and benchmark your program with them. In many cases, you can start with relatively large files like 5 MB, then 10 MB, and together with complexity analysis, deduce what will happen at the 200 GB level. This will allow you to optimize those cases much faster and cheaper.

> Typically it would be too difficult and inefficient to attempt to exactly reproduce a specific workload. A benchmark is usually an abstraction of a workload. It is necessary, in this process of abstracting a workload into a benchmark, to capture the essential aspects of the workload and represent them in a way that maps accurately.
>
> —Alexander Carlton, "Lies, Damn Lies, and Benchmarks"

To sum up, when trying to assess the efficiency or reproduce efficiency regressions, be mindful of the differences between your testing setup and production. Not all of them are worth reproducing, but the first step is to know about those differences and how they can impact the reliability of our benchmarks! Let's now look at what else we can do to improve the confidence of our benchmarking experiments.

Performance Nondeterminism

Perhaps the biggest challenge with efficiency optimizations is the "nondeterministic performance" of modern computers. It means so-called noise, so the variance in our experiment results is because of the high complexity of all layers that impacts the efficiency we learned about in Chapters 4 and 5. As a result, efficiency characteristics are often unpredictable and highly fragile to environmental side effects.

For example, let's consider a single statement in the Go code, an `a += 4`. No matter what conditions this code is executed in, assuming we are the only user of memory used by the `a` variable, the result of `a += 4` is always deterministic—a value of `a` plus 4. This is because, in almost all cases, it is hard to impact correctness. You can put the computer in extreme heat or cold, you can shake it, you can schedule millions of simultaneous processes in the OS, and you can use any version of CPU that exists with any supported type of operating system that supports that hardware. Unless you do something extreme like influencing the electric signal in the memory, or you put the computer out of power, that `a += 4` operation will always give us the same result.

Now let's imagine we are interested to learn how our `a += 4` operation contributes to the latency in the bigger program. At first glance, the latency assessment should be simple—this requires a single CPU instruction (e.g., `ADDQ` (*https://oreil.ly/Vv83D*)) and a single CPU register, so the amortized cost should be as fast as your CPU frequency, so, for example, an average of 0.3 ns for 3 GHz CPU.

In practice, however, overheads are never amortized and never static within a single run, making that statement latency highly nondeterministic. As we learned in Chapter 4, if we don't have the data in the registers, the CPU has to fetch it from L-caches, which might take one nanosecond. If L-caches contain data the CPU needs, our single statement might take 50 ns. Suppose the OS is busy running millions of other processes; our single statement might take milliseconds. Notice that we are talking about a single instruction! On a larger scale, if this noise builds, we can accumulate variance measurable in seconds.

Be mindful. Almost everything can impact the latency of our operations. Busy OS, different versions of hardware elements, and even differences in manufactured CPUs from the same company might mean different latency measurements. Ambient temperature near a laptop's CPU or battery modes can trigger thermal scaling of our CPU frequency up and down. In extreme cases, even screaming at your computer can impact the efficiency![16] The more complexity and layers we have when running our programs, the more fragile our efficiency measurements. Similar problems apply to remote devices, personal computers, and public cloud providers (e.g., AWS or Google) that use shared infrastructure with virtualization like containers or virtual machines.[17]

16 The engineer Brendan Gregg demonstrated (*https://oreil.ly/vI8Rl*) how screaming at server hard drive disks severely impacts their I/O latency due to vibrations.

17 The situation where one workload from a totally different virtual machine impacts our workload is commonly called a noisy neighbor situation (*https://oreil.ly/cLRrD*). It is a serious issue that cloud providers continuously fight, with better or worse results depending on the offering and provider.

Compressible Versus Noncompressible Resources

All efficiency aspects have some nondeterminism, but some resources are more predictive than others. Typically, it is correlated to the categorization known as how compressible resources are. Compression refers to the consequences of the saturation of certain resources (what happens when you don't have enough of the resource).

- The latency and I/O throughput of CPU time, memory or disk access, and network bandwidth are compressible. So if we have too many processes demanding CPU time, we can slow down execution, but eventually, we will execute all the scheduled work. This means we won't see machines crashing due to CPU saturation, but it also results in highly dynamic latency results.
- The space and allocation aspect of the resource, like memory or disk space used, is noncompressible on its own. As we learned in Chapter 5, if the program needs more memory space than the OS has, it has to crash the process or the whole system in most cases. There are mitigations like using space of different mediums instead (OS swap) and compressing the data we want to save, but used space can't compress automatically. This might feel like a challenge, but it is beneficial for benchmarking and measurement purposes—behavior is more deterministic.

The fragility of efficiency assessment is so common that we have to expect it in every benchmarking attempt. Therefore, we have to embrace it and embed mitigations to those risks into our tools.

The first thing you might want to do before mitigating nondeterministic performance is to check if this problem impacts your benchmarks. Verify the repeatability of your test by calculating the variance of your results (e.g., using standard deviation). I will explain a good tool for that in "Understanding the Results" on page 284, but often you can see it in plain sight.

For example, if you run the experiment once and see it finish in 4.05 seconds, and other runs vary from 3.01 to 6.5 seconds, your efficiency assessment might not be accurate. On the other hand, if the variance is low, you can be more confident about the relevance of your benchmarks. Thus, check the repeatability of your benchmark first.

Don't Overuse the Statistics

It is tempting to accept high variance and either remove the extreme results (outliers) or take the mean (average) of all your results. You can apply very complex statistics to find some efficiency numbers with some probability (*https://oreil.ly/594nD*). Increasing benchmark runs can also make your average numbers more stable, thus giving you a bit more confidence.

In practice, there are better ways to try first to mitigate stability. Statistics are great where we can't perform a stable measurement, or we can't verify all samples (e.g., we cannot poll all humans on Earth to find out how many smartphones are used). While benchmarking, we have more control over stability than we might initially think.

There are many best practices we can follow to ensure our efficiency measurements will be more reliable by reducing the potential nondeterministic performance effects:

Ensure the stable state of the machine you benchmark on.
: For most benchmarks that rely on comparisons, it matters less what conditions we benchmark in as long as they are stable (the state of the machine does not change during or between benchmarks). Unfortunately, three mechanics typically get in the way of machine stability:

  Background threads
  : As you learned in Chapter 4, it's hard to isolate processes on machines. Even a single, seemingly small process can make your OS and hardware busy enough to change your efficiency measurements. For example, you might be surprised how much memory and CPU time one browser tab or Slack application might use. On public clouds, it's even more hidden as we might see processes impacting us from different virtual OSes we don't own.

  Thermal scaling
  : The temperature of high-end CPUs increases significantly under load. The CPUs are designed to sustain relatively hot temperatures like 80–110°C, but there are limits. If the fans cannot cool the hardware fast enough, the OS or the firmware will limit the CPU cycles to avoid component meltdown. Especially with remote devices like laptops or smartphones, it's easy to trigger thermal scaling when the ambient temperature is high, your device is in the sunlight, or something is obstructing the cooling fans.

  Power management
  : Similarly, devices can limit the hardware speed to reduce power consumption. This is typically seen on laptops and smartphones with battery-saving modes.

For Most Cases, It's Enough to Maintain Simple Stability Best Practices

To reduce machine instability, you could go extreme and buy a dedicated bare-metal server that only runs OS and your benchmarks. In addition, you could turn off all software updates and all advanced thermal and power management components and keep your server specially cooled. However, for practical efficiency benchmarking, following a few reasonable practices is usually enough to avoid those problems, all while still using your developer device for testing for the quick feedback loop. For example, when benchmarking:

- Try to keep your machine relatively idle, don't actively browse the internet, and avoid running multiple benchmarks at the same time.[18] Close your messaging apps like Slack or Discord or any other programs that might become active during the benchmark. Literally just typing on characters in my IDE editor while performing tests usually impacts my benchmarking results 10%!
- If you use a laptop as your benchmarking machine, keep your laptop connected to power during benchmarks.
- Similarly, don't keep the laptop on your lap or your bed (e.g., on the pillow) when benchmarking. This blocks the fans from pulling the hot air out, which can trigger thermal scaling!

Be extra vigilant on shared infrastructure.

Buying a dedicated virtual machine on a stable cloud provider for benchmarking is not a bad idea. We mentioned noisy neighbor problems, but if done right, the cloud can be sometimes more durable than your desktop machine running various interactive software during benchmarks.

When using cloud resources, ensure you choose the best possible, strict Quality of Service (QoS) contract with the provider. For example, avoid cheaper burstable (*https://oreil.ly/Nu5C6*) or preemptible virtual machines, which by design are prone to infrastructure instabilities and noisy neighbors.

Avoid Continuous Integration (CI) pipelines, especially those from free tiers like GitHub Action (*https://oreil.ly/RcKXR*) or other providers. While they remain a convenient and cheap option, they are designed for correctness testing that has to eventually finish (not as fast as physically possible) and scale dynamically to the user demands to minimize costs. This doesn't provide strict and stable resource allocations required for benchmarks.

18 This is why you won't see me explaining the microbenchmark options like RunParallel (*https://oreil.ly/S74VY*). In general, running multiple benchmark functions in parallel can distort the results. Therefore, I recommend avoiding this option.

Be mindful of benchmark machine limits.

Be aware of your machine spec. For example, if your laptop has only 6 CPU cores (12 virtual cores with Hyper-Threading), don't implement benchmark cases that require the `GOMAXPROCS` to be larger than the CPUs you have available for test. Furthermore, it might make sense to benchmark with only four CPUs for six physical core CPUs on your general-purpose machine to ensure spare room for OS and background processes.[19]

Similarly, be mindful of the limits of other resources, like memory. For example, don't run benchmarks that use close to a maximum capacity of RAM, as memory pressure, faster garbage collection, and memory trashing might slow down all threads on the machine, including the OS!

Run the experiment longer.

One of the easiest ways to reduce variance between benchmark runs is to run the benchmark a bit longer. This allows us to minimize the benchmarking overhead that we might see at the beginning of our benchmarks (e.g., CPU cache warm-up phase). This also statistically gives us more confidence that the average latency or resource consumption metric shows the authentic pattern of the current efficiency level. This method takes time and depends on nontrivial statistics, prone to statistical fallacies, so use it with care and ideally try the suggestions mentioned before.

Avoid Comparing Efficiency with Older Experiment Results!

Put an expiration date on all benchmark results. It is tempting to save benchmarking results after testing one version of your code for later. Then we switch our work focus for a few days, perhaps go on holiday, and get back to optimization flow after a few days or weeks. Resist resuming your benchmarking flow by benchmarking a version with optimization and comparing it with days- or weeks-old benchmarking results stored somewhere in your filesystem.

Chances are that things have changed. For example, your system got upgraded, different processes run on your machine, or there is a different load in your clusters. You also risk other human errors, as it's easy to forget all the past details and environmental conditions you ran in. Solution? Repeat your past benchmarks on demand or invest in continuous benchmarking practices that will do that for you.[20]

---

19 You can also fully dedicate CPU cores to your benchmark; consider the `cpuset` tool (*https://oreil.ly/dCLzw*).

20 I had this problem when writing Chapter 10. I ran some benchmarks in one go on a relatively cold day. Next week there was a heat wave in the UK. I could not continue my optimization effort while reusing the past benchmarking results on such a hot day, as all my code was running 10% slower! I had to redo all the experiments to compare the implementations fairly.

To sum up, be mindful of potential human errors that can lead to confusion. Do care about the relevance of your experiments to the production end goal you and your development team have. Finally, measure the repeatability of your experiments to assess if you can rely on their results. Of course, there will always be some discrepancy between benchmark runs or between benchmark runs and production setup. Still, with these recommendations, you should be able to reduce them to a safe 2–5% variance level.

Perhaps you came to this chapter to learn how to perform Go benchmarks. I can't wait to explain to you step-by-step how to perform those in the next chapter! However, the Go benchmarks are not all we have in our empirical assessment arsenal. Therefore, it's essential to learn when to choose the Go benchmarks and when to fall back on different benchmarking methods. I will outline that in the next section.

Benchmarking Levels

In Chapter 6, we discussed finding latency and resource usage metrics that will allow us reliable measurements. But in the previous section, we learned that this might be only half of the success. By definition, benchmarking requires an experimentation stage that will trigger a certain situation or state of the application, which is valuable to measure.

There is something simpler worth mentioning before we start with experiments. The naive and probably simplest solution to assess the efficiency of, e.g., a new release of our software, is to give it to our customers and collect our metrics during the "production" use. This is great because we don't need to simulate or reproduce anything. Essentially the customer is performing the "experiment" part on our software, and we just measure their experience. We could call it "monitoring" at the source or "production monitoring." Unfortunately, there are some challenges:

- Computer systems are complex. As we learned in "Reproducing Production" on page 258, the efficiency depends on many environmental factors. To truly assess whether our new software versions have better or worse efficiency, we must know about all those "measurement" conditions. However, it is not economical to gather all this information when it runs on client machines.[21] Without it, we cannot derive any meaningful conclusions. On top of that, many users would opt out of any reporting capabilities, meaning we are even more unaware of what happened.

21 In some way, this is why selling your product as a SaaS is so appealing in software. Your "production" is on your premises, making it easier to control the experience of the users and validate some efficiency optimizations.

- Even if we gather that observability information, it isn't guaranteed that a situation causing problems will ever occur again. There is no guarantee that the customer will perform all the steps to reproduce the old problem. Statistically, all meaningful situations will happen at some point, but that eventual timing is too long in practice. For example, imagine that one HTTP request to a particular `/compute` path was causing efficiency problems. We fixed it and deployed it to production. What if no one used this particular path for the next two weeks? The feedback loop can be very long here.

Feedback Loop

The feedback loop is a cycle that starts from the moment of making changes to our code and ends with observations around these changes.

The longer this loop is, the more expensive development is. The frustration of developers is also often underestimated. In extreme cases, it will inevitably result in developers taking shortcuts by ignoring important testing or benchmarking practices.

To overcome this, we must invest in practices that will give us as much reliable feedback as possible in the shortest time.

- Finally, it is often too late if we rely on our users to "benchmark" our software. If it's too slow, we might have already lost their trust. This can be mitigated by canary rollouts (*https://oreil.ly/seUXz*) and feature flags,[22] but still, ideally, we catch efficiency issues before releasing our software to production.

Production monitoring is critical, especially when your software runs 24 hours, 7 days a week. Even more, manual monitoring, like observing efficiency trends and user feedback in your bug tracker, is also useful for the last step of efficiency assessment. Things do slip through the testing strategies we are discussing here, so it makes sense to keep production monitoring as a last verification resort. But as a standalone efficiency assessment, production monitoring is quite limited.

22 Feature flags are configuration options that can be changed dynamically without restarting the service—typically through an HTTP call. This allows reverting new functionality quicker, which helps with testing or benchmarking in production. For feature flags I rely on the excellent `go-flagz` (*https://oreil.ly/rfuh2*) library. I would also pay close attention to the new CNCF project OpenFeature (*https://oreil.ly/7Bsiw*), which is meant to provide more standard interface in this space.

Fortunately, we have more testing options that help to verify efficiency. Without further ado, let's go through the different levels of efficiency testing. If we would put all of them on a single graph that compares them based on the required effort to implement and maintain and the effectiveness of the individual test, it could look like Figure 7-2.

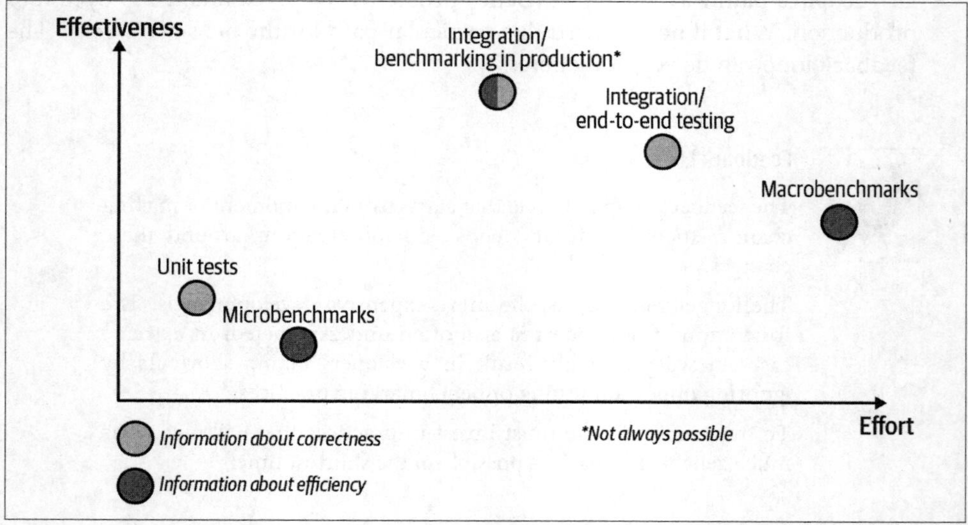

Figure 7-2. Types of efficiency and correctness test methods with respect to difficulty to set up and maintain them (horizontal axis) versus how effective a singular test of a given type is in practice (vertical axis)

Which of the methods presented in Figure 7-2 are used by mature software projects and companies? The answer is all of them. Let me explain.

Benchmarking in Production

Following testing in production practice (*https://oreil.ly/5NUiw*), we could use a live production system to assess efficiency. It might mean hiring "test drivers" (beta users) who will run our software on their devices and create real usage and report issues. Benchmarking in production is also very useful when your company sells the software you develop as a SaaS. For these cases, it is as easy as creating automation (e.g., a batch job or microservice) that periodically or after every rollout benchmarks the cluster using a predefined set of test cases that mimic real user functionalities (e.g., HTTP requests that simulate user traffic). Especially since you control the production environment, you can mitigate the downsides of production monitoring. You can be aware of environmental conditions, revert quickly, use feature flags, perform canary deployments, and so on.

Benchmarking in Production Has Limited Use

Unfortunately, there are many challenges to this testing practice:

- It's easier when you run your software as a SaaS. Otherwise, it's much harder as the developers can't quickly revert or fix potential impacts.
- You have to ensure Quality of Service (QoS). This means you cannot do benchmarking with extreme payloads, as you need to ensure you don't impact—e.g., cause Denial of Service (DoS)—your production environment.
- The feedback loop is quite long for developers in such a model. For example, you need to release your software fully to benchmark it.

On the other hand, if you are fine with those limitations, as presented in Figure 7-2, benchmarking in production might be the most effective and reliable testing strategy. It is ultimately the closest we can get to real production usage, which reduces the risk of inaccurate results. The effort of creating and maintaining such tests is relatively small, assuming we already have production monitoring. We don't need to simulate data, environment, dependencies, etc. We can reuse the existing monitoring tools you need to keep the cluster up.

Macrobenchmarks

Testing or benchmarking in production is reliable, but spotting problems at that point is expensive. That's why the industry introduced testing in earlier stages of development. The benefit is that we can assess the efficiency with just prototypes, which can be produced much quicker. We call the tests on this level "macrobenchmarks."

Macrobenchmarks provide a great balance between good reliability of such tests and faster feedback loop compared to benchmarking in production. In practice, it means building your Go program and benchmarking it in a simulated environment with all required dependencies. For example, for client-side applications, it might mean buying some example client devices (e.g., smartphones if we build the mobile application). Then for some application releases, reinstall your Go program on those devices and thoroughly benchmark it (ideally with some automated suite).

For SaaS-like use cases, it might mean creating copies of production clusters, commonly called "testing" or "staging" environments. Then, to assess efficiency, build your Go program, deploy how you would in production, and benchmark it. We will also discuss more straightforward methods like using an `e2e` framework (*https://oreil.ly/f0IJo*) that you can run on a single development machine without complex orchestration systems like Kubernetes. I will explain those two methods briefly in "Macrobenchmarks" on page 306.

There are many benefits of macrobenchmarking:

- They are highly reliable and effective (yet not as much as benchmarking in production).
- You can delegate such macrobenchmarking to independent QA engineers because you can treat your Go program as a "closed box" (previously known as a "black box"—no need to understand how it is implemented).
- You don't impact production with anything you do.

The downside of this approach, as shown in Figure 7-2, is the effort of building and maintaining such a benchmark suite. Typically, it means complex configuration or code to automate all of it. Additionally, in many cases, any functional changes to our Go program mean we must rebuild parts of the complex macrobenchmarking system. As a result, such macrobenchmarks are viable for more mature projects with stable APIs. On top of that, the feedback loop is still quite long. We also must limit how many benchmarks we can do at once. Naturally, we have a limited number of those testing clusters that we share with other team members for cost efficiency. This means we have to coordinate those benchmarks.

Microbenchmarks

Fortunately, there is a way to have more agile benchmarks! We can follow the pattern of divide and conquer (*https://oreil.ly/ZFxiG*) for optimizations. Instead of looking at the efficiency of the whole system or the Go program, we treat our program in an open box (previously known as a "white box") manner and divide program functionality into smaller parts. We can then use the profiling we will learn in Chapter 9 to identify parts that contribute the most to the efficiency of the whole solution (e.g., use the most CPU or memory resource or add the most to the latency). We can then assess the efficiency of the program's most "expensive" part by writing small unit tests like microbenchmarks just for this small part in isolation. The Go language provides a native benchmarking framework that you can run with the same tool as unit tests: `go test`. We will discuss using this practice in "Microbenchmarks" on page 275.

Microbenchmarks are probably the most fun to write because they are very agile and provide rapid feedback about the efficiency of our Go function, algorithm, or structure. You can quickly run those benchmarks on your (even small!) developer machine, often without going out of your favorite IDE. You can implement such a benchmark test in 10 minutes, execute it in the next 20 minutes, and then tear it down or change it entirely. It is cheap to make, cheap to iterate, like a unit test. You can also treat it as a more reusable development tool—write more complex microbenchmarks that will work as acceptance benchmarks for a small part of the code the whole team can use.

Unfortunately, with agility comes many trade-offs. For example, suppose you wrongly identify the efficiency bottleneck of your program. In that case, you might be celebrating that your local microbenchmarks for some parts of the program take only 200 ms. However, when your program is deployed, it might still cause efficiency problems (and violate the RAER). On top of that, some problems are only visible when you run all the code components together (similar to integration tests). The choice of test data is also nontrivial. In many cases, it is impossible to mimic dependencies in a way that makes sense to reproduce certain efficiency problems, so we have to make some assumptions.

When Microbenchmarking, Don't Forget About the Big Picture

It is not uncommon to perform easy, deliberate optimizations on the part of code that is a bottleneck and see a major improvement. For example, after optimization, our microbenchmarks might indicate that instead of 400 MB, our function now allocates only 2 MB per operation. After thinking about that part of the code, you might have plenty of other ideas about optimizations for that 2 MB of allocations! So you might be tempted to learn and optimize that.

This is a risk. It's easy to fixate on raw numbers from a single microbenchmark and go into the optimization rabbit hole, introducing more complexity and spending valuable engineering time.

In this case, we should most likely be happy with the massive, 200x improvement, and do all it takes to get it deployed. If we want to further improve the performance of the path we were looking at, it's not unlikely that the bottleneck of the code path we were testing has now moved somewhere else!

What Level Should You Use?

As you might have already noticed, there is no "best" benchmark type. Each stage has its purpose and is needed. Every solid software project should eventually have some microbenchmarks, have some macro ones, and potentially benchmark some portion of functionalities in production. This can be confirmed by just looking at some open source projects. There are many examples, but just to pick two:

- The Prometheus project (*https://oreil.ly/FwnBN*) has dozens of microbenchmarks and a semiautomated, dedicated macrobenchmark suite (*https://oreil.ly/QqwrL*) that deploys instances of the Prometheus program in Google Cloud and benchmarks them. Many Prometheus users also test and gather efficiency data directly from production clusters.

- The Vitess project (*https://oreil.ly/tcGNV*) uses microbenchmarks written in Go (*https://oreil.ly/cLr6f*) as well. On top of that, the Vitess project maintains macrobenchmarks (*https://oreil.ly/pxtPO*). Amazingly, it builds automation that runs both types of benchmarks nightly, with results reported on the dedicated website (*https://oreil.ly/8RMw6*). This is an exceptional best-practice example.

What benchmarks to add to the software projects you work on, and when, depends on needs and maturity. Be pragmatic with adding benchmarks. No software needs numerous benchmarks in the early development cycle. When APIs are unstable and detailed requirements are changing, the benchmark will need to change as well. In fact, it can be harmful to the project if we spend time on writing (and later maintaining) benchmarks for a project that hasn't yet functionally proven its usefulness.

Follow this (intelligently) lazy approach instead:

1. If the stakeholder is unhappy with visible efficiency problems, perform the bottleneck analysis explained in Chapter 9 on production and add microbenchmarks (see "Microbenchmarks" on page 275) to the part that is a bottleneck. When optimized, another part will likely be a bottleneck, so new tests must be added. Do this until you are happy with the efficiency, or it's too difficult or expensive to optimize the program further. It will grow organically.
2. When a formal RAER is established, it might be useful to ensure that you test efficiency more end to end. Then you might want to invest in the manual, then automatic, macrobenchmarks (see "Macrobenchmarks" on page 306).
3. If you truly care about accurate and pragmatic tests, and you control your "production" environment (applicable for SaaS software), consider benchmarking in production.

Don't Worry About "Benchmark" Code Coverage!

For functional testing, it's popular to measure the quality of the project by ensuring the test code coverage (*https://oreil.ly/Sfde9*) is high.[23]

Never try to measure how many parts of your program have benchmarks! Ideally, you should only implement benchmarks for the critical places you want to optimize because the data indicates they are (or were) the bottleneck.

23 I am personally not a big fan of this approach. Not every part of the code is equally important to test, and not everything is worth testing. On top of that, engineers tend to gamify this system (*https://oreil.ly/NnjCD*) by writing tests only to improve the coverage, instead on focusing on finding potential problems with the code in the fastest possible way (reducing cost of development).

With this theory, you should know what benchmarking levels are available to you and why there is no silver bullet. Still, benchmarks are in the code of our software efficiency story, and the Go language is no different here. We can't optimize without experimenting and measuring. However, be mindful of the time spent in this phase. Writing, maintaining, and performing benchmarks takes time, so follow the lazy approach and add benchmarks on an appropriate level on demand and only if needed.

Summary

The reliability issues of these tests are perhaps one of the biggest reasons developers, product managers, and stakeholders de-scope efficiency efforts. Where do you think I found all those little best practices to improve reliability? At the beginning of my engineering career, I spent numerous hours on careful load testing and benchmarks with my team, only to realize it meant nothing as we missed a critical element of the environment. For example, our synthetic workloads were not providing a realistic load.

Such cases can discourage even professional developers and product managers. Unfortunately, this is where we typically prefer to pay more for waste computing rather than invest in optimization efforts. That's why it's critically important to ensure the experiment, load tests, and scale tests we do are as reliable as possible to achieve our efficiency goals faster!

In this chapter, you learned the foundations behind reliable efficiency assessment through empirical experiments we call benchmarks.

We discussed the basic complexity analysis that can help optimize our journey. I mentioned the difference between benchmark testing and functional testing and why benchmarks lie if we misinterpret them. You learned common reliability problems that I found truly important during experimentation cycles and the levels of benchmarks commonly spotted in the industry.

We are finally ready to learn how to implement those benchmarks on all levels mentioned above, so let's jump right into it!

CHAPTER 8
Benchmarking

Hopefully, your Go IDE is ready and warmed up for some action! It's time to stress our Go code to find its efficiency characteristics on the micro and macro levels mentioned in Chapter 7.

In this chapter, we will start with "Microbenchmarks", where we will go through the basics of microbenchmarking and introduce Go native benchmarking. Next, I will explain how to interpret the output with tools like `benchstat`. Then I will go through the microbenchmark aspects and tricks that I learned that are incredibly useful for the practical use of microbenchmarks.

In the second half of this chapter, we'll go through "Macrobenchmarks" on page 306, which is rarely in the scope of programming books due to its size and complexity. In my opinion, macrobenchmarking is as critical to Go development as microbenchmarking, so every developer caring about efficiency should be able to work with that level of testing. Next, in "Go e2e Framework" on page 310 we will go through a complete example of a macro test written fully in Go using containers. We will discuss results and common observability in the process.

Without further ado, let's jump into the most agile way of assessing the efficiency of smaller parts of the code, namely microbenchmarking.

Microbenchmarks

A benchmark can be called a microbenchmark if it's focused on a single, isolated functionality on a small piece of code running in a single process. You can think of microbenchmarks as a tool for efficiency assessment of optimizations made for a single component on the code or algorithm level (discussed in "Optimization Design Levels" on page 98). Anything more complex might be challenging to benchmark on the micro level. By more complex, I mean, for example, trying to benchmark:

- Multiple functionalities at once.
- Long-running functionalities (over 5–10 seconds long).
- Bigger multistructure components.
- Multiprocess functionalities. Multigoroutine functionalities are acceptable if they don't spin too many goroutines (e.g., over one hundred) during our tests.
- Functionalities that require more resources to run than a moderate development machine (e.g., allocating 40 GB of memory to compute an answer or prepare a test dataset).

If your code violates any of those elements, you might consider splitting it into smaller microbenchmarks or consider using macrobenchmarks on ones with different frameworks (see "Macrobenchmarks" on page 306).

Keep Microbenchmarks Micro

The more we are benchmarking at once on a micro level, the more time it takes to implement and perform such benchmarks. This results in cascading consequences—we try to make benchmarks more reusable and spend even more time building more abstractions over them. Ultimately, we try to make them stable and harder to change.

This is a problem because microbenchmarks were designed for agility. We change code often, so we want benchmarks to be updated quickly and not get in our way. So you write them quickly, keep them simple, and change them.

On top of that, Go benchmarks do not have (and should not have!) sophisticated observability, which is another reason to keep them small.

The benchmark definition means that it's very rare for the microbenchmark to validate if your program matches the high-level user RAER for certain functionality, e.g., "The p95 of this API should be under one minute." In other words, it is usually not well suited to answer questions requiring absolute data. Therefore, while writing microbenchmarks, we should instead focus on answers that relate to a certain baseline or pattern, for example:

Learning about runtime complexity
: Microbenchmarks are a fantastic way to learn more about the Go function or method efficiency behavior over certain dimensions. For example, how is latency impacted by different shares and sizes of the input and test data? Do allocations grow in an unbounded way with the size of input? What are the constant factors and the overhead of the algorithm you chose?

Thanks to the quick feedback loop, it's easy to manually play with test inputs and see what your function efficiency looks like for various test data and cases.

A/B testing

A/B tests are defined by performing the same test on version A of your program and then on version B, which is different (ideally) only by one thing (e.g., you reused one slice). They can tell us the relative impact of our changes.

Microbenchmarks are a great way to assess if a new change of the code, configuration, or hardware can potentially affect the efficiency. For example, suppose we know that the absolute latency of some requests is two minutes, and we know that 60% of that latency is caused by a certain Go function in a code we develop. In this case, we can try optimizing this function and perform a microbenchmark before and after. As long as our test data is reliable, if after optimization, our microbenchmark shows our optimization makes our code 20% faster, the full system will also be 18% faster.

Sometimes the absolute numbers on microbenchmarking for latency might matter less. For example, it doesn't tell us much if our microbenchmark shows 900 ms per operation on our machine. On a different laptop, it might show 500 ms. What matters is that on the same machine, with as few changes to the environment as possible and running one benchmark after another, the latency between version A and B is higher or lower. As we learned in "Reproducing Production" on page 258, there are high chances that this relation is then reproducible in any other environment where you will benchmark those versions.

The best way to implement and run microbenchmarks in Go is through its native benchmarking framework built into the `go test` tool. It is battle tested, integrated into testing flows, has native support for profiling, and you can see many benchmark examples in the Go community. I already mentioned the basics around the Go benchmark framework with Example 6-3, and we saw some preprocessed results in Example 7-2 outputs, but it's now time to dive into details!

Go Benchmarks

Creating microbenchmarks in Go (*https://oreil.ly/0h0y0*) starts by creating a particular function with a specific signature. Go tooling is not very picky—a function has to satisfy three elements to be considered a benchmark:

- The file where the function is created must end with the *_test.go* suffix.[1]
- The function name must start with the case-sensitive `Benchmark` prefix, e.g., `BenchmarkSum`.
- The function must have exactly one function argument of the type `*testing.B`.

In "Complexity Analysis" on page 240, we discussed the space complexity of the Example 4-1 code. In Chapter 10, I will show you how to optimize this code with a few different requirements. I wouldn't be able to optimize those successfully without Go benchmarks. I used them to obtain estimated numbers for the number of allocations and latency. Let's now see how that benchmarking process looks.

The Go Benchmark Naming Convention

I try to follow the consistent naming pattern[2] for the `<NAME>` part on all types of functions in the Go testing framework, like benchmarks (`Benchmark<NAME>`), tests (`Test<NAME>`), fuzzing tests (`Fuzz<NAME>`), and examples (`Example<NAME>`). The idea is simple:

- Calling a test `BenchmarkSum` means it tests the `Sum` function efficiency. `BenchmarkSum_withDuplicates` means the same, but the suffix (notice it starts with a lowercase letter) tells us a certain condition we test in.
- `BenchmarkCalculator_Sum` means it tests a method `Sum` from the `Calculator` struct. As above, we can add a suffix if we have more tests for the same method to distinguish between cases, e.g., `BenchmarkCalculator_Sum_withDuplicates`.
- Additionally, you can put an input size as yet another suffix e.g., `BenchmarkCalculator_Sum_10M`.

Given that `Sum` in Example 4-1 is a single-purpose short function, one good microbenchmark should suffice to tell its efficiency. So I created a new function in the *sum_test.go* file with the name `BenchmarkSum`. However, before I did anything else, I added the raw template of the small boilerplate required for most benchmarks, as presented in Example 8-1.

1 For bigger projects, I would suggest adding the *_bench_test.go* suffix for an easier way of discovering benchmarks.

2 It is well explained in the testing package's Example documentation (*https://oreil.ly/PRrlW*).

Example 8-1. Core Go benchmark elements

```
func BenchmarkSum(b *testing.B) {
    b.ReportAllocs() ❶

    // TODO(bwplotka): Add any initialization that is needed.

    b.ResetTimer() ❷
    for i := 0; i < b.N; i++ { ❸
        // TODO(bwplotka): Add tested functionality.
    }
}
```

❶ Optional method (*https://oreil.ly/ootGE*) that tells the Go benchmark to provide the number of allocations and the total amount of allocated memory. It's equivalent to setting the `-benchmem` flag when running the test. While it might, in theory, add a tiny overhead to measured latency, it is only visible in very fast functions. I rarely need to remove allocation tracing in practice, so I always have it on. Often, it's useful to see a number of allocations even if you expect the job to be only CPU sensitive. As mentioned in "Memory Relevance" on page 150, some allocations can be surprising!

❷ In most cases, we don't want to benchmark the resources required to initialize the test data, structure, or mocked dependencies. To do this "outside" of the latency clock and allocation tracking, reset the timer (*https://oreil.ly/5et2N*) right before the actual benchmark. If we don't have any initialization, we can remove it.

❸ This exact `for` loop sequence with `b.N` is a mandatory element of any Go benchmark. Never change it or remove it! Similarly, never use `i` from the loop for your function. It can be confusing at the start, but to run your benchmark, `go test` might run `BenchmarkSum` multiple times to find the right `b.N`, depending on how we run it. By default, `go test` will aim to run this benchmark for at least 1 second. This means it will execute our benchmark once with `b.N` that equals 1 m only to assess a single iteration duration. Based on that, it will try to find the smallest `b.N` that will make the whole `BenchmarkSum` execute at least 1 second.[3]

The `Sum` function I wanted to benchmark takes one argument—the filename containing a list of the integers to sum. As we discussed in "Complexity Analysis" on page

---

3 If we would remove `b.N` completely, the Go benchmark will try to increase a number of `N` until the whole `BenchmarkSum` will take at least 1 second. Without the `b.N` loop, our benchmark will never exceed 1 second as it does not depend on `b.N`. Such a benchmark will stop at `b.N` being equal to 1 billion iterations, but with just a single iteration being executed, the benchmark results will be wrong.

240, the algorithm used in Example 4-1 depends on the number of integers in the file. In this case, space and time complexity are `O(N)`, where `N` is a number of integers. This means that `Sum` with a single integer will be faster and allocate less memory than `Sum` with thousands of integers. As a result, the choice of input will significantly change the efficiency results. But how do we find the correct test input for our benchmark? Unfortunately, there is no single answer.

The Choice of Test Data and Conditions for Our Benchmarks

Generally, we want the smallest possible (thus quickest and cheapest to use!) dataset, which will give us enough knowledge and confidence in our program efficiency characteristic patterns. On the other hand, it should be big enough to trigger potential limits and bottlenecks that users might experience. As we mentioned in "Reproducing Production" on page 258, the test data should simulate the production workload as much as possible. We aim for "typicality."

However, if our functionality has a massive problem for specific input, we should also include that in our benchmarks!

To make things more difficult, we are additionally constrained with the data size for microbenchmarks. Typically, we want to ensure those benchmarks can run at maximum within a matter of minutes and in our development environments for the best agility and shortest feedback loop possible. On the bright side, there are ways to find some efficiency pattern of your program, run benchmarks with a couple of times smaller dataset than the potential production dataset, and extrapolate the possible results.

For example, on my machine it takes Example 4-1 about 78.4 ms to sum 2 million integers. If I benchmark with 1 million integers, it takes 30.5 ms. Given these two numbers, we could assume with some confidence[4] that our algorithm, on average, requires around 29 nanoseconds to sum a single integer.[5] If our RAER specifies, for example, that we have to sum 2 billion integers under 30 seconds, we can assume our implementation is too slow as 29 ns * 2 billion is around 58 seconds.

For those reasons, I decided to stick with 2 million integers for the Example 4-1 benchmark. It is a big enough number to show some bottlenecks and efficiency patterns but small enough to keep our program relatively quick (on my machine, it can

4 As mentioned earlier, microbenchmarks are always based on some amount of assumptions; we cannot simulate everything in such a small test.

5 Note that it definitely will not take 29 nanoseconds for a benchmark with a single integer. This number is a latency we see for a larger number of integers.

perform around 14 operations within 1 second.)[6] For now, I created a *testdata* directory (excluded from the compilation) and manually created a file called *test.2M.txt* with 2 million integers. With the test data and Example 8-1, I added the functionality I want to test, as presented in Example 8-2.

Example 8-2. Simplest Go benchmark for assessing efficiency of the `Sum` function

```
func BenchmarkSum(b *testing.B) {
    for i := 0; i < b.N; i++ {
        _, _ = Sum("testdata/test.2M.txt")
    }
}
```

To run this benchmark, we can use the `go test` command, which is available when we install Go (*https://oreil.ly/dQ57t*) on our machine. `go test` allows us to run all specified tests, fuzzing tests, or benchmarks. For benchmarks, `go test` has many options that allow us to control how it will execute our benchmark and what artifacts it will produce after a run. Let's go through example options, presented in Example 8-3.

Example 8-3. Example commands we can use to run Example 8-2

```
$ go test -run '^$' -bench '^BenchmarkSum$' ❶
$ go test -run '^$' -bench '^BenchmarkSum$' -benchtime 10s ❷
$ go test -run '^$' -bench '^BenchmarkSum$' -benchtime 100x ❸
$ go test -run '^$' -bench '^BenchmarkSum$' -benchtime 1s -count 5 ❹
```

❶ This command executes a single benchmark function with the explicit name `BenchmarkSum`. You can use the RE2 regex language (*https://oreil.ly/KDIL9*) to filter the tests you want to run. Notice the `-run` flag that strictly matches no functional test. This is to make sure no unit test will be run, allowing us to focus on the benchmark. Empty `-run` flags mean that all unit tests will be executed.

❷ With `-benchtime`, we can control how long or how many iterations (functional operations) our benchmark should execute. In this example, we choose to have as many iterations as can fit in a 10-second interval.[7]

6 Note that it is acceptable to change test data in future versions of our program and benchmark. Usually, our optimizations over time make our test dataset "too small," so we can increase it over time to spot different problems if we need to optimize further.

7 As explained previously, note that the full benchmarking process can take longer than 10 seconds because the Go framework will try to find a correct number of iterations. The more variance in the test results—potentially the longer the test will last.

❸ We can choose to set `-benchtime` to the exact amount of iterations. This is used less often because, as a microbenchmark user, you want to focus on a quick feedback loop. When iterations are specified, we don't know when the test will end and if we need to wait 10 seconds or 2 hours. This is why it's often preferred to limit the benchmark time, and if we see too few iterations, increase the number in `-benchtime` a little, or change the benchmark implementation or test data.

❹ We can also repeat the benchmark cycle with the `-count` flag. Doing so is very useful, as it allows us to calculate the variance between runs (with tools explained in "Understanding the Results" on page 284).

The full list of options is pretty long, and you can list them anytime using `go help testflag` (*https://oreil.ly/F2wTM*).

Running Go Benchmarks Through IDE

Almost all modern IDEs allow us to simply click on the Go benchmark function and execute it from the IDE. So feel free to do it. Just set up the correct options, or at least be aware of what options are there by default!

I use the IDE to trigger initial, one-second benchmark runs, but I prefer good old CLI commands for more complex cases. They are easy to use and it's easy to share the test run configuration with others. In the end, use what you feel the most comfortable with!

For my `Sum` benchmark, I created a helpful one-liner with all the options I need, presented in Example 8-4.

Example 8-4. One-line shell command to benchmark Example 4-1

```
$ export ver=v1 && \ ❶
    go test -run '^$' -bench '^BenchmarkSum$' -benchtime 10s -count 5 \
        -cpu 4 \ ❷
        -benchmem \ ❸
        -memprofile=${ver}.mem.pprof -cpuprofile=${ver}.cpu.pprof \ ❹
    | tee ${ver}.txt ❺
```

❶ It is very tempting to write complex scripts or frameworks to save the result in the correct place, create automation that compares results for your use, etc. In many cases, that is a trap because Go benchmarks are typically ephemeral and easy to run. Still, I decided to add a tiny amount of bash scripting to ensure the artifacts my benchmark will produce have the same name I can refer to later. When I benchmark a new code version with optimizations, I can manually adjust the `ver` variable to different values like `v2`, `v3`, or `v2-with-streaming` for later comparisons.

❷ Sometimes if we aim to optimize latency via concurrent code, as in "Optimizing Latency Using Concurrency" on page 402, it is important to control the number of CPU cores the benchmarks were allowed to use. This can be achieved with the `-cpu` flag. It sets the correct `GOMAXPROCS` setting. As we mentioned in "Performance Nondeterminism" on page 260, the choice of the exact value highly depends on what the production environment looks like and how many CPUs your development machine has.[8]

❸ There is no point in optimizing latency if our optimization allocates an extreme amount of memory which, as we learned in "Memory Relevance" on page 150, might be our first enemy. In my experience, the memory allocations cause more problems than CPU usage, so I always try to pay attention to allocations with `-benchmem`.

❹ If you run your microbenchmark and see results you are not happy with, your first question is probably what caused that slowdown or high memory usage. This is why the Go benchmark has built-in support for profiling, explained in Chapter 9. I am lazy, so I usually keep those options on by default, similar to `-benchtime`. As a result, I can always dive into the profile to find the line of code that contributed to suspicious resource usage. Similar to `-benchtime` and `ReportAllocs`, those are turned off by default because they add a slight overhead to latency measurements. However, it's usually safe to leave them turned on unless you measure ultra-low latency operations (tens of nanoseconds). Especially the `-cpuprofile` option adds some allocations and latency in the background.

❺ By default, `go test` prints results to standard output. However, to reliably compare and not get lost in what results correspond to what runs, I recommend saving them in temporary files. I recommend using `tee` to write both to file and standard output, so you can follow the progress of the benchmark.

---

8 You can also provide multiple numbers after a comma. For example, `-cpu=1,2,3` will run a test with `GOMAXPROCS` set to 1, then to 2, and the third run with 3 CPUs.

With the benchmark implementation, input file, and execution command, it's time to perform our benchmark. I executed Example 8-4 in the directory of the test file on my machine, and after 32 seconds, it finished. It created three files: *v1.cpu.pprof*, *v1.mem.pprof*, and *v1.txt*. In this chapter, we are most interested in the last file, so you can learn how to read and understand the Go benchmark output. Let's do that in the next section.

Understanding the Results

After each run, the `go test` benchmark prints the result in a consistent format.[9] Example 8-5 presents the output runs executed with Example 8-4 on the code presented in Example 4-1.

Example 8-5. The output of the v1.txt file produced by the Example 8-4 command

```
goos: linux ❶
goarch: amd64
pkg: github.com/efficientgo/examples/pkg/sum
cpu: Intel(R) Core(TM) i7-9850H CPU @ 2.60GHz
BenchmarkSum-4    67    79043706 ns/op    60807308 B/op    1600006 allocs/op ❷
BenchmarkSum-4    74    79312463 ns/op    60806508 B/op    1600006 allocs/op
BenchmarkSum-4    66    80477766 ns/op    60806472 B/op    1600006 allocs/op
BenchmarkSum-4    66    80010618 ns/op    60806224 B/op    1600006 allocs/op
BenchmarkSum-4    74    80793880 ns/op    60806445 B/op    1600006 allocs/op
PASS
ok      github.com/efficientgo/examples/pkg/sum    38.214s
```

❶ Every benchmark run captures some basic information about the environment like architecture, operating system type, the package we run the benchmark in, and the CPU on the machine. Unfortunately, as we discussed in "Reliability of Experiments" on page 256, there are many more elements that could be worth capturing[10] that can impact the benchmark.

❷ Every row represents a single run (i.e., if you ran the benchmark with `-count=1`, you would have just a single line). The line consists of three or more columns. The number depends on the benchmark configuration, but the order is consistent. From the left, we have:

9 The internal representation of that format can be explored by looking at `BenchmarkResult` type (*https://oreil.ly/90wO2*).

10 Things like the Go version, Linux kernel version, other processes running at the same time, CPU mode, etc. Unfortunately, the full list is almost impossible to capture.

- Name of the benchmark with the suffix representing the number of CPUs available (in theory[11]) for this benchmark. This tells us what we can expect for concurrent implementations.
- Number of iterations in this benchmark run. Pay attention to this number; if it's too low, the numbers in the other columns might not reflect reality.
- Nanoseconds per operation resulting from `-benchtime` divided by a number of runs.
- Allocated bytes per operation on the heap. As you learned in Chapter 5, remember that this does not tell us how much memory is allocated in any other segments, like manual mappings, caches, and stack! This column is present only if the `-benchmem` flag was set (or `ReportAllocs`).
- Number of allocations per operation on the heap (also only present with the `-benchmem` flag set).
- Optionally, you can report your own metrics per operation using the `b.ReportMetric` method. See this example (*https://oreil.ly/IuwYl*). This will appear as further columns and can be aggregated similarly with the tooling explained later.

If you run Example 8-4 and you see no output for a long time, it might mean that the first run of your microbenchmark is taking that long. If your `-benchtime` is time based, the `go test` quickly checks how long it takes to run a single iteration to find the estimated number of iterations.

If it takes too much time, unless you want to run 30+ minute tests, you might need to optimize the benchmark setup, reduce the data size, or split the microbenchmark into smaller functionality. Otherwise, you won't achieve hundreds or dozens of required iterations.

If you see the initial output (`goos`, `goarch`, `pkg`, and benchmark name), a single iteration run has completed, and a proper benchmark has started.

The results presented in Example 8-5 can be read directly, but there are some challenges. First of all, the numbers are in the base unit—it's not obvious at first glance to see if we allocate 600 MB, 60 MB, or 6 MB. It's the same if we translate our latency to

11 The Go testing framework does not check how many CPUs are free to be used for this benchmark. As you learned in Chapter 4, CPUs are shared fairly across other processes, so with more processes in the system, the four CPUs, in my case, are not fully reserved for the benchmark. On top of that, programmatic changes to `runtime.GOMAXPROCS` are not reflected here.

seconds. Secondly, we have five measurements, so which one do we choose? Finally, how do we compare a second microbenchmark result done for the code with the optimization?

Fortunately, the Go community created another CLI tool, `benchstat` (*https://oreil.ly/PWSN4*), that performs further processing and statistical analysis of one or multiple benchmark results for easier assessment. As a result, it has become the most popular solution for presenting and interpreting Go microbenchmark results in recent years.

You can install `benchstat` using the standard `go install` tooling, for example, `go install golang.org/x/perf/cmd/benchstat@latest`. Once completed, it will be present in your $GOBIN or *$GOPATH/bin* directory. You can then use it to present the results we got in Example 8-5; see the example usage in Example 8-6.

Example 8-6. Running `benchstat` on the results presented in Example 8-5

```
$ benchstat v1.txt ❶
name    time/op
Sum-4   79.9ms ± 1% ❷

name    alloc/op
Sum-4   60.8MB ± 0%

name    allocs/op
Sum-4    1.60M ± 0%
```

❶ We can run `benchstat` with the *v1.txt* containing Example 8-5. The `benchstat` can parse the format of the `go test` tooling from one or multiple benchmarks performed once or multiple times on the same code version.

❷ For each benchmark, `benchstat` calculates the mean (average) of all runs and ± the variance across runs (1% in this case). This is why it's essential to run `go test` benchmarks multiple times (e.g., with the `-count` flag); otherwise, with just a single run, the variance will indicate a misleading 0%. Running more tests allows us to assess the repeatability of the result, as we discussed in "Performance Nondeterminism" on page 260. Run `benchstat --help` to see more options.

Once we have confidence in our test run, we can call it baseline results. We typically want to assess the efficiency of our code with the new optimization by comparing it with our baseline. For example, in Chapter 10 we will optimize the `Sum`, and one of the optimized versions will be twice as fast. I found this by changing the `Sum` function visible in Example 4-1 to `ConcurrentSum3` (the code is presented in Example 10-12). Then I ran the benchmark implemented in Example 8-2 using exactly the same command shown in Example 8-4, just changing `ver=v1` to `ver=v2` to produce *v2.txt* and *v2.cpu.pprof* and *v2.mem.pprof*.

The benchstat helped us calculate variance and provided human-readable units. But there is another helpful feature: comparing results from different benchmark runs. For example, Example 8-7 shows how I checked the difference between the naive and improved concurrent implementation.

Example 8-7. Running benchstat to compare results from v1.txt *and* v2.txt

```
$ benchstat v1.txt v2.txt ❶
name    old time/op    new time/op    delta
Sum-4     79.9ms ± 1%    39.5ms ± 2%  -50.52%  (p=0.008 n=5+5) ❷

name    old alloc/op   new alloc/op   delta
Sum-4     60.8MB ± 0%    60.8MB ± 0%     ~     (p=0.151 n=5+5)

name    old allocs/op  new allocs/op  delta
Sum-4      1.60M ± 0%     1.60M ± 0%   +0.00%  (p=0.008 n=5+5)
```

❶ Running benchstat with two files enables comparison mode.

❷ In comparison mode, benchstat provides a delta column showing the delta between two means in a percentage or ~ if the significance test fails. The significance test is defaulted to the Mann-Whitney U test (*https://oreil.ly/ESCAz*) and can be disabled with `-delta-test=none`. The significance test is an extra statistical analysis that calculates the p-value (*https://oreil.ly/6K0zl*), which by default should be smaller than `0.05` (configurable with `-alpha`). It gives us additional information on top of the variance (after ±) if the results can be safely compared. The `n=5+5` represents the sample sizes in both results (both benchmark runs were done with `-count=5`).

Thanks to benchstat and Go benchmarks, we can tell with some confidence that our concurrent implementation is around 50% faster and does not impact allocations.

Careful readers might notice that the allocation size failed the significance test of benchstat (p is higher than 0.05). I could improve that by running benchmarks with a higher `-count` (e.g., 8 or 10).

I left this significance test failing on purpose to show you that there are cases when you can apply common reasoning. Both results indicate large 60.8 MB allocations with minimal variance. We can clearly say that both implementations use a similar amount of memory. Do we care whether one implementation uses a few KB more or less? Probably not, so we can skip the benchstat significance test that verifies if we can trust the delta. No need to spend more time here than needed!

Analyzing microbenchmarks might be confusing initially, but hopefully, the presented flow using `benchstat` taught you how to assess efficiencies of different implementations without having a degree in data science! Generally, while using `benchstat`, remember to:

- Run more tests than one (`-count`) to be able to spot the noise.
- Check that the variance number after ± is not higher than 3–5%. Be especially vigilant in variance for smaller numbers.
- To rely on an accurate delta across results with higher variance, check the significance test (p-value).

With this in mind, let's go through a few common advanced tricks that you might find very useful in your day-to-day work with Go benchmarks!

Tips and Tricks for Microbenchmarking

The best practices for microbenchmarking are often learned from your own mistakes and rarely shared with others. Let's break that up by mentioning some of the common aspects of Go microbenchmarks that are worth being aware of.

Too-High Variance

As we learned in "Performance Nondeterminism" on page 260, knowing the variance of our tests is critical. If the difference between microbenchmarks is more than, let's say, 5%, it indicates potential noise, and we might not be able to rely on those results entirely.

I had this case when preparing "Optimizing Latency Using Concurrency" on page 402. When benchmarking, my results had way too large a variance as the `benchstat` result suggested. The results from that run are presented in Example 8-8.

Example 8-8. `benchstat` indicating large variance in latency results

```
name   time/op
Sum-4  45.7ms ±19% ❶

name   alloc/op
Sum-4  60.8MB ± 0%

name   allocs/op
Sum-4   1.60M ± 0%
```

❶ Nineteen percent variance is quite scary. We should ignore such results and stabilize the benchmark before making any conclusions.

What can we do in this case? We already mentioned a few things in "Performance Nondeterminism" on page 260. We should consider running the benchmark longer, redesigning our benchmark, or running it in different environmental conditions. In my case I had to close my browser and increase `-benchtime` from 5 s to 15 s to achieve the 2% variance run in Example 8-7.

Find Your Workflow

In "Go Benchmarks" on page 277, you followed me through my efficiency assessment cycle on a micro level. Of course, this can vary, but it is generally based on `git` branches, and can be summarized as follows:

1. I check for any existing microbenchmark implementation for what I want to test. If none exists, I will create one.
2. In my terminal, I execute a command similar to Example 8-4 to run the benchmark several times (5–10). I save results to something like *v1.txt*, save profiles, and assume that as my baseline.
3. I assess the *v1.txt* results to check if the resource consumption is roughly what I expect from my understanding of the implementation and the input size. To confirm or reject, I perform the bottleneck analysis explained in Chapter 9. I might perform more benchmarks for different inputs at this stage to learn more. This tells me roughly if there is room for some easy optimizations, should I invest in more dangerous and deliberate optimization, or should I move to optimizations on a different level.
4. Assuming room for some optimizations, I create a new `git` branch (*https://oreil.ly/AcM1D*) and implement it.
5. Following the TFBO flow, I test my implementation first.
6. I commit the changes, run the benchmarking function with the same command, and save it to, e.g., *v2.txt*.
7. I compare the results with `benchstat` and adjust the benchmark or optimizations to achieve the best results.
8. If I want to try a different optimization, I create yet another `git` branch or build new commits on the same branch and repeat the process (e.g., produce *v3.txt*, *v4.txt*, and so on). This allows me to get back to previous optimizations if an attempt makes me pessimistic.
9. I jot findings in my notes, commit message, or repository change set (e.g., pull requests), and discard my *.txt* results (expiration date!).

This flow works for me, but you might want to try a different one! As long as it's not confusing for you, is reliable, and follows the TFBO pattern we discussed in

"Efficiency-Aware Development Flow" on page 102, use it. There are many other options, for example:

- You can use your terminal history to track benchmarking results.
- You can create different functions for the same functionality with different optimizations. Then you can swap what function you use in your benchmark functions if you don't want to use `git` here.
- Use `git stash` instead of commits.
- Finally, you can follow the Dave Cheney flow (*https://oreil.ly/1MJNT*) that uses the `go test -c` command to build the testing framework and your code into a separate binary. You can then save this binary and perform benchmarks without rebuilding source code or saving your test results.[12]

I would propose trying different flows and learning what helps you the most!

I would suggest avoiding writing too complex automation for our local microbenchmarking workflow (e.g., complex bash script to automate some steps). Microbenchmarks are meant to be more interactive, where you can manually dig information you care for. Writing complex automation might mean more overhead and a longer feedback loop than needed. Still, if this is working for you, do it!

Test Your Benchmark for Correctness!

One of the most common mistakes we make in benchmarking is assessing the efficiency of the function that does not provide correct results. Due to the nature of deliberate optimizations, it is easy to introduce a bug that breaks the functionality of our code. Sometimes, optimizing failed executions is important,[13] but it should be an explicit decision.

The "Testing" part in TFBO, explained in "Efficiency-Aware Development Flow" on page 102, is not there by mistake. Our priority should be to write a unit test for the same functionality we will benchmark. An example unit test for our `Sum` function can look like Example 8-9.

12 Make sure to strictly control the Go version you use to build those binaries. Testing binaries built using a different Go version might create misleading results. For example, you can build a binary and add a suffix to its name with the `git` hash of the version of your source code.

13 This is especially important for distributed systems and user-facing applications that handle errors very often, and it's part of the normal program life cycle. For example, I often worked with code that was fast for database writes, but was allocating an extreme amount of memory on failed runs, causing cascading failures.

Example 8-9. Example unit test to assess the correctness of the `Sum` function

```
// import "github.com/efficientgo/core/testutil"

func TestSum(t *testing.T) {
    ret, err := Sum("testdata/input.txt")
    testutil.Ok(t, err)
    testutil.Equals(t, 3110800, ret)
}
```

Having the unit test ensures that with the right CI configured, when we propose our change to the main repository (perhaps via a pull request (*https://oreil.ly/r24MR*) [PR]), we will notice if our code is correct or not. So this already improves the reliability of our optimization job.

However, there are still things we could do to improve this process. If you only test as the last development step, you might have already performed all the effort of benchmarking and optimizing without realizing that the code is broken. This can be mitigated by manually running the unit test in Example 8-10 before each benchmarking run, e.g., the Example 8-2 code. This helps, but there are still some slight problems:

- It is tedious to run yet another thing after our changes. So it's too tempting to skip that manual process of running functional tests after the change to save time and achieve an even quicker feedback loop.
- The function might be well tested in the unit test, but there are differences between how you invoke your function in the unit test and the benchmark.
- Additionally, as you learned in "Comparison to Functional Testing" on page 252, for benchmarks we need different inputs. A new thing means a new place for making an error! For example, when preparing the benchmark for this book in Example 8-2, I accidentally made a typo in the filename (*testdata/test2M.txt* instead of *testdata/test.2M.txt*). When I ran my benchmark, it passed with very low latency results. Turns out the `Sum` did not work other than failing with the file does not exist error. Because in Example 8-2 I ignored all errors for simplicity, I missed that information. Only intuition told me that my benchmark ran a bit too quickly to be true, so I double-checked what `Sum` actually returned.
- During benchmarking at higher load, new errors might appear. For example, perhaps we could not open another file due to the limit of file descriptors on the machine, or our code does not clean files on disk, so we can't write changes to the file due to a lack of disk space.

Fortunately, an easy solution to that problem is adding a quick error check to the benchmark iteration. It could look like Example 8-10.

Example 8-10. Go benchmark for assessing the efficiency of the `Sum` function with error check

```
func BenchmarkSum(b *testing.B) {
    for i := 0; i < b.N; i++ {
        _, err := Sum("testdata/test.2M.txt")
        testutil.Ok(b, err) ❶
    }
}
```

❶ Asserting `Sum` does not return an error on every iteration loop.

It's important to notice that the efficiency metrics we get after the benchmark will include the latency contributed by the `testutil.Ok(b, err)` invocation,[14] even if there is no error. This is because we invoke this function in our `b.N` loop, so it adds a certain overhead.

Should we accept this overhead? This is the same question we have about including `-benchmem` and profile generation for tests, which also can add small noise. Such overhead is unacceptable if we try to benchmark very fast operations (let's say under milliseconds fast). For the majority of benchmarks, however, such an assertion will not change your benchmarking results. One would even argue that such error assertion will exist in production, so it should be included in the efficiency assessment.[15] Similar to `-benchmem` and profiles, I add that assertion to almost all microbenchmarks I work with.

In some ways, we are still prone to mistakes. Perhaps with the large input, the `Sum` function does not provide a correct answer without returning an error. As with all testing, we will never stop all mistakes—there has to be a balance between the effort of writing, executing, and maintaining extra tests and confidence. It's up to you to decide how much you trust your workflow.

If you want to choose the preceding case for more confidence, you can add a check that compares the returned sum with the expected result. In our case, it will not be a big overhead to add `testutil.Equals(t, <expected number>, ret)`, but usually it is more expensive and thus inappropriate to add for microbenchmarks. For those purposes, I created a small `testutil.TB` object (*https://oreil.ly/wMX6O*) that allows you to run a single iteration of your microbenchmark for unit test purposes. This allows it to be always up-to-date in terms of correctness, which is especially

14 In my benchmarks, on my machine, this instruction alone takes 244 ns and allocates zero bytes.

15 Profiling, explained in "Profiling in Go" on page 331, can also help determine how much your benchmark affects those overheads.

challenging in bigger shared code repositories. For example, continuous testing of our Sum benchmark could look like Example 8-11.[16]

Example 8-11. Testable Go benchmark for assessing the efficiency of the Sum *function*

```
func TestBenchSum(t *testing.T) {
    benchmarkSum(testutil.NewTB(t))
}

func BenchmarkSum(b *testing.B) {
    benchmarkSum(testutil.NewTB(b))
}

func benchmarkSum(tb testutil.TB) { ❶
    for i := 0; i < tb.N(); i++ { ❷
        ret, err := Sum("testdata/test.2M.txt")
        testutil.Ok(tb, err)
        if !tb.IsBenchmark() {
            // More expensive result checks can be here.
            testutil.Equals(tb, int64(6221600000), ret) ❸
        }
    }
}
```

❶ `testutil.TB` is an interface that allows running a function as both benchmarks and a unit test. Furthermore, it allows us to design our code, so the same benchmark is executed by other functions, e.g., with extra profiling, as shown in Example 10-2.

❷ The `tb.N()` method returns `b.N` for the benchmark, allowing normal microbenchmark execution. It returns `1` to perform one test run for unit tests.

❸ We can now put the extra code that might be more expensive (e.g., more complex test assertions) in the space unreachable for benchmarks, thanks to the `tb.IsBenchmark()` method.

To sum up, please test your microbenchmark code. It will save you and your team time in the long run. On top of that, it can provide a natural countermeasure against unwanted compiler optimizations, explained in "Compiler Optimizations Versus Benchmark" on page 301.

16 Note that TB is my own invention and it's not common or recommended by the Go community, so use with care!

Sharing Benchmarks with the Team (and Your Future Self)

Once you finish your TFBO cycle and are happy with your next optimization iteration, it's time to commit to new code. Share what you found or achieved with your team for more than your small one-person project. When someone proposes an optimization change, it's not uncommon to see the optimization in the production code and only a small description: "I benchmarked it, and it was 30% faster." This is not ideal for multiple reasons:

- It's hard for the reviewer to validate the benchmark without seeing the actual microbenchmark code you use. It's not that reviewers should not trust that you tell the truth, but rather it's easy to make a mistake, forget a side effect, or benchmark wrongly.[17] For example, the input has to be of a certain size to trigger the problem, or the input does not reflect the expected use cases. This can only be validated by another person looking at your benchmarking code. It's especially important when we work remotely with the team and in open source projects, where strong communication is essential.
- Once merged, it's likely any other change that touches this code might accidentally introduce efficiency regression.
- If you or anyone else wants to try to improve the same part of code, they have no other option than to re-create the benchmark and go through the same effort you did in your pull request because the previous benchmark implementation is gone (or stored on your machine).

The solution here is to provide as much context as possible on your experiment details, input, and implementation of the benchmark. Of course, we can provide that in some form of documentation (e.g., in the description of the pull report), but there is nothing better than committing the actual microbenchmark next to your production code! In practice, however, it isn't so simple. Some extra pieces are worth adding before sharing the microbenchmark with others.

I optimized our `Sum` function and explained my benchmarking process. However, you don't want to write an entire chapter to explain the optimization you made to your team (and your future self)! Instead, you could provide all that is needed in a single piece of code as presented in Example 8-12.

17 In fact, we should not even trust ourselves there! A second careful reviewer is always a good idea.

Example 8-12. Well-documented, reusable Go benchmark for assessing concurrent implementations of the `Sum` function

```
// BenchmarkSum assesses `Sum` function. ❶
// NOTE(bwplotka): Test it with a maximum of 4 CPU cores, given we don't allocate
// more in our production containers.
//
// Recommended run options:
/*
export ver=v1 && go test \
    -run '^$' -bench '^BenchmarkSum$' \
    -benchtime 10s -count 5 -cpu 4 -benchmem \
    -memprofile=${ver}.mem.pprof -cpuprofile=${ver}.cpu.pprof \
  | tee ${ver}.txt ❷
*/
func BenchmarkSum(b *testing.B) {
   // Create 7.55 MB file with 2 million lines.
   fn := filepath.Join(b.TempDir(), "/test.2M.txt")
   testutil.Ok(b, createTestInput(fn, 2e6)) ❸

   b.ResetTimer()
   for i := 0; i < b.N; i++ {
      _, err := Sum(fn)
      testutil.Ok(b, err) ❹
   }
}
```

❶ It might feel excessive for a simple benchmark, but good documentation significantly increases the reliability of your and your team's benchmarking. Mention any surprising facts around this benchmark, dataset choice, conditions, or prerequisites in the commentary.

❷ I recommend commenting on the benchmark with the suggested way to invoke it. It's not to force anything but rather to describe how you envisioned running this benchmark (e.g., for how long). Future you or your team members will thank you!

❸ Provide the exact input you intend to run your benchmark with. You could create a static file for unit tests and commit it to your repository. Unfortunately, the benchmarking inputs are often too big to be committed to your source code (e.g., `git`). For this purpose, I created a small `createTestInput` function that can generate a dynamic number of lines. Notice the use of `b.TempDir()` (*https://oreil.ly/elBJa*), which creates a temporary directory and cares about cleaning it manually afterward.[18]

---

18 Note that the `t.TempDir` and `b.TempDir` methods create a new, unique directory every time they are invoked!

❹ Because you want to reuse this benchmark in the future, and it will also be used by other team members, it makes sense to ensure others do not measure the wrong thing, thus testing for basic error modes even in the benchmark.

Thanks to `b.ResetTimer()`, even if the input file creation is relatively slow, latency and resource usage won't be visible in the benchmarking results. However, it might not be very pleasant for you while repeatedly running that benchmark. Even more, you will experience that slowness more than once after. As we learned in "Go Benchmarks" on page 277, Go can run the benchmark multiple times to find the correct `N` value. If the initialization takes too much time and impacts your feedback loop, you can add the code that will cache test the input on the filesystem. See Example 8-13 for how you can add a simple `os.Stat` to achieve this.

Example 8-13. Example of the benchmark with input creation executed only once and cached on disk

```
func lazyCreateTestInput(tb testing.TB, numLines int) string {
    tb.Helper() ❶

    fn := fmt.Sprintf("testdata/test.%v.txt", numLines)
    if _, err := os.Stat(fn); errors.Is(err, os.ErrNotExist) { ❷
        testutil.Ok(tb, createTestInput(fn, numLines))
    } else {
        testutil.Ok(tb, err)
    }
    return fn
}

func BenchmarkSum(b *testing.B) {
    // Create a 7.55 MB file with 2 million lines if it does not exist.
    fn := lazyCreateTestInput(tb, 2e6)

    b.ResetTimer()
    for i := 0; i < b.N; i++ {
        _, err := Sum(fn)
        testutil.Ok(b, err)
    }
}
```

❶ `t.Helper` tells the testing framework to point out the line that invokes `lazyCreateTestInput` when a potential error happens.

❷ `os.Stat` stops executing `createTestInput` if the file exists. Be careful when changing the characteristics or size of the input file. If you don't change the filename, the risk is that people who ran those tests will have a cached old version of the input. However, that small risk is worth it if the creation of the input is slower than a few seconds or so.

Such a benchmark provides elegant and concise information about the benchmark implementation, purpose, input, run command, and prerequisites. Moreover, it allows you and your team to replicate or reuse the same benchmark with little effort.

Running Benchmarks for Different Inputs

It's often helpful to learn how the efficiency of our implementation changes for different sizes and types of input. Sometimes it's fine to manually change the input in our code and rerun our benchmark, but sometimes we would like to program benchmarks for the same piece of code against different inputs in our source code (e.g., for our team to use later). Table tests are perfect for such use cases. Typically, we see this pattern in functional tests, but we can use it in microbenchmarks, as presented in Example 8-14.

Example 8-14. Table benchmark using a common pattern with `b.Run`

```
func BenchmarkSum(b *testing.B) {
    for _, tcase := range []struct { ❶
       numLines int
    }{
        {numLines: 0},
        {numLines: 1e2},
        {numLines: 1e4},
        {numLines: 1e6},
        {numLines: 2e6},
    } {
        b.Run(fmt.Sprintf("lines-%d", tcase.numLines), func(b *testing.B) { ❷
            b.ReportAllocs() ❸

            fn := lazyCreateTestInput(tb, tcase.numLines)

            b.ResetTimer()
            for i := 0; i < b.N; i++ { ❹
                _, err := Sum(fn)
                testutil.Ok(b, err)
            }
        })
    }
}
```

❶ An inlined slice of anonymous structures works well here because you don't need to reference this type anywhere. Feel free to add any fields here to map test cases as you need.

❷ In the test case loop, we can run `b.Run` that tells `go test` about a subbenchmark. If you put the `""` empty string as the name, `go test` will use numbers as your test case identification. I decided to present a number of lines as a unique description of each test case. The test case identification will be added as a suffix, so `BenchmarkSum/<test-case>`.

❸ For these tests, `go test` ignores any `b.ReportAllocs` and other benchmark methods outside the `b.Run`, so make sure to repeat them here.

❹ A common pitfall here is to accidentally use `b` from the main function, not from the closure created for the inner function. This is common if you try to avoid shadowing the `b` variable and use a different variable name for the inner `*testing.B`, e.g., `b.Run("", func(b2 *testing.B)`. These problems are hard to debug, so I recommend always using the same name, e.g., `b`.

Amazingly, we can use the same recommended `run` command presented in Example 8-4 for a nontable test. The example run output processes by `benchstat` will then look like Example 8-15.

Example 8-15. `benchstat` output on results from the Example 8-14 test

```
name                  time/op
Sum/lines-0-4         2.79µs ± 1%
Sum/lines-100-4       8.10µs ± 5%
Sum/lines-10000-4      407µs ± 6%
Sum/lines-1000000-4   40.5ms ± 1%
Sum/lines-2000000-4   78.4ms ± 3%

name                  alloc/op
Sum/lines-0-4           872B ± 0%
Sum/lines-100-4       3.82kB ± 0%
Sum/lines-10000-4      315kB ± 0%
Sum/lines-1000000-4   30.4MB ± 0%
Sum/lines-2000000-4   60.8MB ± 0%

name                  allocs/op
Sum/lines-0-4           6.00 ± 0%
Sum/lines-100-4         86.0 ± 0%
Sum/lines-10000-4      8.01k ± 0%
Sum/lines-1000000-4     800k ± 0%
Sum/lines-2000000-4    1.60M ± 0%
```

I find the table tests great for quickly learning about the estimated complexity (discussed in "Complexity Analysis" on page 240) of our application. Then, after I learn more, I can trim the number of cases to those that can truly trigger bottlenecks we saw in the past. In addition, committing such a benchmark to our team's source code will increase the chances that other team members (and yourself!) will reuse it and run a microbenchmark with all cases that matter for the project.

Microbenchmarks Versus Memory Management

The simplicity of microbenchmarks has many benefits but also downsides. One of the most surprising problems is that the memory statistics reported in the `go test` benchmarks don't tell a lot. Unfortunately, given how memory management is implemented in Go ("Go Memory Management" on page 172), we can't reproduce all the aspects of memory efficiency of our Go programs with microbenchmarks.

As we saw in Example 8-6, the naive implementation of `Sum` in Example 4-1 allocates around 60 MB of memory on the heap with the 1.6 million objects to calculate a sum for 2 million integers. This tells us less about memory efficiency than we might think. It only tells us three things:

- Some of the latency we experience in microbenchmark results inevitably come from the sole fact of making so many allocations (and we can confirm with profiles how much it matters).
- We can compare that number and size of allocations with other implementations.
- We can compare the number and size of the allocation with expected space complexity ("Complexity Analysis" on page 240).

Unfortunately, any other conclusion based on those numbers is in the realm of estimations, which only can be verified when we run "Macrobenchmarks" on page 269 or "Benchmarking in Production" on page 268. The reason is very simple—there is no special GC schedule for benchmarks because we want to ensure as close to production simulation as possible. They run on a normal schedule like in production code, which means that during our 100 iterations of our benchmark, the GC might run 1,000 times, 10 times, or for fast benchmarks it might not run at all! Therefore, any attempts to manually trigger `runtime.GC()` are also poor options, given that it's not how it will be running in production and might clash with normal GC schedules.

As a result, the microbenchmark will not give us a clear idea and the following memory efficiency questions:

GC latency
: As we learned in "Go Memory Management" on page 172, a bigger heap (more objects in a heap) will mean more work for the GC, which always translates to increased CPU usage or, more often, GC cycles (even with fair 25% CPU usage

mechanisms). Because of nondeterministic GC and quick benchmarking operations, we most likely won't see GC impact on a microbenchmark level.[19]

Maximum memory usage
: If a single operation allocates 60 MB, does it mean that the program performing one such operation at the time will need no more and no less than ~60 MB of memory in our system? Unfortunately, for the same reason mentioned previously, we can't tell with microbenchmarks.

    It might be that our single operation doesn't need all objects for the full duration. This might mean that the maximum usage of memory will be, for example, only 10 MB, despite the 60 MB allocation number, as the GC can do clean-up runs multiple times in practice.

    You might even have the opposite situation too! Especially for Example 4-1, most of the memory is kept during the whole operation (it is kept in the file buffer—we can tell that from profiling, explained in "Profiling in Go" on page 331). On top of that, the GC might not clean the memory fast enough, resulting in the next operation allocating 60 MB on top of the original 60 MB, requiring 120 MB in total from the OS. This situation can be even worse if we do a larger concurrency of our operations.

This is unfortunate, as the preceding problems are often seen in our Go code. If we could verify those problems on microbenchmarks, it would be easier to tell if we can reuse memory better (e.g., through "Memory Reuse and Pooling" on page 449) or if we should straight reduce allocation and to what level. Unfortunately, to tell for sure, we need to move to "Macrobenchmarks" on page 306.

Still, the microbenchmark allocation information is incredibly useful if we assume that, generally, more allocations can cause more problems. This is why simply focusing on reducing the number of allocations or allocated space in our micro-optimization cycle is still very effective. What we need to acknowledge, however, is that those numbers from just microbenchmarking might not give us complete confidence about whether the end GC overhead or maximum memory usage will be acceptable or problematic. We can try to estimate this, but we won't know for sure until we move to the macro level to assess that.

19 For longer microbenchmarks, you might see the GC latency. Some tutorials also recommend running microbenchmarks without GC (*https://oreil.ly/7v3oE*) (using `GOGC=off`), but I found this not useful in practice. Ideally, move to the macro level to understand the full impact.

Compiler Optimizations Versus Benchmark

There is a very interesting "meta" dynamic between microbenchmarking and compiler optimizations, which is sometimes controversial. It is worth knowing about this problem, the potential consequences, and how to mitigate them.

Our goal when microbenchmarking is to assess the efficiency of the small part of our production code with as high confidence as possible (given the amount of time available and problem constraints). For this reason, the Go compiler treats our "Go Benchmarks" on page 277 benchmarking function like any other production code. The same AST conversions, type safety, memory safety, dead code elimination, and optimizations rules discussed in "Understanding Go Compiler" on page 118 are performed by the compiler on all parts of the code—no special exceptions for benchmarks. Therefore, we are reproducing all production conditions, including the compilation stage.

This premise is great, but what gets in the way of this philosophy is that microbenchmarks are a little special. From the runtime process perspective, there are three main differences between how this code is executed on production and when we want to learn about production code efficiency:

- No other user code is running at the same time in the same process.[20]
- We are invoking the same code in a loop.
- We typically don't use the output or return arguments.

Those three elements might not seem like a big difference, but as we learned in "CPU and Memory Wall Problem" on page 126, modern CPUs can already run differently in those cases due to, e.g., different branch prediction and L-cache locality. On top of that, you can imagine a smart enough compiler that will adjust the machine code differently based on those cases too!

This problem is especially visible when programming in Java because some compilation phases are done in runtime, thanks to the mature just-in-time (JIT) compiler. As a result, Java engineers must be very careful when benchmarking (*https://oreil.ly/OJKNS*) and use special frameworks (*https://oreil.ly/Cil2Z*) for Java to ensure simulating production conditions with warm-up phases and other tricks to increase the reliability of benchmarks.

20 Unless you run with the parallel option I discouraged in "Performance Nondeterminism" on page 260.

In Go, things are simpler. The compiler is less mature than Java's, and no JIT compilation exists. While JIT is not even planned, some form of runtime profile-guided compiler optimization (PGO) (*https://oreil.ly/yFYut*) is being considered for Go (*https://oreil.ly/jDYqF*), which might make our microbenchmark more complex in future. Time will tell.

However, even if we focus on the current compiler, it sometimes can apply unwanted optimizations to our benchmarking code. One of the known problems is called dead code elimination (*https://oreil.ly/QG1y1*). Let's consider a low-level function representing `population count` instruction (*https://oreil.ly/lnuMl*) and the naive microbenchmark in Example 8-16.[21]

Example 8-16. `popcnt` function with the naive implementation of microbenchmark impacted by compiler optimizations

```
const m1 = 0x5555555555555555
const m2 = 0x3333333333333333
const m4 = 0x0f0f0f0f0f0f0f0f
const h01 = 0x0101010101010101

func popcnt(x uint64) uint64 {
   x -= (x >> 1) & m1
   x = (x & m2) + ((x >> 2) & m2)
   x = (x + (x >> 4)) & m4
   return (x * h01) >> 56
}

func BenchmarkPopcnt(b *testing.B) {
   for i := 0; i < b.N; i++ {
      popcnt(math.MaxUint64) ❶
   }
}
```

❶ In the original issue #14813, the input for the function was taken from `uint64(i)`, which is a huge anti-pattern. You should never use `i'` from the `b.N` loop! I want to focus on the surprising compiler optimization risk in this example, so let's imagine we want to assess the efficiency of `popcnt` working on the largest unsigned integer possible (using `math.MaxInt64` to obtain it). This also will expose us to an unexpected behavior mentioned below.

If we execute this benchmark for a second, we will get slightly concerning output, as presented in Example 8-17.

21 The idea behind this function comes from amazing Dave's tutorial (*https://oreil.ly/BKZfr*) and issue 14813 (*https://oreil.ly/m3Yiy*), with some modifications.

Example 8-17. The output of the `BenchmarkPopcnt` benchmark from Example 8-16

```
goos: linux
goarch: amd64
pkg: github.com/efficientgo/examples/pkg/comp-opt-away
cpu: Intel(R) Core(TM) i7-9850H CPU @ 2.60GHz
BenchmarkPopcnt
BenchmarkPopcnt-12      1000000000          0.2344 ns/op ❶
PASS
```

❶ Every time you see your benchmark making a billion iterations (maximum number of iterations `go test` will do), you know your benchmark is wrong. It means we will see a loop overhead rather than the latency we are measuring. This can be caused by the compiler optimizing away your code or by measuring something too fast to be measured with a Go benchmark (e.g., single instruction).

What is happening? The first problem is that the Go compiler inlines the `popcnt` code, and further optimization phases detected that no other code is using the result of the inlined calculation. The compiler detects that no change in observable behavior would occur if we remove this code, so it elides that inlined code part. If we would list assembly code using `-gcflags=-S` on `go build` or `go test`, you would notice there is no code responsible for performing statements behind `popcnt` (we run an empty loop!). This can also be confirmed by running `GOSSAFUNC=BenchmarkPopcnt go build` and opening *ssa.html* in your browser, which also lists the generated assembly more interactively. We can verify this problem by running a test with `-gcflags=-N`, which turns off all compiler optimizations. Executing or looking at the assembly will show you the large difference.

The second problem is that all the iterations of our benchmark run `popcnt` with the same constant number—the largest unsigned integer. Even if code elimination did not happen, with inlining, the Go compiler is smart enough to precompute some logic (sometimes referred to as `intrinsic` (*https://oreil.ly/NEOyQ*)). The result of `popcnt(math.MaxUint64)` is always 64, no matter how many times and where we run it; thus, the machine code will simply use `64` instead of calculating `popcnt` in every iteration.

Generally, there are three practical countermeasures against compiler optimization in benchmarks:

Move to the macro level.
On a macro level, there is no special code within the same binary, so we can use the same machine code for both benchmarks and production code.

Microbenchmark more complex functionality.

If compiler optimizations impact, you might be optimizing Go on a too low level.

I personally haven't been impacted by compiler optimization, because I tend to microbenchmark on higher-level functionalities. If you benchmark really small functions like Example 8-16, typically inlined and a few nanoseconds fast, expect the CPU and compiler effect to impact you more. For more complex code, the compiler typically is not as clever to inline or adjust the machine code for benchmarking purposes. The number of instructions and data on bigger macrobenchmarks will also more likely break the CPU branch predictor and cache locality like it would at production.[22]

Outsmart compiler in microbenchmark.

If you want to microbenchmark such a tiny function like Example 8-16, there is no other way to obfuscate the compiler code analysis. What typically works is using exported global variables. They are hard to predict given the current per-package Go compilation logic[23] or using `runtime.KeepAlive`, which is a newer way to tell compile that "this variable is used" (which is a side effect of telling the GC to keep this variable on the heap). The `//go:noinline` directive that stops the compiler from inlining function might also work, but it's not recommended as on production, your code might be inlined and optimized, which we want to benchmark too.

If we would like to improve the Go benchmark shown in Example 8-16, we could add the `Sink` pattern[24] and global variable for input, as presented in Example 8-18. This works in Go 1.18 with the `gc` compiler, but it's not prone to future improvements in the Go compiler.

Example 8-18. `Sink` pattern and variable input countermeasure unwanted compiler optimization on microbenchmarks

```
var Input uint64 = math.MaxUint64 ❶
var Sink uint64 ❷

func BenchmarkPopcnt(b *testing.B) {
    var s uint64
```

22 I am not discouraging microbenchmarks on super low-level functions. You can still compare things, but be mindful that production numbers might surprise you.

23 This does not mean that the future Go compiler won't be able to be smarter and consider optimization with global variables.

24 The `sink` pattern is also popular in C++ for the same reasons (*https://oreil.ly/UpGFo*).

```
    b.ResetTimer()
    for i := 0; i < b.N; i++ {
        s = popcnt(Input) ❸
    }
    Sink = s
}
```

❶ The global `Input` variable masks the fact that `math.MaxUint64` is constant. This forces the compiler to not be lazy and do the work in our benchmark iteration. This works because the compiler can't tell if anyone else will change this variable in runtime before or during experiments.

❷ `Sink` is a similar global variable to `Input`, but it hides from the compiler that the value of our function is never used, so the compiler won't assume it's a dead code.

❸ Notice that we don't assign a value directly to the global variable as it's more expensive (*https://oreil.ly/yvNAi*), thus potentially adding even more overhead to our benchmark.

Thanks to the techniques presented in Example 8-18, I can assess that such an operation on my machine takes around 1.6 nanoseconds. Unfortunately, although I got a stable result that (one would hope) is realistic, assessing efficiency for such low-level code is fragile and complicated. Outsmarting the compiler or disabling optimizations are quite controversial techniques—they go against the philosophy that benchmarked code should be as close to production code as possible.

Don't Put Sinks Everywhere!

This section might feel scary and complicated. Initially, when I learned about these complex compilation impacts, I was putting a sink to all my microbenchmarks or assert errors only to avoid potential elision problems.

That is unnecessary. Be pragmatic, be vigilant of benchmarking results you can't explain (as mentioned in "Human Errors" on page 256), and add those special countermeasures.

> Personally, I'd rather not see sinks appear everywhere until they are needed. In many cases they won't be, and the code is clearer without them. My advice is to wait until the benchmark is clearly optimized away and only then put them in. The details of the sink can depend on the context. If you have a function returning an int, it's fine to sum them up and then assign the result to a global, for example.
>
> —Russ Cox (rsc), "Benchmarks vs Dead Code Elimination," email thread (*https://oreil.ly/xGDYr*)

In summary, be mindful of how the compiler can impact your microbenchmark. It does not happen too often, especially if you are benchmarking on a reasonable level, but when it happens, you should now know how to mitigate those problems. My recommendation is to avoid relying on a microbenchmark at such a low level. Instead, unless you are an experienced engineer interested in the ultra-high performance of your Go code for a specific use case, move to a higher level by testing more complex functionality. Fortunately, most of the code you will work with will likely be too complex to trigger such a "battle" with the Go compiler.

Macrobenchmarks

Programming books that cover performance and optimization topics don't usually describe benchmarking on a larger level than micro. This is because testing on a macro level is a gray area for developers. Typically, it is the responsibility of dedicated tester teams or QA engineers. However, for backend applications and services, such macrobenchmarking involves experience, skills, and tools to work with many dependencies, orchestration systems, and generally bigger infrastructure. As a result, such activity used to be the domain of operation teams, system administrators, and DevOps engineers.

However, things are changing a bit, especially for the infrastructure software, which is my area of expertise. The cloud-native ecosystem makes infrastructure tools more accessible for developers, with standards and technologies like Kubernetes (*https://kubernetes.io*), containers, and paradigms like Site Reliability Engineering (SRE) (*https://sre.google*). On top of that, the popular microservice architecture allows breaking functional pieces into smaller programs with clear APIs. This allows developers to take more responsibility for their areas of expertise. Therefore, in the last decades, we are seeing the move toward making testing (and running) software on all levels easier for developers.

Participate in Macrobenchmarks That Touch Your Software!

As a developer, it is extremely insightful to participate in testing your software, even on a macro level. Seeing your software's bugs and slowdowns gives crystal clarity to the priority. Additionally, if you catch those problems on the setup you control or are familiar with, it is easier to debug the problem or find the bottleneck, ensuring a quick fix or optimization.

I would like to break the mentioned convention and introduce you to some basic concepts required for effective macrobenchmarking. Especially for backend applications, developers these days have much more to say when it comes to accurate efficiency assessment and bottleneck analysis at higher levels. So let's use this fact and discuss some basic principles and provide a practical example of running a macrobenchmark via `go test`.

Basics

As we learned in "Benchmarking Levels" on page 266, macrobenchmarks focus on testing your code at the product level (application, service, or system) close to your functional and efficiency requirements (as described in "Efficiency Requirements Should Be Formalized" on page 83). As a result, we could compare macrobenchmarking to integration or end-to-end (e2e) functional testing.

In this section, I will mostly focus on benchmarking server-side, multicomponent Go backend applications. There are three reasons why:

- That's my speciality.
- It's the typical target environment of applications written in the Go language.
- This application typically involves working with nontrivial infrastructure and many complex dependencies.

Especially the last two items make it beneficial for me to focus on backend applications, as other types of programs (CLI, frontend, mobile) might require less-complex architecture. Still, all types will reuse some patterns and learnings from this section.

For instance, in "Microbenchmarks" on page 275, we assessed the efficiency of the `Sum` function (Example 4-1) in our Go code, but that function might have been a bottleneck for a much bigger product or service. Imagine that our team's task is to develop and maintain a bigger microservice called `labeler` that uses the `Sum`.

The `labeler` will run in a container and connect to an object storage[25] with various files. Each file has potentially millions of integers in each new line (the same input as in our `Sum` problem). The `labeler` job is to return a label—the metadata and some statistics of the specified object when the user calls the HTTP `GET` method `/label_object`. The returned label contains attributes like the object name, object size, checksum, and more. One of the key label fields is the sum of all numbers in the object.[26]

You learned first how to assess the efficiency of the smaller `Sum` function on a micro level because it's simpler. On the product level the situation is much more complex. That's why to perform reliable benchmarking (or bottleneck analysis) on a macro

25 Object storage is cheap cloud storage with simple APIs for uploading objects and reading them or their byte ranges. It treats all data in the form of objects with a certain ID that typically looks similar to the file path.

26 You can find simplified microservice code in the `labeler` package (*https://oreil.ly/myFWw*).

level, there are a few differences to notice and extra components to have. Let's go through them, as presented in Figure 8-1.

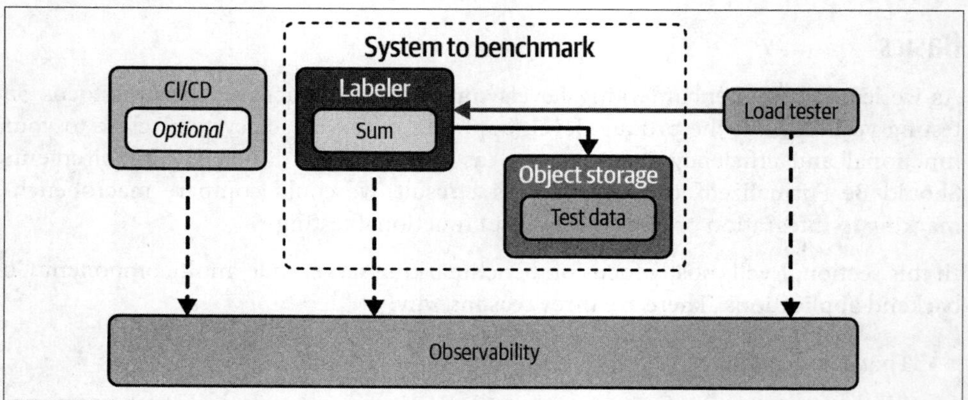

Figure 8-1. Common elements required for the macrobenchmark, for example, to benchmark the `labeler` service

The specific differences from our `Sum` microbenchmark can be outlined as follows:

Our Go program as a separate process
: Thanks to "Go Benchmarks" on page 277, we understand the efficiency of the `Sum` function and can optimize it. But what if another part of the code is now a bigger bottleneck in our flow? This is why we typically want to benchmark our Go program with its full user flow on a macro level. This means running the process in a similar fashion and configuration as in production. But unfortunately, this also means we can't run the `go test` benchmarking framework anymore as we benchmark on the process level.

Dependencies, e.g., object storage
: One of the key elements of macrobenchmarks is that we typically want to analyze the efficiency of the full system, including all key dependencies. This is especially important when our code might rely on certain efficiency characteristics of the dependency. In our `labeler` example, we use object storage, which usually means transferring bytes over the network. There might be little point in optimizing `Sum` if the object storage communication is the main bottleneck in latency or resource consumption. There are generally three ways of handling dependencies on a macro level:

  - We can try to use realistic dependency (e.g., in our example, the exact object storage provider that will be used on production, with a similar dataset size). This is typically the best idea if we want to test the end-to-end efficiency of the whole system.

- We can try to implement or use a fake (*https://oreil.ly/06UmC*) or adapter that will simulate production problems. However, this often takes too much effort and it's hard to simulate the exact behavior of, for example, a slow TCP connection or server.
- We could implement the simplest fake for our dependency and assess the isolated efficiency of our program. In our example, this might mean running local, open source object storage like Minio (*https://min.io*). It will not reflect all the problems we might have with production dependencies, but it will give us some estimates on the problems and overhead for our program. We will use this in "Go e2e Framework" on page 310 for simplicity.

Observability
: We can't use "Go Benchmarks" on page 277 on a macro level, so we don't have built-in support for latency, allocations, and custom metrics. So we have to provide our observability and monitoring solution. Fortunately, we already discussed instrumentation and observability for Go programs in Chapter 6, which we can use on a macro level. In "Go e2e Framework" on page 310, I will show you a framework that has built-in support for the open source Prometheus (*https://prometheus.io*) project, which allows gathering latency, usage, and custom benchmarking metrics. You can enrich this setup with other tools like tracing, logging, and continuous profiling to debug the functional and efficiency problems even easier.

Load tester
: Another consequence of getting out of the Go benchmark framework is the missing logic of triggering the experiment cases. Go benchmark was executing our code the desired amount of times with desired arguments. On the macro level, we might want to use this service as the user would use the HTTP REST API for web services like `labeler`. This is why we need some load-tester code that understands our APIs and will call them the desired amount of times and arguments.

  You can implement your own to simulate the user traffic, which unfortunately is prone to errors.[27] There are ways to "fork" or replay production traffic to the testing product using more advanced solutions like Kafka. Perhaps the easiest solution is to pick an off-the-shelf framework like an open source k6 (*https://k6.io*) project, which is designed and battle-tested for load-testing purposes. I will present an example of using k6 in "Go e2e Framework".

27 One common pitfall is to implement inefficient load-testing code. There is a risk that your application does not allow the throughput you want only because the client is not sending the traffic fast enough!

Continuous Integration (CI) and Continuous Deployment (CD)
: Finally, we rarely run macrobenchmarks on local development machines for more complex systems. This means we might want to invest in automation that schedules the load test and deploys required components with the desired version.

With such architecture, we can perform the efficiency analysis on a macro level. Our goals are similar to what we have for "Microbenchmarks" on page 275, just on a more complex system, such as A/B testing and learning the space and runtime complexity of your system functionality. However, given that we are closer to how users use our system, we can also treat it as an acceptance test that will validate efficiency with our RAER.

The theory is important, but how does it look in practice? Unfortunately, there is no consistent way of performing macrobenchmarks with Go, as it highly depends on your use case, environment, and goals. However, I would like to provide an example of a pragmatic and fast macrobenchmark of `labeler` that we can perform on our local development machine using Go code! So let's dive into the next section.

Go e2e Framework

Backend macrobenchmarking does not necessarily always mean using the same deployment mechanism we have in production (e.g., Kubernetes). However, to reduce the feedback loop, we can try macrobenchmarking with all the required dependencies, dedicated load tester, and observability on our developer machine or small virtual machine (VM). In many cases, it might give you reliable enough results on a macro level.

For experiments, you can manually deploy all the elements mentioned in "Basics" on page 307 on your machine. For example, you can write a bash script or Ansible (*https://oreil.ly/x9LTf*) runbook. However, since we are Go developers looking to improve the efficiency of our code, what about implementing such a benchmark in Go code and saving it next to your benchmarked code?

For this purpose, I would like to introduce you to the `e2e` (*https://oreil.ly/f0IJo*) Go framework that allows running interactive or automated experiments on a single machine using Go code and Docker containers. The container (*https://oreil.ly/aMXxz*) is a concept that allows running processes in an isolated, secure sandbox environment while reusing the host's kernel. In this concept, we execute software inside predefined container images. This means we must build (or download) a required image of the software we want to run beforehand. Alternatively, we can build our container image and add required software like pre-build binary of our Go program, e.g., `labeler`.

A container is not a first-class citizen on any OS. Instead, it can be constructed with existing Linux mechanisms like `cgroups`, `namespaces`, and Linux Security Modules (LSMs (*https://oreil.ly/C4h3z*)). Docker provides one implementation of the container engine, among others.[28] Containers are also heavily used for large cloud-native infrastructure thanks to orchestration systems like Kubernetes.

Benefits of Benchmarking in Containers

There are many reasons why on a macro level, I prefer using containers, even for single-node local tests:

- They allow isolating our processes, enabling more reliable observability and limitation facilities. This allows us to constraint certain resources to simulate different production aspects and account for resource usage to a given process (e.g., network usage or CPU usage).
- If you use containers on production, you can use the same container images in your macrobenchmarks. This ensures higher reliability—no unknowns are introduced by building, packaging, or installing phases.
- Similarly, for analyzing the benchmarking situation, we can use the same instrumentation and observability as we use for production.[29]
- The isolation of containers has little overhead compared to heavier virtualization like virtual machines (VMs) (*https://oreil.ly/HEtBk*) that have to fully virtualize hardware resources like memory and CPU.
- Easier installation and use of dependencies (portability!).

To leverage all benefits of containers, run only one process per container! Putting more processes (e.g., local database) into one container is tempting. But that defies the point of observing and isolating containers. Tools like Kubernetes or Docker are designed for singular processes per container, so put auxiliary processes in sidecar containers.

Let's go through a complete macrobenchmark implementation divided into two parts, Examples 8-19 and 8-20, that assess latency and memory usage of our

---

28 This space expanded quite quickly with two separate specifications (CRI and OCI) and various implementations of various parts of the container ecosystem. Read more about it here (*https://oreil.ly/yKSL8*).

29 This is often underestimated. Creating reusable dashboards, learning about your instrumentation, and what metrics mean takes a nontrivial amount of work. If our local testing and production environment share the same metrics and other signals, it saves us a lot of time and increases the chances our observability is high quality.

`labeler` service introduced in "Basics" on page 307. For convenience, our implementation can be scripted and executed as a normal `go test` guarded by `t.Skip` or build tag (*https://oreil.ly/tyue6*) to execute it manually or in a different cadence than functional tests.[30]

Example 8-19. Go test running the macrobenchmark in interactive mode (part 1)

```
import (
    "testing"

    "github.com/efficientgo/e2e"
    e2edb "github.com/efficientgo/e2e/db"
    e2einteractive "github.com/efficientgo/e2e/interactive"
    e2emonitoring "github.com/efficientgo/e2e/monitoring"
    "github.com/efficientgo/core/testutil"
    "github.com/thanos-io/objstore/providers/s3"
)

func TestLabeler_LabelObject(t *testing.T) {
    e, err := e2e.NewDockerEnvironment("labeler") ❶
    testutil.Ok(t, err)
    t.Cleanup(e.Close)

    mon, err := e2emonitoring.Start(e) ❷
    testutil.Ok(t, err)
    testutil.Ok(t, mon.OpenUserInterfaceInBrowser()) ❸

    minio := e2edb.NewMinio(e, "object-storage", "test") ❹
    testutil.Ok(t, e2e.StartAndWaitReady(minio))

    labeler := e2e.NewInstrumentedRunnable(e, "labeler"). ❺
        WithPorts(map[string]int{"http": 8080}, "http").
        Init(e2e.StartOptions{
            Image: "labeler:test", ❻
            LimitCPUs: 4.0,
            Command: e2e.NewCommand(
                "/labeler",
                "-listen-address=:8080",
                "-objstore.config="+marshal(t, client.BucketConfig{
                    Type: client.S3,
                    Config: s3.Config{
                        Bucket:    "test",
                        AccessKey: e2edb.MinioAccessKey,
                        SecretKey: e2edb.MinioSecretKey,
                        Endpoint:  minio.InternalEndpoint(e2edb.AccessPortName),
                        Insecure:  true,
```

30 You can run this code yourself or explore the e2e framework to see how it configures all components here (*https://oreil.ly/ftAY1*).

```
                },
            }),
        ),
    })
testutil.Ok(t, e2e.StartAndWaitReady(labeler))
```

❶ The e2e project is a Go module that allows the creation of end-to-end testing environments. It currently supports running the components (in any language) in Docker containers (*https://oreil.ly/iXrgX*), which allows clean isolation for both filesystems, network, and observability. Containers can talk to each other but can't connect with the host. Instead, the host can connect to the container via mapped `localhost` ports printed at the container start.

❷ The `e2emonitoring.Start` method starts Prometheus and cadvisor (*https://oreil.ly/v9gEL*). The latter translates cgroups related to our containers to Prometheus metric format so it can collect them. Prometheus will also automatically collect metrics from all containers started using `e2e.NewInstrumentedRunnable`.

❸ For an interactive exploration of resource usage and application metrics, we can invoke `mon.OpenUserInterfaceInBrowser()` that will open the Prometheus UI in our browser (if running on a desktop).

❹ `Labeler` uses object storage dependency. As mentioned in "Basics" on page 307, I simplified this benchmark by focusing on `labeler` Go program efficiency without the impact of remote object storage. For that purpose, local `Minio` container is suitable.

❺ Finally, it's time to start our `labeler` Go program in the container. It is worth noticing that I set the container CPU limit to `4` (enforced by Linux `cgroups`) to ensure our local benchmark is not saturating all the CPUs my machines have. Finally, we inject object storage configuration to connect with the local `minio` instance.

❻ I used the `labeler:test` image that is built locally. I often add a script in `Makefile` to produce such an image, e.g., `make docker`. You risk forgetting to build the image with the desired Go program version you want to benchmark, so be mindful of what you are testing!

Example 8-20. Go test running the macrobenchmark in interactive mode (part 2)

```
testutil.Ok(t, uploadTestInput(minio, "object1.txt", 2e6)) ❶

k6 := e.Runnable("k6").Init(e2e.StartOptions{
```

```
        Command: e2e.NewCommandRunUntilStop(),
        Image: "grafana/k6:0.39.0",
    })
    testutil.Ok(t, e2e.StartAndWaitReady(k6))

    url := fmt.Sprintf(
        "http://%s/label_object?object_id=object1.txt",
        labeler.InternalEndpoint("http"),
    )
    testutil.Ok(t, k6.Exec(e2e.NewCommand(
        "/bin/sh", "-c", `cat << EOF | k6 run -u 1 -d 5m - ❷
import http from 'k6/http'; ❸
import { check, sleep } from 'k6';

export default function () {
    const res = http.get('`+url`');
    check(res, { ❹
        'is status 200': (r) => r.status === 200,
        'response': (r) =>
            r.body.includes(
    '{"object_id":"object1.txt","sum":6221600000,"checksum":"SUUr'
            ),
    });
    sleep(0.5)
}
EOF`)))

    testutil.Ok(t, `e2einteractive.RunUntilEndpointHit()`) ❺
}
```

❶ We have to upload some test data. In our simple test, we upload a single file with two million lines, using a similar pattern we used in "Go Benchmarks" on page 277.

❷ I choose `k6` as my load tester. `k6` works as a batch job, so I first have to create a long-running empty container. I can then execute new processes in the `k6` environment to put the desired load on my `labeler` service. As a shell command, I pass the load-testing script as an input to the `k6` CLI. I also specify the number of virtual users (`-u` or `--vus`) I want. VUS represents the workers or threads running load-test functions specified in the script. To keep our tests and results simple, let's stick to one user for now to avoid simultaneous HTTP calls. The `-d` (short flag for `--duration`) is similar to the `-benchtime` flag in our "Go Benchmarks" on page 277. See more tips about using `k6` here (*https://oreil.ly/AbLOD*).

❸ `k6` accepts load-testing logic programmed in simple JavaScript code. My load test is simple. Make an HTTP `GET` call to the `labeler` path I want to benchmark. I choose to sleep 500 ms after each HTTP call to give the `labeler` server time to clean resources after each call.

❹ Similar to "Test Your Benchmark for Correctness!" on page 290, we have to test the output. If we trigger a bug in the `labeler` code or macrobenchmark implementation, we might be measuring the wrong thing! Using the `check` JavaScript functions allows us to assert the expected HTTP code and output.

❺ We might want to add here the automatic assertion rules that pass these tests when latency or memory usage is within a certain threshold. However, as we learned in "Comparison to Functional Testing" on page 252, finding reliable assertion for efficiency is difficult. Instead, I recommend learning about our `labeler` efficiency in a more interactive way. The `e2einteractive.RunUntilEndpointHit()` stops the `go test` benchmark until you hit the printed HTTP URL. It allows us to explore all outputs and our observability signals, e.g., collected metrics about `labeler` and the test in Prometheus.

The code snippet might be long, but it's relatively small and readable compared to how many things it orchestrates. On the other hand, it has to describe quite a complex macrobenchmark to configure and schedule five processes in one reliable benchmark with rich instrumentation for containers and internal Go metrics.

Keep Your Container Images Versioned!

It is important to ensure you benchmark against a deterministic version of dependencies. This is why you should avoid using `:latest` tags, as it is very common to update them without noticing them transparently. Furthermore, it's quite upsetting to realize after the second benchmark that you cannot compare it to the result of the first one because the dependency version changed, which might (or might not!) potentially impact the results.

You can start the benchmark in Example 8-19 either via your IDE or a simple `go test . -v -run TestLabeler_LabelObject` command. Once the `e2e` framework creates a new Docker network, start Prometheus, cadvisor, `labeler`, and `k6` containers, and stream their output to your terminal. Finally, the `k6` load test will be executed. After the specified five minutes, we should have results printed with summarized statistics around correctness and latency for our tested functionality. The test will stop when we hit the printed URL. If we do that, the test will remove all containers and the Docker network.

Duration of Macrobenchmarks

In "Go Benchmarks" on page 277, it was often enough to run a benchmark for 5–15 seconds. Why do I choose to run the macro load test for five minutes? Two main reasons:

- Generally, the more complex functionality we benchmark, the more time and iterations we want to repeat to stabilize all the system components. For example, as we learned in "Microbenchmarks Versus Memory Management" on page 299, microbenchmarks do not give us an accurate impact that GC might have on our code. With macrobenchmarks, we run a full `labeler` process, so we want to see how the Go GC will cope with the `labeler` work. However, to see the frequency, the impact of GC, and maximum memory usage, we need to run our program longer under stress.
- For sustainable and cheaper observability and monitoring in production, we avoid measuring the state of our application too often. This is how the recommended Prometheus collection (scrape) interval is around 15 to 30 s. As a result, we might want to run our test through a couple of collection periods to obtain accurate measurements while also sharing the same observability as production.

In the next section, I will go through the outputs this experiment gives us and potential observations we can make.

Understanding Results and Observations

As we saw in "Understanding the Results" on page 284, experimenting is only half of the success. The second half is to correctly interpret the results. After running Example 8-19 for around seven minutes, we should see k6 output[31] that might look like Example 8-21.

Example 8-21. Last 24 lines of the macrobenchmark output from a 7-minute test with one virtual user (VUS) using k6

```
running (5m00.0s), 1/1 VUs, 476 complete and 0 interrupted iterations
default   [ 100% ] 1 VUs  5m00.0s/5m0s
running (5m00.4s), 0/1 VUs, 477 complete and 0 interrupted iterations
default ✓ [ 100% ] 1 VUs  5m0s
✓ is status 200
✓ response
```

31 There is also a way to push those results directly to Prometheus (*https://oreil.ly/1UdNR*).

```
checks.....................: 100.00% ✓ 954       ✗ 0 ❶
data_received..............: 108 kB  359 B/s
data_sent..................: 57 kB   191 B/s
http_req_blocked...........: avg=9.05µs  min=2.48µs  med=8.5µs     max=553.13µs
    p(90)=11.69µs p(95)=14.68µs
http_req_connecting........: avg=393ns   min=0s      med=0s        max=187.71µs
http_req_duration..........: avg=128.9ms min=92.53ms med=126.05ms max=229.35ms ❷
    p(90)=160.43ms p(95)=186.77ms ❷
{ expected_response:true }: avg=128.9ms min=92.53ms med=126.05ms max=229.35ms
    p(90)=160.43ms p(95)=186.77ms
http_req_failed............: 0.00%   ✓ 0         ✗ 477
http_req_receiving.........: avg=60.17µs min=30.98µs med=46.48µs  max=348.96µs
    p(90)=95.05µs  p(95)=124.73µs
http_req_sending...........: avg=35.12µs min=11.34µs med=36.72µs  max=139.1µs
    p(90)=59.99µs  p(95)=67.34µs
http_req_waiting...........: avg=128.81ms min=92.45ms med=125.97ms max=229.22ms
    p(90)=160.24ms p(95)=186.7ms
http_reqs..................: 477     1.587802/s ❸
iteration_duration.........: avg=629.75ms min=593.8ms med=626.51ms max=730.08ms
    p(90)=661.23ms p(95)=687.81ms
iterations.................: 477     1.587802/s ❸
vus........................: 1       min=1      max=1
vus_max....................: 1       min=1      max=1
```

❶ Check this line to ensure you measure successful calls!

❷ `http_req_duration` is the most important measurement if we want to track the latency of the total HTTP request latency.

❸ It's also important to note the total number of calls we made (the more iterations we have, the more reliable it will be).

From the client's perspective, the `k6` results can tell us much about the achieved throughput and latencies of different HTTP stages. It seems that with just one "worker" calling our method and waiting 500 ms, we reached around 1.6 calls per second (`http_reqs`) and the average client latency of 128.9 ms (`http_req_duration`). As we learned in "Latency" on page 221, tail latency might be more relevant for latency measurements. For that, `k6` calculates the percentiles as well, which indicates that 90% of requests (`p90`) were faster than 160 ms. In "Go Benchmarks" on page 277, we learned that the `Sum` function involved in the process is taking 79 ms on average, which means it accounts for most of the average latency or even total `p90` latency. If we care about optimizing latency in this case, we should try to optimize `Sum`. We will learn how to verify that percentage and identify other bottlenecks in Chapter 9 with tools like profiling.

Another important result we should check is the variance of our runs. I wish `k6` provided out-of-the-box variance calculation because it's hard to tell how repeatable our iterations were without it. For example, we see that the fastest request took 92 ms, while the slowest took 229 ms. This looks concerning, but it's normal to have first requests take longer. To tell for sure, we would need to perform the same test twice and measure the average and percentile values variance. For example, on my machine, the next run of the same 5-minute test gave me an average of 129 ms and a `p90` of 163 ms, which suggests the variance is small. Still, it's best to gather those numbers in some spreadsheet and calculate the standard deviation to find the variance percentage. There might be room for a quick CLI tool like `benchstat` that would give us a similar analysis. This is important, as the same "Reliability of Experiments" on page 256 aspects apply to macrobenchmarks. If our results are not repeatable, we might want to improve our testing environment, reduce the number of unknowns, or test longer.

The `k6` output is not everything we have! The beauty of macrobenchmarks with good usage monitoring and observability, like Prometheus, is that we can assess and debug many efficiency problems and questions. In the Example 8-19 setup, we have instrumentation that gives us `cgroup` metrics about containers and processes thanks to `cadvisor`, built-in process and heap metrics from the `labeler` Go runtime, and application-level HTTP metrics I manually instrumented in `labeler` code. As a result, we can check the usage metrics we care for based on our goals and the RAER (see "Efficiency-Aware Development Flow" on page 102), for example, the metrics we discussed in "Efficiency Metrics Semantics" on page 220 and more.

Let's go through some metric visualizations I could see in Prometheus after my run.

Server-side latency

In our local tests, we use a local network, so there should be almost no difference between server and client latency (we talked about this difference in "Latency" on page 221). However, more complex macro tests that may load test systems from different servers or remote devices in another geolocation might introduce network overhead that we may want or don't want to account for in our results. If we don't, we can query Prometheus for the average request duration server handled for our `/label_object` path, as presented in Figure 8-2.

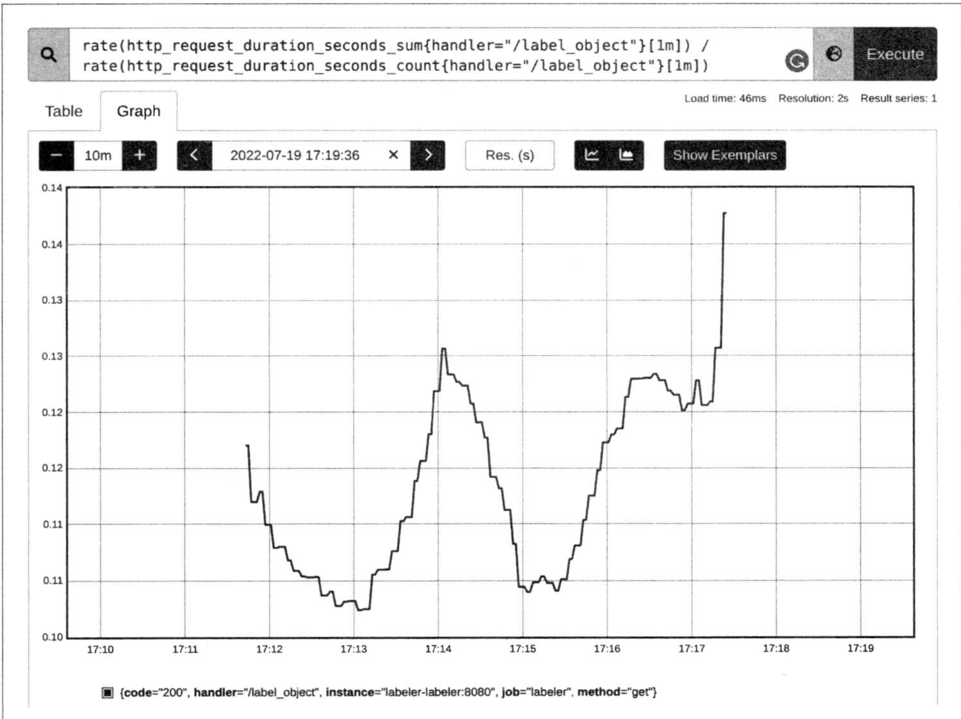

Figure 8-2. Dividing `http_request_duration_seconds` histogram sum by count rates to obtain server-side latency

The results confirm what we saw in Example 8-21. The observed average latency is around 0.12–0.15 seconds, depending on the moment. The metric comes from manually created HTTP middleware I added in Go using the `prometheus/client_golang` library (*https://oreil.ly/j1k4E*).[32]

Prometheus Rate Duration

Notice I am using `[1m]` range vectors for Prometheus counters in queries for this macrobenchmark. This is because we only run our tests for 5 minutes. With a 15-second scrape, 1 minute should have enough samples for `rate` to make sense, but also I can see more details in my metric value with one-time minute window granularity.

32 See the example code (*https://oreil.ly/22YQp*) that `labeler` uses.

When it comes to the server-side percentile, we rely on a bucketed histogram. This means that the accuracy of the result is up to the nearest bucket. In Example 8-21, we saw that results are 92 ms to 229 ms, with `p90` equal to 136 ms. At the moment of benchmark, the buckets were defined in `labeler` as follows: `0.001, 0.01, 0.1, 0.3, 0.6, 1, 3, 6, 9, 20, 30, 60, 90, 120, 240, 360, 720`. As a result, we can only tell that 90% of requests were faster than 300 ms, as presented in Figure 8-3.

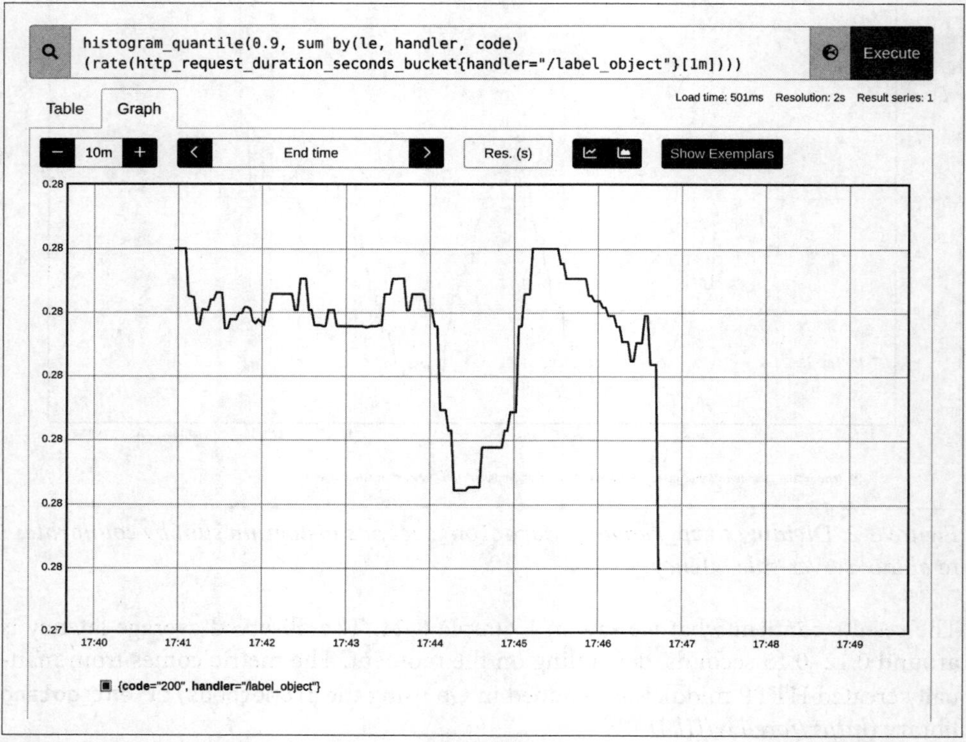

Figure 8-3. Using the `http_request_duration_seconds` histogram to calculate the `p90` quantile of the `/label_object` request

To find more accurate results, we might need to adjust buckets manually or use a new sparse histogram feature in the upcoming Prometheus 2.40 version. The default buckets work well in cases when we don't care if the request was handled in 100 ms or 300 ms, but we care if it was suddenly 1 second.

CPU time

Latency is one thing, but CPU time can tell us how much time the CPU needs to fulfill its job, how much concurrency can help, and if our process is CPU or I/O bound. We can also tell if we gave enough CPU for the current process load. As we learned in Chapter 4, higher latency of our iterations might be a result of the CPU saturation—our program using all available CPU cores (or close to the limit), in effect slowing the execution of all goroutines.

In our benchmark we can use either the Go runtime `process_cpu_seconds_total` counter or the `cadvisor` `container_cpu_usage_seconds_total` counter to find that number. This is because `labeler` is the only process in its container. Both metrics look similar, with the latter presented in Figure 8-4.

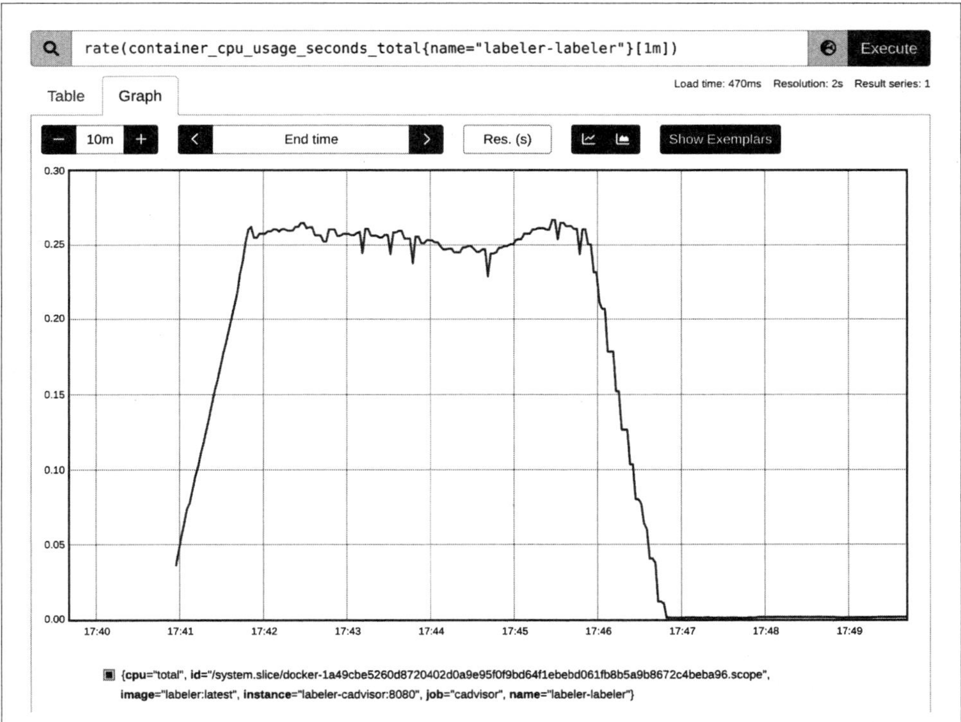

Figure 8-4. Using the `container_cpu_usage_seconds_total` counter to assess `labeler` CPU usage

The value oscillates between 0.25–0.27 CPU seconds, which represents the amount of CPU time the `labeler` needed for this load. I limited `labeler` to 4 CPU cores, but it used a maximum of 27% of a single CPU. This means that, most likely, the CPUs are not saturated (unless there are a lot of noisy neighbors running at the same moment, which we would see in the latency numbers). The 270 ms of CPU time per second seems like a sane value given that our requests take, on average, 128.9 ms, and after that, `k6` was waiting for 500 ms. This gives us 20%[33] of load-testing time, so the `k6` was actually demanding some work from `labeler`, which might not all be used on CPU, but also on I/O time. The `labeler /label_object` execution in our current version is sequential, but there are some background tasks, like listening to signal, metric collection, GC, and HTTP background goroutines. Again, see "Profiling in Go" on page 331 as the best way to tell exactly what's taking the CPU here.

Memory

In "Microbenchmarks" on page 275, we learned how much memory `Sum` allocates, but `Sum` is not the only logic `labeler` has to perform. Therefore, if we want to assess the memory efficiency of `labeler`, we need to look at the process or container level memory metrics we gathered during our benchmark. On top of that, we mentioned in "Microbenchmarks Versus Memory Management" on page 299 that only on the macro level do we have a chance to learn more about GC impact and maximum memory usage of our `labeler` process.

Looking at the heap metric presented in Figure 8-5, we can observe that a single `/label_object` is using the nontrivial amount of memory. This is not unexpected after seeing the `Sum` function microbenchmarks results in Example 8-7 showing 60.8 MB per iteration.

This observation shows us the eventuality of GC that might cause problems. Given a single "worker" (VUS) in `k6`, the `labeler` should never need more than ~61 MB of live memory if the `Sum` is the main bottleneck. However, we can see that for durations of 2 scrapes (30 seconds) and then 1 scrape, the memory got bumped to 118 MB. Most likely, GC had not released memory from the previous HTTP `/label_object` call before the second call started. If we account for spikes, the overall maximum heap size is stable at around 120 MB, which should tell us there are no immediate memory leaks.[34]

---

33 128.9 ms divided by 128.9+500 milliseconds to tell what portion of time the load tester was actively load-testing.

34 Looking on `go_goroutines` also helps. If we see a visible trend, we might forget to close some resources.

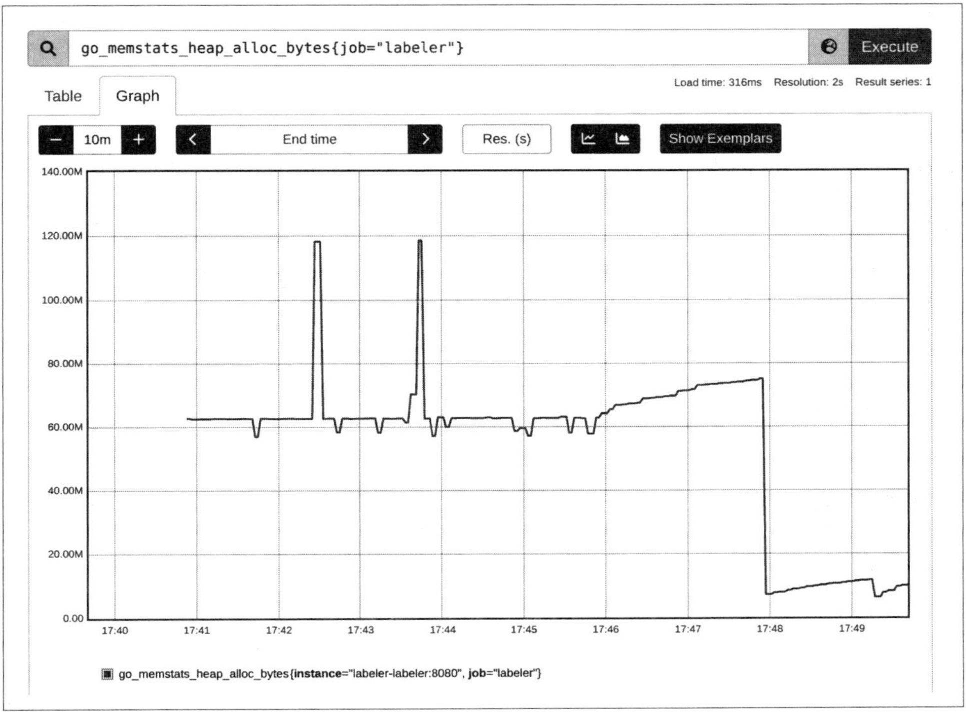

Figure 8-5. Using the `go_memstats_heap_alloc_bytes` gauge to assess `labeler` heap usage

go_memstats_heap_alloc_bytes Gauge and Temporary Changes

Be careful with any Prometheus gauges that monitor changes that occur more often than the scrape interval. For example, our Go program might have more spikes like the two we see in Figure 8-5, but they were too short to be observed by Prometheus in the `go_memstats_heap_alloc_bytes` metric.[35]

Something similar can happen when querying a gauge metric over a long period, like a dozen hours or days. The UI resolution (so-called `step`) is adjusted for longer periods and can potentially hide interesting moments. Ensure lower resolution or use `max_over_time` to know for sure what were the observed maximums (or `min_over_time` for minimums).

35 The solution is to use counters. For memory, it would mean using the existing `rate(go_memstats_alloc_bytes_total[1m])` and dividing it by the rate of bytes released by the GC. Unfortunately, the Prometheus Go collector does not expose such metrics. Go allows us to get this information (*https://oreil.ly/Noqnp*), so it is possible to get it added in the future.

This is rarely the problem in terms of memory as the GC and OS react very slowly with lazy memory release mechanisms, explained in "OS Memory Management" on page 156.

Unfortunately, as we learned in "OS Memory Management" on page 156 and "Memory Usage" on page 234, the memory used by the heap is only a portion of the RAM space that is used by the Go program. The space allocated for goroutine stacks, manually created memory maps, and kernel cache (e.g., for file access) requires the OS to reserve more pages on the physical memory. We can see that when we look at our container-level RSS metric presented in Figure 8-6.

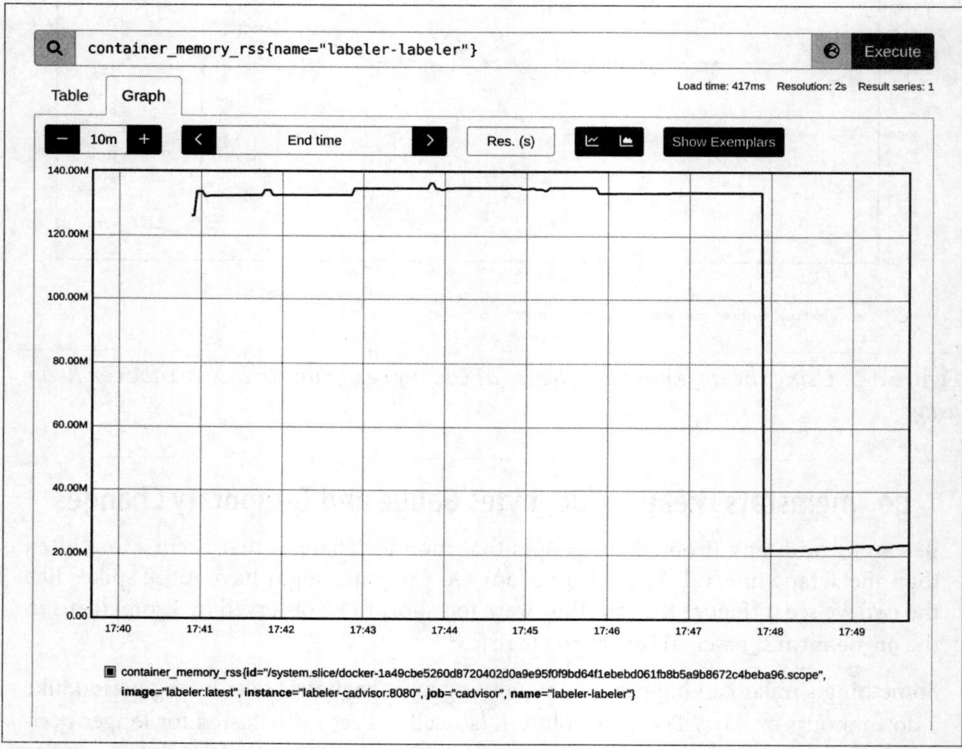

Figure 8-6. Using the `container_memory_rss` gauge to assess `labeler` physical RAM usage

Fortunately, nothing unexpected on the RSS side as well. The active memory pages were more or less the size of the heap and returned to a smaller level as soon as the test finished. So we can assess that `labeler` requires around 130 MB of memory for this load.

To sum up, we assessed the efficiency of latency and resources like CPU and memory on a macro level. In practice, we can assess much more, depending on our efficiency goals like disk, network, I/O devices, DB usage, and more. The `k6` configuration was straightforward in our test—single worker and sequential calls with a pause. Let's explore other variations and possibilities in the next section.

Common Macrobenchmarking Workflows

The example test in "Go e2e Framework" on page 310 should give you some awareness of how to configure the example load-testing tool, hook in dependencies, and set up and use pragmatic observability for efficiency analysis. On top of that, you can expand such local `e2e` tests in the direction you and your project need based on the efficiency goals. For example:

- Load test your system with more than one worker to assess how many resources it takes to sustain a given request per second (RPS) rate while sustaining a desired `p90` latency.[36]
- Run `k6` or other load-testing tools to simulate realistic client traffic in a different location.
- Deploy the macrobenchmark on remote servers, perhaps with the same hardware as your production.
- Deploy dependencies in a remote location; e.g., in our `labeler` example, use the AWS S3 service (*https://oreil.ly/pzeua*) instead of the local object storage instance.
- Scale out your macro test and services to multiple replicas to check if the traffic can be load balanced properly, so the system's efficiency stays predictable.

Similar to "Find Your Workflow" on page 289, you should find the workflow for performing such experiments and analysis that suits you the most. For example, for myself and the teams I worked with, the process of designing and using the macrobenchmark like in "Go e2e Framework" on page 310 might look as follows:

1. As a team, we plan the macrobenchmark elements, dependencies, what aspects we want to benchmark, and what load we want to put on it.
2. I ensure a clean code state for `labeler` and macrobenchmark code. I commit all the changes to know what I am testing and with what benchmark. Let's say we end up with a benchmark as in "Go e2e Framework" on page 310.

36 For bigger tests, consider making sure your load tester has enough resources. For `k6`, see this guide (*https://oreil.ly/v4DGs*).

3. Before starting the benchmark, I create a shared Google Document[37] and note all the experiment details like environmental conditions and software version.
4. I perform the benchmark to assess the efficiency of a given program version:
   - I run my macrobenchmarks, e.g., by starting the `go test` with the Go e2e framework (see "Go e2e Framework" on page 310) in Goland IDE and waiting until the load test finishes.
   - I confirm no functional errors are present.
   - I save the `k6` results to Google Documents.
   - I gather interesting observations of the resources I want to focus on, for example, heap and RSS to assess memory efficiency. I capture screenshots and paste them to my Google document.[38] Finally, I note all conclusions I made.
   - Optionally, I gather profiles for the "Profiling in Go" on page 331 process.
5. If the findings allowed me to find the optimization in my code, I implement it and save it as a new `git` commit. Then I benchmark again (see step 5) and save the new results to the same Google Doc under a different version, so I can compare my A/B test later on.

The preceding workflow allows us to analyze the results and conclude an efficiency assessment given the assumptions that can be formulated thanks to the document I create. Linking the exact benchmark, which ideally is committed to the source code, allows others to reproduce the same test to verify results or perform further benchmarks and tests. Again, feel free to use any practice you need as long as you care for the elements mentioned in "Reliability of Experiments" on page 256. There is no single consistent procedure and framework for macrobenchmarking, and it all highly depends on the type of software, production conditions, and price you want to invest in to ensure your product's efficiency.

It's also worth mentioning that macrobenchmarking is not so far from "Benchmarking in Production" on page 268. You can reuse many elements for macrobenchmarks like load tester and observability tooling in benchmarking against production (and vice versa). Such interoperability allows us to save time on building and learning new tools. The main difference in performing benchmarks in a production environment

---

37 Any other medium like Jira ticket comments or GitHub issue works too. Just ensure you can easily paste screenshots so it's less fuss and there are fewer occasions to make mistakes on what screenshot was for what experiment!

38 Don't just make it all screenshots first and delay describing them until later. Try to iterate on each observation in Google Documents, as it's easy to forget later what situation you were capturing. Additionally, I saw many incidents of thinking screenshots were saved in my laptop's local directory, then losing all benchmarking results.

is to assure the quality of the production users—either by ensuring basic qualities of a new software version on different testing and benchmarking levels, or by leveraging beta testers or canary deployments.

Summary

Congratulations! With this chapter, you should now understand how to practically perform micro- and macrobenchmarks, which are core ways to understand if we have to optimize our software further, what to optimize if we have to, and how much. Moreover, both micro- and macrobenchmarks are also invaluable in other aspects of software development connected to efficiency like capacity planning and scalability.[39]

In my daily career in software development, I lean heavily on micro- and macrobenchmarks. Thanks to the micro-level fast feedback loop, I often do them for smaller functions in the critical path to decide how the implementation should go. They are easy to write and easy to delete.

Macrobenchmarks require more investment, so I especially recommend creating and doing such benchmarks:

- As an acceptance test against the RAER assessment of the entire system after a bigger feature or release.
- When debugging and optimizing regressions or incidents that trigger efficiency problems.

The experimentation involved in both micro- and macrobenchmarks is useful for efficiency assessment and in "6. Find the main bottleneck" on page 107. However, during that benchmark, we can also perform profiling of our Go program to deduce the main efficiency bottlenecks. Let's see how to do that in action in the next chapter!

39 Explained well in Martin Kleppmann's book *Designing Data-Intensive Applications: The Big Ideas Behind Reliable, Scalable, and Maintainable Systems* (*https://oreil.ly/M9RYQ*) (O'Reilly).

[illegible]

[illegible]

CHAPTER 9

Data-Driven Bottleneck Analysis

> Programmers are usually notoriously bad at guessing which parts of the code are the primary consumers of the resources. It is all too common for a programmer to modify a piece of code expecting see a huge time savings and then to find that it makes no difference at all because the code was rarely executed.
>
> — Jon Louis Bentley, *Writing Efficient Programs*

One of the key steps to improving the efficiency of our Go programs is to know where is the main source of the latency or resource usage you want to improve. Therefore, we should make a conscious effort to first focus on the code parts that contribute the most (the bottleneck or hot spot) to get the biggest value for our optimizations.

It is very tempting to use our experience in software development to estimate what part of the code is the most expensive or too slow to compute. We might have already seen similar code fragments causing efficiency problems in the past. For example, "Oh, I worked with linked lists in Go, it was so slow, this must be it!" or "We create a lot of new slices here, I think this is our bottleneck, let's reuse some." We might still remember the pain or stress it might have caused. Unfortunately, those feelings-based conclusions are often wrong. Every program, use case, and environment is different. The software might struggle in other places. It's essential to uncover that part quickly and reliably so we know where to spend our optimization efforts.

Fortunately, we don't need to guess. We can gather appropriate data! Go provides and integrates very rich tools we can use for bottleneck analysis. We will start our journey with the "Root Cause Analysis, but for Efficiency" on page 330 that introduces some of them. Then, I will introduce you to "Profiling in Go" on page 331, where you will learn about the `pprof` ecosystem. This profiling foundation is quite popular, yet it isn't easy to understand its results if you don't know the basics. The tooling, reports, and views are poorly documented, so I will spend a few sections describing the principles and common representations. In "Capturing the Profiling Signal" on page 355, you will learn how to instrument and collect profiles. In "Common Profile Instrumentation" on page 360, I will explain a few important existing profiles we can use right now in Go. Finally, we go through some "Tips and Tricks" on page 373, including the recently popular technique called "Continuous Profiling" on page 373!

This is one of those chapters where I learned a lot while researching and preparing the content. This is why I am even more excited to share that knowledge with you! Let's start with root cause analysis and its connection to bottleneck analysis.

Root Cause Analysis, but for Efficiency

The bottleneck analysis process is no different from the causal analysis (*https://oreil.ly/3MhUA*) or root cause analysis (*https://oreil.ly/KNqVV*) engineers perform after system incidents or failed tests. In fact, efficiency problems cause many of those incidents, e.g., HTTP requests timing out as the CPUs were saturated. As a result, it's best if we equip ourselves with similar mindsets and tools during bottleneck analysis of our system or program.

For more complex systems with multiple processes, the investigation might be quite involved with many symptoms,[1] red herrings,[2] or even multiple bottlenecks.

The tools in Chapter 6 are always invaluable for bottleneck analysis. With metrics around resource usage, we can narrow down when and which process allocated or used the most memory or CPU time, etc. With detailed logging, we could provide extra latency measurements for each stage. With tracing, we can analyze the request path and find which process and sometimes program function[3] contribute the most to the latency of the whole operation.

The other naive way is trial-and-error flow. We can always manually experiment by disabling certain code parts one by one to check if we can reproduce that efficiency

1 A symptom is an effect we see caused by some underlying situation, e.g., OOM is a symptom of the Go program requiring more memory than allowed. The problem with symptoms is that they often look like a root cause, but there might be an underlying bottleneck causing them. For example, the high memory usage of a process that caused the OOM might look like a root cause, but it can as well be just a symptom of a different issue if it was caused by a dependency not processing requests fast enough.

2 A red herring (*https://oreil.ly/5AKbS*) is an unexpected behavior that turns out to not be a problem to the general topic of our investigation. For example, while investigating the higher latency of our requests, it might be concerning to see the debug log "started handling request" in our application and not see a "finished request" for hours. It often turns out that the "finish" log message we might expect was not implemented, or we just dropped it in our logging system. Things often can mislead us; that's why we should be clean and explicit without observability and program flows to mislead us when we need to find the problem fast.

3 Usually, tracing does not provide a full stack trace, just the most important functionalities. This is to limit overhead and cost of tracing.

error or not. However, for large systems, this is likely to be infeasible in practice. There might be a better way to determine the main contributor to the extensive resource usage or high latency. Something that, in seconds, can tell us the exact code line responsible for it.

That convenient signal is called *profiling*, and it's often described as the fourth pillar of observability. Let's explore profiling in detail in the next section.

Profiling in Go

> Profiling is a form of dynamic code analysis. You capture characteristics of the application as it runs, and then you use this information to identify how to make your application faster and more efficient.
>
> —"Profiling Concepts," Google Cloud Documentation (*https://oreil.ly/okyge*)

Profiling is a perfect concept for representing the exact usage of something (e.g., elapsed time, CPU time, memory, goroutines, or rows in the database) caused by a specific code line in a program. Depending on what we look for, we can compare the contribution of something for different code lines or grouped by functions[4] or files.

In my experience, profiling is one of the most mature debugging methods in the Go community. It's rich, efficient, and accessible to everyone, with the Go standard library providing six profile implementations out of the box, community-created ones, and easy-to-build custom ones. What's amazing is that all these profiles might have different meanings and are related to different resources, but their representation follows the same convention and format. This means that no matter if you want to explore heap (see "Heap" on page 360), goroutine (see "Goroutine" on page 365), or CPU (see "CPU" on page 367), you can use the same visualization and analysis tools and patterns.

Without a doubt, many thanks should go to the `pprof` project (*https://oreil.ly/jBj18*) ("pprof" stands for performance profiles). There are many profilers out there. We have `perf_events` (`perf` tool) (*https://oreil.ly/M08S8*) for Linux, `hwpmc` (*https://oreil.ly/JJ8Gp*) for FreeBSD, DTrace (*https://oreil.ly/hUm9r*), and much more. What's special about `pprof` is that it establishes a common representation, file format, and visualization tooling for profiling data. This means you can use any of the preceding tools, or implement a profiler in Go from scratch and use the same tooling and semantics for analyzing those profiles.

4 Or methods, but that is treated in Go in the same way. Especially in this chapter, I will use the term *function* very often, and I mean both Go functions and methods.

Profiler

A profiler is a piece of software that can collect the stack traces and usage of a certain resource (or time) and then save it into a profile. Configured, installed, or instrumented profiler can be called profiling instrumentation.

Let's dive into `pprof` in the next section.

pprof Format

> The original `pprof` tool was a Perl script developed internally at Google. Based on the copyright header, development might go back to 1998. It was first released in 2005 as part of `gperftools`, and added to the Go project in 2010. In 2014 the Go project replaced the Perl based version of the pprof tool with a Go implementation by Raul Silvera that was already used inside of Google at this point. This implementation was re-released as a standalone project in 2016. Since then the Go project has been vendoring a copy of the upstream project, updating it on a regular basis.
>
> —Felix Geisendörfer, "Go's pprof Tool and Format" (*https://oreil.ly/FmOz8*)

Many programming languages like Go and C++ (*https://oreil.ly/0maaM*), and tools like Linux `perf` (*https://oreil.ly/PTJFN*) can leverage the `pprof` format, so it's worth learning about it more. To truly understand profiling, let's quickly create our custom profiling to track currently opened files in our Go program. There is a limit to how many file descriptors the program can hold simultaneously. If our program encounters such a problem, the file descriptor profiling might be beneficial to find what part of the program is responsible for opening the largest number of descriptors.[5]

For such basic profiling, we don't need to implement any `pprof` encoding or tracking code. Instead, we can use a simple `runtime/pprof.Profile` struct (*https://oreil.ly/f2OkA*) that the standard library implements. It allows for creating profiles that record counts and sources of the currently used objects of the desired type. `pprof.Profile` is very simple and a bit limited,[6] but it's perfect to start our journey with profiling.

---

5 Such a profiler was already proposed in the Go community to be included in the standard library. However, for now, the idea was rejected (*https://oreil.ly/YZoiR*) by the Go team as you can, in theory, track opened files thanks to the memory profile focused on allocations from `os.Open`.

6 With `pprof.Profile`, we can only track objects. We cannot profile advanced things like past object creation, I/O usage, etc. We also can't customize what is in the resulted `pprof` file, like extra labels, custom sampling, other value types, etc. Such custom profiling requires more code, but it is still relatively easy to implement thanks to Go packages like `github.com/google/pprof/profile` (*https://oreil.ly/DgeqN*).

The basic profiler example is presented in Example 9-1.

Example 9-1. Implementing file descriptor profiling using `pprof.Profile` functionality

```
package fd

import (
    "os"
    "runtime/pprof"
)

var fdProfile = pprof.NewProfile("fd.inuse") ❶

// File is a wrapper on os.File that tracks file descriptor lifetime.
type File struct {
    *os.File
}

// Open opens a file and tracks it in the `fd` profile`.
func Open(name string) (*File, error) {
    f, err := os.Open(name)
    if err != nil {
        return nil, err
    }
    fdProfile.Add(f, 2) ❷
    return &File{File: f}, nil
}

// Close closes files and updates profile.
func (f *File) Close() error {
    defer fdProfile.Remove(f.File) ❸
    return f.File.Close()
}

// Write saves the profile of the currently open file
// descriptors into a file in pprof format.
func Write(profileOutPath string) error {
    out, err := os.Create(profileOutPath)
    if err != nil {
        return err
    }
    if err := fdProfile.WriteTo(out, 0); err != nil { ❹
        _ = out.Close()
        return err
    }
    return out.Close()
}
```

❶ `pprof.NewProfile` is designed to be used as a global variable. It registers profiles with the provided name, which has to be unique. In this example, I use the `fd.inuse` name to indicate the profile tracks in-use file descriptors.

Unfortunately, this global registry convention has a few downsides. If you import two packages that create profiles you don't want to use, or they register profiles with common names, our program will panic. On the other hand, the global pattern allows us to use `pprof.Lookup("fd.inuse")` to get the created profile from different packages. It also automatically works with the `net/http/pprof` handler, explained in "Capturing the Profiling Signal" on page 355. For our example, it works fine, but I would usually not recommend using global conventions for any serious custom profiler.

❷ To record living file descriptors, we offer an `Open` function that mimics the `os.Open` function. It opens a file and records it. It also wraps the `os.File`, so we know when it's closed. The `Add` method records the object. The second argument tells how many calls to skip in the stack trace. The stack trace is used to record the location of the profile in the further `pprof` format.

I decided to use the `Open` function as the reference to sample creation, so I have to skip two stack frames.

❸ We can remove the object when the file is closed. Note I am using the same inner `*os.File`, so the `pprof` package can track and find the object I opened.

❹ Standard Go profiles offer a `WriteTo` method that writes bytes of a full `pprof` file into a provided writer. However, we typically want to save it to the file, so I added the `Write` method.

Many standard profiles, like those mentioned later in "Common Profile Instrumentation" on page 360, are transparently instrumented. For example, we don't have to allocate memory differently to see it in the heap profile (see "Heap" on page 360). For custom profiles like ours, a profiler has to be manually instrumented in our program. For example, I created `TestApp` that simulates an app that opens exactly 112 files. The code using Example 9-1 is presented in Example 9-2.

Example 9-2. `TestApp` code instrumented with `fd.inuse` profiling saves the profile at the end to the `fd.pprof` file

```
package main

// import "github.com/efficientgo/examples/pkg/profile/fd"

type TestApp struct {
```

```
    files []io.ReadCloser
}

func (a *TestApp) Close() {
    for _, cl := range a.files {
        _ = cl.Close() // TODO: Check error. ❷
    }
    a.files = a.files[:0]
}

func (a *TestApp) open(name string) {
    f, _ := fd.Open(name) // TODO: Check error. ❶
    a.files = append(a.files, f)
}

func (a *TestApp) OpenSingleFile(name string) {
    a.open(name)
}

func (a *TestApp) OpenTenFiles(name string) {
    for i := 0; i < 10; i++ {
        a.open(name)
    }
}

func (a *TestApp) Open100FilesConcurrently(name string) {
    wg := sync.WaitGroup{}
    wg.Add(10)
    for i := 0; i < 10; i++ {
        go func() {
            a.OpenTenFiles(name)
            wg.Done()
        }()
    }
    wg.Wait()
}

func main() {
    a := &TestApp{}
    defer a.Close()

    // No matter how many files we opened in the past...
    for i := 0; i < 10; i++ {
        a.OpenTenFiles("/dev/null") ❸
        a.Close()
    }

    // ...after the last Close, only files below will be used in the profile.
    f, _ := fd.Open("/dev/null") // TODO: Check error.
    a.files = append(a.files, f)

    a.OpenSingleFile("/dev/null")
```

```
    a.OpenTenFiles("/dev/null")
    a.Open100FilesConcurrently("/dev/null")

    if err := fd.Write("fd.pprof"); err != nil { ❹
        log.Fatal(err)
    }
}
```

❶ We open the file using our `fd.Open` function, which starts recording it in the profile as a side effect of opening the file.

❷ We always need to ensure the file will be closed when we don't need it anymore. This saves resources (like file descriptor) and more importantly, flushes any buffered writes and records that the file is no longer used.

❸ To demonstrate our profiling works, we first open 10 files and close them, repeated 10 times. We use */dev/null* as our dummy file for testing purposes.

❹ Finally, we create 110 files using methods that are chained in some way. Then we take a snapshot of this situation in the form of our `fd.inuse` profile. I use the *.pprof* file extension for this file (Go documentation uses *.prof*), but technically it's a gzipped (compressed using `gzip` program) protobuf file, so the *.pb.gz* file extension is often used. Use whatever you find more readable.

What's happening in the code in Example 9-2 might seem straightforward. In practice, however, the complexity of our Go program might cause us to wonder what piece of code creates so many files that are not closed. The data saved in the created *fd.pprof* should give us an answer to this question. We refer to the `pprof` format in the Go community as simply a gzipped protobuf (*https://oreil.ly/2Lgbl*) (binary format) file. The format is typed with the schema defined in the `.proto` language and officially defined in *google/pprof* project's *proto* file (*https://oreil.ly/CiEKb*).

To learn the `pprof` schema and its primitives quickly, let's look at what the *fd.pprof* file produced in Example 9-2 could store. The high-level representation of the open (in use) and total file descriptors diagram is presented in Figure 9-1.

Figure 9-1 shows what objects are stored in `pprof` format and a few core fields those objects contain (there are more). As you might notice, this format is designed for efficiency, with many indirections (referencing other things via integer IDs). I skipped that detail on the diagram for simplicity, but all strings are also referenced as integers with the string table for interning (*https://oreil.ly/KT4UY*).

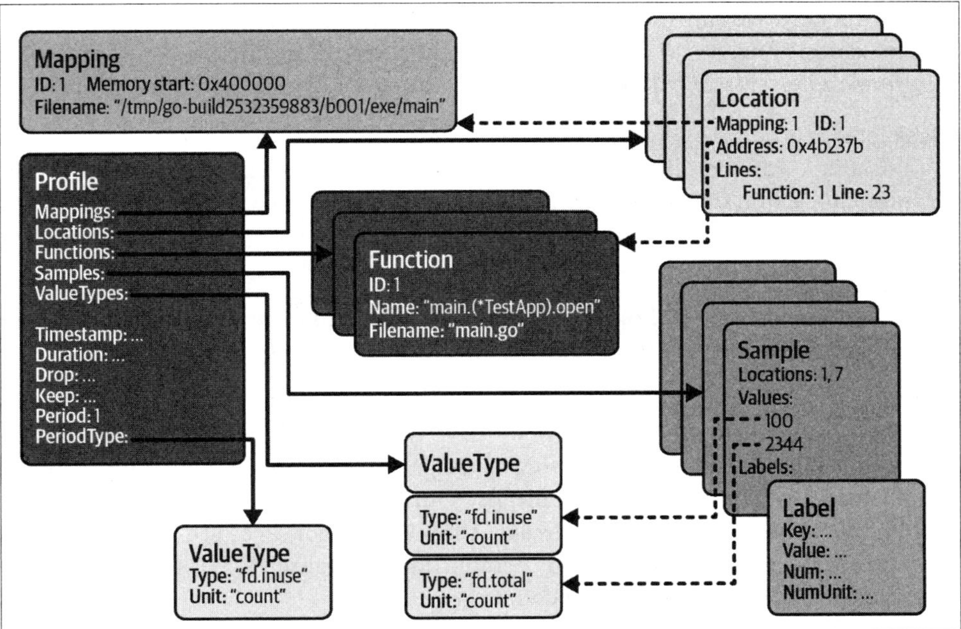

Figure 9-1. The high-level representation of open (in use) and total file descriptors in `pprof` format

`pprof` format starts with the single root object called `Profile`, which contains the following child objects:

`Mappings`
: Not every program has debugging symbols inside the binary. For example, in "Understanding Go Compiler" on page 118, we mentioned that Go has them by default to provide human-readable stack traces that refer to source code. However, someone compiling binary might remove this information to make the binary size much smaller. If there are no symbols, the `pprof` file can be used with addresses of stack frames (locations). Those addresses will then be dynamically translated to the exact source code line by further tooling in a process called symbolization (*https://oreil.ly/zcZKa*). Mapping allows specifying how addresses are mapped to the binary if it's dynamically provided in a later step.

  Unfortunately, if you need a binary file, it has to be built from the same source code version and architecture from which we gathered profiles. This is usually very tricky. For example, when we obtain profiles from remote services (more on that in "Capturing the Profiling Signal" on page 355), we most likely won't have the same binary on the machine where we analyze the profiles.

Fortunately, we can store all required metadata in the `pprof` profile, so no symbolization is needed. This is what's used for standard profiles in Go from Go 1.9 (*https://oreil.ly/qONe8*), so I will skip explaining the symbolization techniques.

`Locations`
: Locations are code lines (or their addresses). For convenience, a location can point to a function it was defined in and the source code filename. Location essentially represents a stack frame.

`Functions`
: Functions structures hold metadata about functions in which locations are defined. They are only filled if debug symbols were present in the binary.

`ValueTypes`
: This tells how many dimensions we have in our profiles. Each location can be responsible for using (contributing to usage of) some values. Value types define the unit and what that value means. Our Example 9-1 profile has only the `fd.inuse` type, because the current, simplistic `pprof.Profile` does not allow putting more dimensions; but for demonstration, Figure 9-1 has two types representing total count and current count.

Contributions

The `pprof` format profile does not limit what the profile value means. It's up to the implementation to define the measured value semantics. For example, in Example 9-1, I defined it as the number of open files present at the moment of the profile snapshot. For other "Common Profile Instrumentation" on page 360, the value means something else: the time spent on CPU, allocated bytes, or the number of goroutines executing in a specific location. Always clarify what your profile values mean!

Generally, most profile values tell us how much each part of our code uses a certain resource or time. That's why I stick to the *contribution* verb when explaining profile values on samples.

`Samples`
: The measurement or measured contribution by a given stack trace of some value for a given value type. To represent a stack trace (call sequence), a sample lists all location IDs starting from the top of the stack trace. The important detail is that the sample has to have the exact number of values equal to the number (and order) of value types we have defined. We can also attach labels to samples. For

example, we could attach the example filename that was open in that stack trace. "Heap" on page 360 uses it to show average allocation size.

Further metadata
: Information like when the profile was captured, data tracking duration (if applicable), and some filtering information can also be in the profile object. One of the most important fields is the `period` field, which tells us if the profile was sampled or not. We track all the instrumented `Open` calls in Example 9-2, so we have `period` equal to one.

With all those components, the `pprof` data model is very well designed with the profiling data that describes any aspect of our software. It also works well with statistical profiles, which capture the data from a small portion of all the things that happened.

In Example 9-2, tracking opened files does not pose too much overhead to the application. Perhaps in extreme production cases calling `Add` and `Remove`, and mapping objects on every file open and closed, might slow down some critical paths. However, the situation is much worse with complex profiles like "CPU" on page 367 or "Heap" on page 360. For the CPU profile that profiles the use of the CPU by our program, it's impractical (and impossible) to track what exact instruction was executed in every single cycle. This is because, for every cycle, we would need to capture a stack trace and record it in memory, which, as we learned in Chapter 4, can take hundreds of CPU cycles alone.

This is why the CPU profile has to be sampled. This is similar to other profiles, like memory. As you will learn in "Heap" on page 360, we sample it because tracking all individual allocations would add significant overhead and slow down all allocations in our program.

Fortunately, even with highly sampled profiles, profiling is extremely useful. By design, profiling is primarily used for bottleneck analysis. By definition, the bottleneck is something that uses most of some resources or time. This means that no matter if we capture 100%, 10%, or even 1% of events that use, e.g., the CPU time, statistically, the code that uses the most CPU should still be at the top with the largest usage number. This is why the more expensive profiles will always be sampled in some way, which allows Go developers to safely pre-enable profiles in almost all our programs. It also enables the continuous profiling practices discussed in "Continuous Profiling" on page 373.

Statistical Profiles Are Not 100% Precise

In the sampled profile, you can miss some portion of the contributions.

Profilers like Go have a sophisticated scaling mechanism that attempts to find (*https://oreil.ly/DrfIA*) the probability of missing the allocations and adjust for it, which usually is precise enough.

Yet, those are only approximations. We can sometimes miss some code locations with smaller allocations on our profiles. Sometimes the real allocation is a little larger or smaller than estimated.

Make sure to check the `period` information in the `pprof` profiles (explained in "go tool pprof Reports"), and be aware of the sampling in your profiles to reach the right conclusions. Don't be surprised and worried that your benchmarked allocation numbers do not exactly match the numbers in the profile. We can be entirely certain about absolute numbers only when we obtain a profile with a period equal to one (100% samples).

With the fundamentals of the `pprof` standard explained, let's look at what we can do with such a *.pprof* file. Fortunately, we have plenty of tools that understand this format and help us analyze the profiling data.

go tool pprof Reports

There are many tools (and websites!) out there you can use to parse and analyze `pprof` profiles. Thanks to a clear schema, you can also easily write your own tool. However, the most popular one out there is the `google/pprof` project, which implements the `pprof` CLI tool (*https://oreil.ly/lGZJG*) for this purpose. The same tool is also vendored in the Go project (*https://oreil.ly/pbDk3*), which allows us to use it through the Go CLI. For example, we can report all the `pprof` relevant fields in semi-human readable format using the `go tool pprof -raw fd.pprof` command, as presented in Example 9-3.

Example 9-3. Raw debug output of the .pprof file using the Go CLI

```
go tool pprof -raw fd.pprof
PeriodType: fd.inuse count
Period: 1 ❶
Time: 2022-07-29 15:18:58.76536008 +0200 CEST
Samples:
fd.inuse/count
        100: 1 2
         10: 1 3 4
          1: 5 4
          1: 6 4
```

```
Locations
1: 0x4b237b M=1 main.(*TestApp).open example/main.go:23 s=0
    main.(*TestApp).OpenTenFiles example/main.go:33 s=0
2: 0x4b25cd M=1 main.(*TestApp).Open100FilesConcurrently.func1 (...)
3: 0x4b283a M=1 main.main example/main.go:64 s=0
4: 0x435b51 M=1 runtime.main /go1.18.3/src/runtime/proc.go:250 s=0
5: 0x4b26f2 M=1 main.main example/main.go:60 s=0
6: 0x4b2799 M=1 main.(*TestApp).open example/main.go:23 s=0
    main.(*TestApp).OpenSingleFile example/main.go:28 s=0
    main.main example/main.go:63 s=0
Mappings
1: 0x400000/0x4b3000/0x0 /tmp/go-build3464577057/b001/exe/main  [FN]
```

❶ The `-raw` output is currently the best way (*https://oreil.ly/juE75*) to discover what sampling (`period`) was used when capturing the profile. Using it with the `head` utility lets us see the first few rows containing that information, which is useful for large profiles, for example, `go tool pprof -raw fd.pprof | head`.

The raw output can reveal some basic information about the data contained by the profile, and it helped create the diagram in Figure 9-1. However, there are much better ways to analyze bigger profiles. For example, if you run `go tool pprof fd.pprof`, it will enter an interactive mode that lets you inspect different locations and generate various reports. We won't cover this mode in this book because there is a much better way these days that does almost all the interactive mode can—the web viewer!

The most common way to run a web viewer is to run a local server on your machine via the Go CLI. Use the `-http` flag to specify the address with the port to listen on. For example, running the `go tool pprof -http :8080 fd.pprof`[7] command will open the web viewer website[8] in your browser showing the profile obtained in Example 9-2. The first page you would see is a directed graph rendered based on the given `fd.pprof` profile (see "Graph" on page 347). But before we get there, let's get familiar with the top navigation menu available[9] in the web interface, shown in Figure 9-2.

7 The `:8080` is shorthand for `0.0.0.0:8080`, so listening on all network interfaces of your machine.

8 To run this command or generate graphs, you need to install the `graphviz` tool (*http://www.graphviz.org*) on your machine.

9 This guide is for the web interface from Go 1.19. There are no hints that it will change, but the `pprof` tool may be enhanced or updated in subsequent versions of Go.

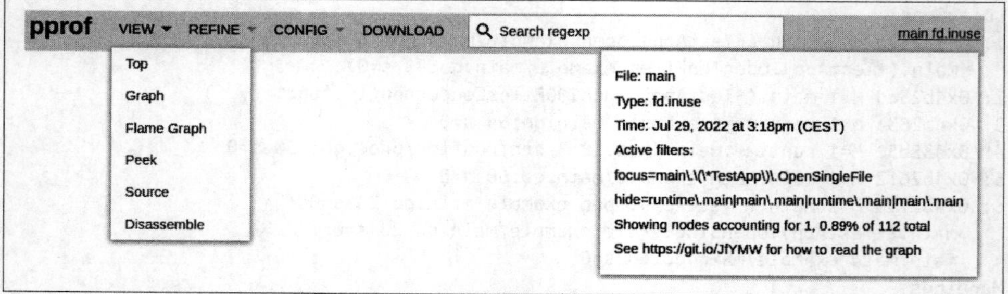

Figure 9-2. The top navigation on the `pprof` web interface

From the left, the top gray overlay menu has the following buttons and inputs:[10]

VIEW
: Allows you to choose different views (reports) of the same profiling data. We will go through all six view types in the subsections below. They all show profiles from a slightly different angle and have a purpose; you might favor different ones. They are generated from the location hierarchy (stack trace) that can be reconstructed from the samples in Figure 9-1.

SAMPLE
: This menu option is not present in Figure 9-2 because we only have one sample value type (`fd.inuse` type with `count` unit), but for profiles with more types, the SAMPLE menu allows us to choose what sample type we want to use (we can use one at a time). This is commonly present on heap profiles.

REFINE
: This menu works only in the Graph and Top views (see "Graph" on page 347 and "Top" on page 345). It allows filtering the Graph or Top views to certain locations of interest: nodes in the graph and rows in the top table. It is especially useful for very complex profiles with hundreds or more locations. To use it, click on one or more Graph nodes or rows in the Top table to select the locations. Then click REFINE and choose if you want to focus, ignore, hide, or show them.

Focus and Ignore control the visibility of samples that go through a selected node or row, allowing you to focus on or ignore full stack traces. Hide and Show control only the node or row's visibility without impacting samples.

The same filtering can be applied using `-focus` and other flags in the `go tool pprof` CLI (*https://oreil.ly/OVQLC*). Additionally, the REFINE > Reset option

10 You can also hover over each menu item, and after three seconds a short help pop-up will appear.

brings us back to a nonfiltered view, and if you change to a view that does not support refined options, it only persists in the Focus value.

Focus and Ignore are incredibly useful when you want to find the exact contribution of a certain code path. On the other hand, you can use Hide and Show when you want to present the graph to somebody or as documentation for a clearer picture.

Don't use those options if you're trying to mentally correlate your code with the profile, as you can get easily confused, especially at the start of your profiling journey.

CONFIG
: The refinement settings you used from the REFINE option are saved in the URL. However, you can save these settings to a special, named configuration (as well as a zoom option for the Graph view). Click CONFIG > Save As ..., then choose the configuration you will be using. The `Default` configuration works like REFINE > Reset. The configuration is saved under *<os.UserConfigDir>/pprof/settings.json* (*https://oreil.ly/nWfnq*). On my Linux machine, it is in *~/.config/pprof/settings.json*. This option also works only on the Top and Graph views and automatically changes to Default if you change to any other view.

DOWNLOAD
: This option downloads the same profile you used in `go tool pprof`. It is useful if someone exposes the web viewer on the remote server and you want to save the remote profile.

Search regexp
: You can search for samples of interest using the RE2 regular expression (*https://oreil.ly/cOvAq*) syntax by the location's function name, filename, or object name. This sets the Focus option in the REFINE menu. In some views, like Top, Graph, and os.ReadFile, the interface also highlights matched samples as you write the expression.

The binary name and sample type
: In the right-hand corner is a link with the chosen binary name and sample value type. You can click this menu item to open a small pop-up with quick statistics about the profile, view, and options we are running with. For example, Figure 9-2 shows what you see when you click on that link with some REFINE options on.

Before diving into the different views available in the `pprof` tool, we have to understand important concepts of Flat and Cumulative (Cum for short) values for certain location granularity.

Every pprof view shows Flat and Cumulative values for one or more locations:

- Flat represents a certain node's *direct* responsibility for resource or time usage.
- Cumulative is a sum of *direct* in *indirect* contributions. Indirect means that the locations did not create any resource (or were not used anytime) directly, but may have invoked one or more functions that did.

Using code examples is best to explain those definitions in detail. Let's use part of the main() function from Example 9-2 presented in Example 9-4.

Example 9-4. Snippet of Example 9-2 explaining Flat and Cumulative values

```
func main() { ❶
    // ...

    f, _ := fd.Open("/dev/null") // TODO: Check error. ❷
    a.files = append(a.files, f) ❸

    a.OpenSingleFile("/dev/null")
    a.OpenTenFiles("/dev/null") ❹

    // ...
}
```

❶ Profiling is tightly coupled with a stack trace representing a call sequence that led to a certain sample, so in our case, opening files. However, we could aggregate all samples going through the main() function to learn more. In this case, the main() function Flat number of open files is 1, Cum is 12. This is because, in the main function, we directly open only one file (via fd.Open);[11] the rest were opened via chained (descendant) functions.

❷ From our *fd.pprof* profile, we could find that this code line Flat value is 1 and Cum is 1. It directly opens one file and does not contribute indirectly to any more file descriptor usage.

❸ append does not contribute to any sample. Therefore, no sample should include this code line.

11 From the perspective of the profiling, the direct (Flat) contribution is decided by instrumentation implementation. Our custom code in Example 9-1 treats the fd.Open function as the moment the file descriptor was opened. Different profiling implementations might define the moment of "use" differently (moment of the allocation, use of CPU time, waiting for lock opening, etc.).

❹ The code line that invokes the `a.OpenSingleFile` method has a Flat value of 0 and a Cum of 1. Similarly, the `a.OpenTenFiles` method Flat value is 0 and Cum is 10. Both directly in the moment of the CPU touching this program line do not create (yet) any files.

I find the Flat and Cum names quite confusing, so I will use the direct and cumulative terms in further content. Both numbers are beneficial to compare what parts of the code contribute to the resource usage (or time used). The cumulative number helps us understand what flow is more expensive, whereas the direct value tells us the source of the potential bottleneck.

Let's walk through the different views and see how we can use them to analyze the *fd.pprof* file obtained in Example 9-2.

Top

First on the VIEW list, the Top report shows a table of statistics per location grouped by functions. The view for the *fd.pprof* file is presented in Figure 9-3.

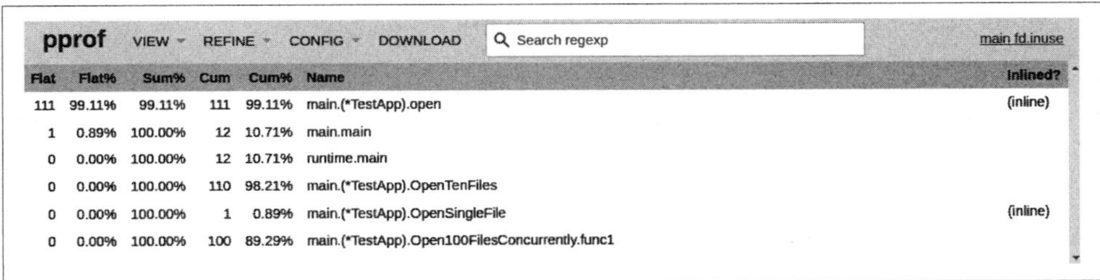

Flat	Flat%	Sum%	Cum	Cum%	Name	Inlined?
111	99.11%	99.11%	111	99.11%	main.(*TestApp).open	(inline)
1	0.89%	100.00%	12	10.71%	main.main	
0	0.00%	100.00%	12	10.71%	runtime.main	
0	0.00%	100.00%	110	98.21%	main.(*TestApp).OpenTenFiles	
0	0.00%	100.00%	1	0.89%	main.(*TestApp).OpenSingleFile	(inline)
0	0.00%	100.00%	100	89.29%	main.(*TestApp).Open100FilesConcurrently.func1	

Figure 9-3. The Top view is sorted by the direct value

Each row represents direct and cumulative contributions of open files for the single function, which, as we learned from Example 9-4, aggregates the usage of one or multiple lines within that function. This is called function granularity, which can be configured by URL or CLI flag.

Choose Your Granularity

Certain views like Top, Graph, and Flame Graph allow us to group locations by file, function, or not group at all (grouping by line or address). This means that one entry, row, or graph node will group contributions from all lines within a single function or file.

You can choose the granularity by using one of the following flags in the `go tool pprof` command: `-functions` (default option), `-files`, `-lines`, or `-address`. Similarly, you can set this using the URL parameter `?g=<granularity>`.

Usually, function granularity is low enough, especially given low sampling rates (e.g., in the CPU profile). However, switching to line granularity effectively tells us exactly which code line contributes to the resource we profile for and where to find it. If the function has a nonzero direct contribution value, you might want to check what exact part of the function is a bottleneck!

We already defined the values represented by the Flat and Cum columns. Other columns in this view are:

Flat%
: The percentage of the row's direct contributions to the program's total contributions. In our case, 99.11% of the open file descriptors were created directly by the `open` method (111 out of 112).

Sum%
: The third column is the percentage of all direct values from the top to the current flow to the total contributions. For instance, the 2 top rows are directly responsible for all 112 file descriptors. This statistic allows us to narrow down to the functions that might matter the most for our bottleneck analysis.

Cum%
: The percentage of the cumulative contribution of the row to the total contributions.

Be Careful When Goroutines Are Involved

The cumulative value can be misleading in some cases with goroutines. For example, Figure 9-3 indicated that `runtime.main` cumulatively opened 12 files. However, from Example 9-2 you can find that it also executes the `Open100FilesConcurrently` method, which then executes `Open100FilesConcurrently.func1` (anonymous function) as a new goroutine. I would expect a link from `runtime.main` to `Open100FilesConcurrently.func1` in the Graph, and the cumulative value of `runtime.main` to be 112.

The problem is that stack traces of each goroutine in Go are always separate. Therefore, there is no relation between goroutines in which goroutine created which one, which will be clear when we look at the goroutine profiles in "Goroutine" on page 365. We must keep this in mind while analyzing our program's bottleneck.

Name and Inlined
: The function name for the location and whether it was inlined during compilation. In Example 9-2, both `open` and `OpenSingleFile` were simple enough for compiler to inline them to the parent functions. You can represent the situation from the binary (after inline) by adding the `-noinlines` flag to the `pprof` command or by adding the `?noinlines=t` URL parameter. Seeing the situation before inlining is still recommended to map what happened to the source code more easily.

The sorting order of rows in our Top table is by direct contribution, but we can change it with the `-cum` flag to order by cumulative values. We can also click on each header in the table to trigger different sorting in this view.

The Top view might be the simplest and fastest way to find the functions (or files or lines, depending on the chosen granularity) directly or cumulatively responsible for using resources or time you are profiling for. The downside is that it does not tell us the exact link between those rows, which would tell us which code flow (full stack trace) might have triggered the usage. For such cases, it might be worth using the Graph view explained in the next section.

Graph

The Graph view is the first thing you see when opening the `pprof` tool web interface. This is not without reason—humans work better if things are visualized (*https://oreil.ly/VElUH*) than if we have to parse and visualize all in our brain from the text report. This is my favorite view as well, especially for profiles obtained from less familiar code bases.

To render the Graph view, the `pprof` tool generates a graphical directed acyclic graph (DAG) (*https://oreil.ly/hzglQ*) from the provided profile in the DOT (*https://oreil.ly/HiRV9*) format. We can then use the `-dot` flag with `go tool pprof`, and use other rendering tools or render it to the format we want with the `-svg`, `-png`, `-jpg`, `-gif`, or `-pdf` formats. On the other hand, we have the `-http` option that generates a temporary graphic using the `.svg` format and starts the web browser from it. From the browser, we can see the `.svg` visualization in the Graph view and use the interactive REFINE options explained before: zoom in, zoom out, and move around through the graph. The example Graph view from our *fd.pprof* format is presented in Figure 9-4.

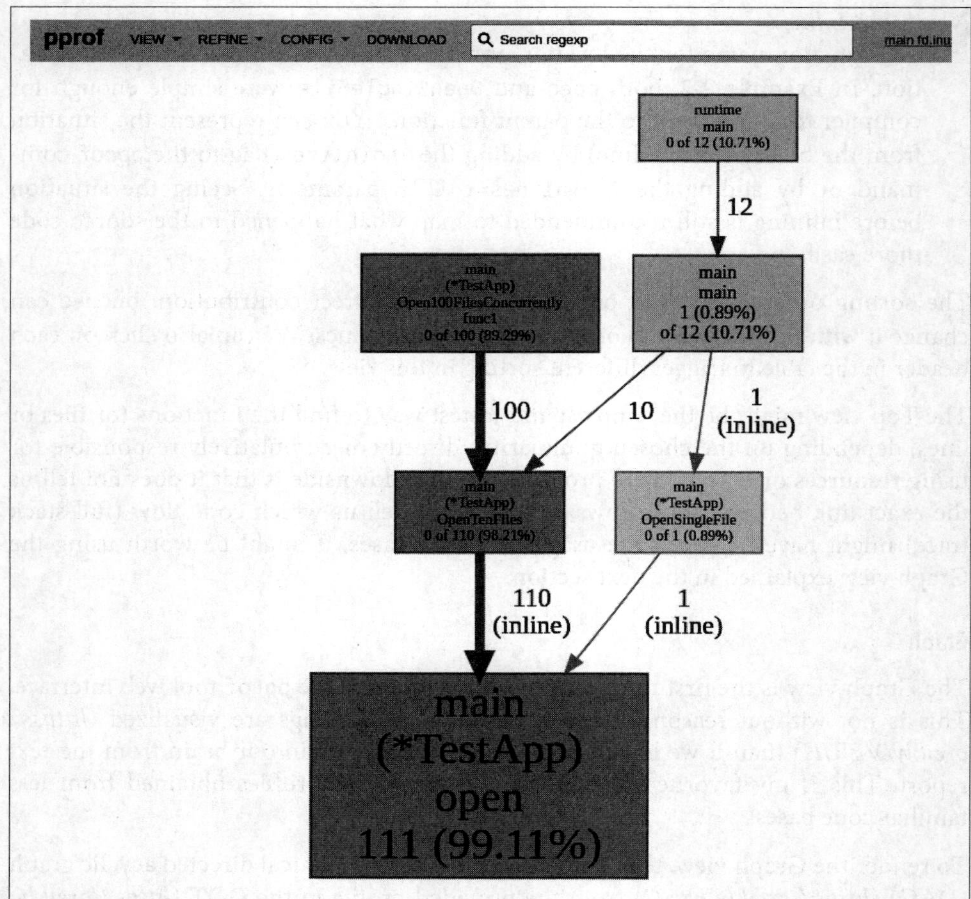

Figure 9-4. The Graph view of Example 9-2 with function granularity

What I love about this view is that it clearly represents the relation (hierarchy) of different execution parts of your program regarding resource or time usage. While it might be tempting, you cannot move nodes around. You can only hide or show them using the REFINE options. Hovering over a node also shows the full package name or code line.

On top of that, every aspect of this graph has its meaning (*https://oreil.ly/GQbNn*), which helps to find the most expensive parts. Let's go through the graph attributes:

Node

Each node represents the contribution of a function for the currently opened files. This is why the first part of the text in the node shows the Go package and function (or method). We would see the code line or file if we chose a different granularity. The second part of the node shows the direct and cumulative values. If any of the values are nonzero, we see that the percentage of that value to the total contributions. For example, in Figure 9-4 we see the `main.main()` node (on the right) confirms the number we found in Example 9-4. Using `pprof`, we recorded 1 direct contribution and 12 cumulative ones in that function. The color and size tell us something too:

- The size of the node represents direct contributions. The bigger the node, the more resource or time it used directly.
- The border and fill color represent cumulative values. The normal color is gold. Large positive cumulative numbers make the node red. Cumulative values close to zero cause the node to be gray.

Edge

Each edge represents the call path between functions (files or lines). The call does not need to be direct. For example, if you use the `REFINE` option, you can hide multiple nodes that were called between two, causing the edge to show an indirect link. The value on the edge represents the cumulative contributions of that code path. The `inline` word next to the number tells us that the call pointed to by edge was inlined into the caller. Other characteristics matter as well:

- The weight of the edge indicates cumulative contributions by a path. The thicker the edge, the more resources were used.
- The color shows the same. Normally an edge is gold. Larger positive values color an edge red, close to zero to gray.
- A dashed edge indicates that some connected locations were removed, e.g., because of a node limit.[12]

12 The REFINE hidden option keeps the line solid.

Some Nodes Might Be Hidden!

Don't be surprised if you don't see every contribution to the resource you profile in the Graph view. As I mentioned before, most of the profiles are sampled. This means that statistically, the locations that contribute a little might be missed in the resulting profile.

The second reason is the node limit in the `pprof` viewer. By default, it does not show more than 80 nodes (*https://oreil.ly/Wcwsu*) for readability. You can change that limit using the `-nodecount` flag.

Finally, the `-edgefraction` and `-nodefraction` settings hide the edges and nodes with the fraction of direct contribution to the total contribution lower than the specified value. By default (*https://oreil.ly/oVfrt*) it is 0.005 (0.5%) for node fraction and 0.001 (0.1%) for edge fraction.

With theory aside, what can we learn from the `pprof` Graph view? This view is perfect for learning about efficiency bottlenecks and how to find their source. From Figure 9-4 we can immediately see that the biggest cumulative contributor is `Open100FilesConcurrently`, which seems to be a new goroutine since it is not connected to the `runtime/main` function. It might be a good idea to optimize that path first. The most open files come from `OpenTenFiles` and `open`. This tells us that it's a critical path for the efficiency of this resource. If some new functionality required creating an additional file on every `open` call, we would see a significant growth in opened file descriptors by our Go program.

The Graph view is an excellent method to understand how your application's different functionalities impact your program's resource usage. It is especially important for more complex programs with large dependencies your team did not create. As it turns out, it is easy to misunderstand the right way of using the library you depend on. Unfortunately, this also means that there will be a lot of function names or code lines you don't recognize or don't understand. See Figure 9-5, taken from the optimized `Sum` we optimize in "Optimizing Latency Using Concurrency" on page 402.

This result also proves the importance of the skill of switching between different granularity. It's as easy as adding to a URL `?g=lines` to switch to line granularity—it's way more effective than reopening `go tool pprof` with the `-lines` flag.

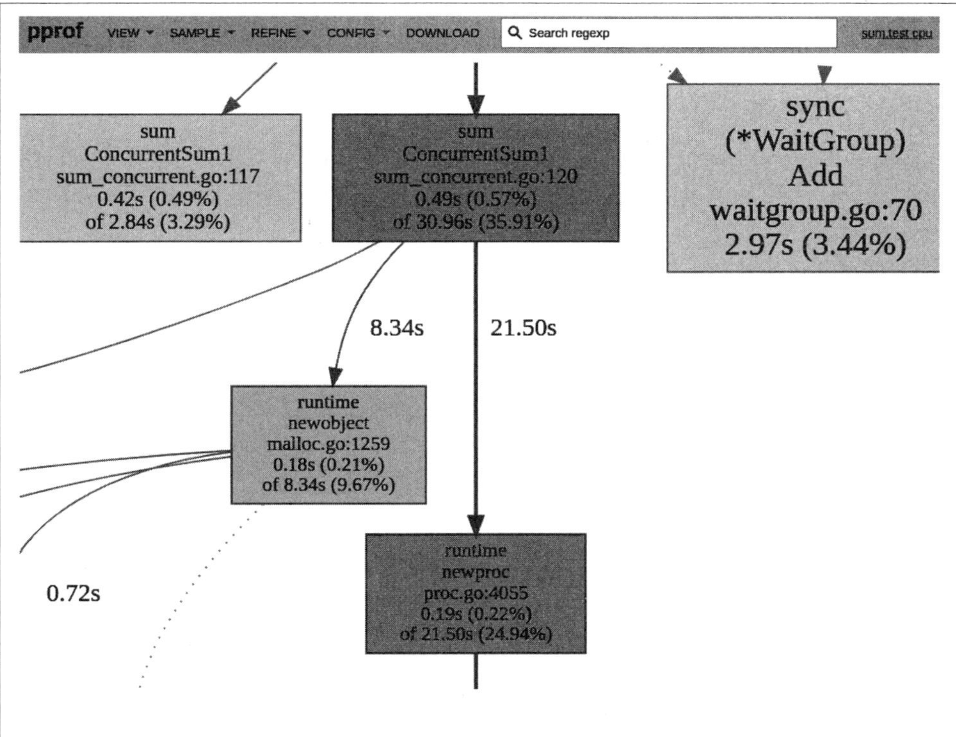

Figure 9-5. Snippet of the Graph view of the CPU profile taken from Example 10-10 with line granularity

Don't Be Afraid of Unknowns!

It's normal to feel a bit anxious if you see your Go program's Graph profile for the first time. For example, it isn't uncommon to see various runtime functions. We don't need to always have an idea of what they do exactly, but we can always find that information if we want to!

Build your confidence in being able to dive into any new function or code line that matters. In Figure 9-5, the `runtime.newproc` function was one of the biggest bottlenecks for CPU time. Instinct might tell us it has something to do with creating a new goroutine (kind of new process), but it's relatively easy to confirm:

- A quick Google search for `runtime.newproc github` or Peek, Source, and Disassemble views gives us the exact code line of `newproc` (*https://oreil.ly/3tgSz*). From this, we can try to read the comment or code and figure out what this function is responsible for (not always trivial).

- The Top view or Graph view with line granularity tells the exact line where this contribution starts. As presented in Figure 9-5, it is triggered by line 120 (*https://oreil.ly/YlASl*), which clearly shows creation of the goroutine!

Following the Graph view, we have the latest addition to the `pprof` tool—the Flame Graph view, which many members of the Go community prefer. So let's dive into it.

Flame Graph

The Flame Graph (sometimes also called the Icicle Graph) view in `pprof` is inspired by Brendan Gregg's work (*https://oreil.ly/sKFbH*), focused initially on CPU profiling.

> A flame graph visualizes a collection of stack traces (aka call stacks), shown as an adjacency diagram with an inverted icicle layout. Flame graphs are commonly used to visualize CPU profiler output, where stack traces are collected using sampling.
>
> —Brendan Gregg, "The Flame Graphs" (*https://oreil.ly/RAsrK*)

The Flame Graph report rendered from *fd.pprof* is presented in Figure 9-6.

Color and Order of Segments Usually Do Not Matter

This depends on the tool that renders the Flame Graph, but for the `pprof` tool, both color and order do not have any meaning here. The segments are typically sorted by the location name or label value.

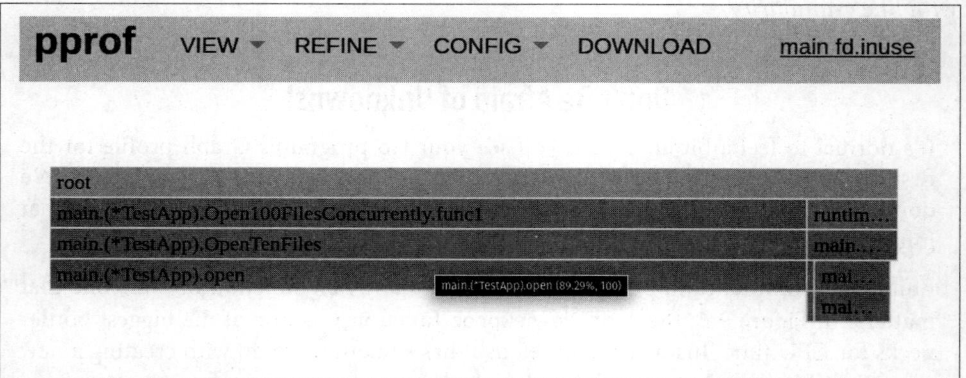

Figure 9-6. The Flame Graph view of Example 9-2 with a function granularity

The `pprof` is an inverted version of the original Flame Graph, where each significant code flow forms a separate icicle. The main attribute that matters here is the width of the rectangular segment, which represents the node from the Graph view—function in our case. The wider the block, the larger the cumulative contribution it is responsible for. You can hover over individual segments to see their absolute and percentage cumulative values. Click on each block to focus the view on the given code path.

Instead of edges, we can follow call hierarchy by looking at what's above the current segment. Don't focus too much on the height of the icicle—it only shows how complex (deep) the call stack is. It's the width that matters here.

In some way, a Flame Graph is often favored by more advanced engineers because it's more compact. It allows a pragmatic insight into the biggest bottlenecks of the system. It immediately shows the percentage of all resources that each code path contributed. At a glance, in Figure 9-6 we can quickly tell without any interactivity that `Open100FilesConcurrently.func1` is the major bottleneck of opened files with approximately 90% of resources used by it. The Flame Graph is also excellent to show if there is any major bottleneck. On some occasions, a lot of small contributors might together generate a large usage. A Flame Graph will tell us about this situation immediately. Note that similar to the Figure 9-4 view, it can drop many nodes from the view. The number of dropped nodes is presented if you click the binary name at the top right corner.

Any of the three views we discussed—Top, Graph, or Flame Graph—should be the first point of interest to find the biggest bottleneck in our program efficiency. Remember about sampling, switching granularity to learn more, and focusing your time on the biggest bottlenecks first. However, three more views are worth briefly mentioning: Peek, Source, and Disassemble. Let's look at them in the next section.

Peek, Source, and Disassemble

The other three views—Peek, Source, and Disassemble—are not affected by the granularity option. They all show the raw line or address level of locations, which is especially useful if you want to go back to your source code to focus on your code optimization inside your favorite IDE.

The Peek view provides a table similar to the Top view. The only difference is that each code line shows all direct callers and the usage distribution in the Call and Calls % columns. It helps in cases with many callers where you want to narrow down the code path that contributes the most.

One of my favorite tools is the Source view. It shows the exact code line in the context of the program source code. In addition, it shows the few lines before and after. Unfortunately, the output is not ordered, so you have to use previous views to know what function or code line you want to focus on, and use the Search feature to focus on what you want. For example, we could see direct and cumulative contributions of `Open100FilesConcurrently` directly mapped to the code line in our code, as presented in Figure 9-7.

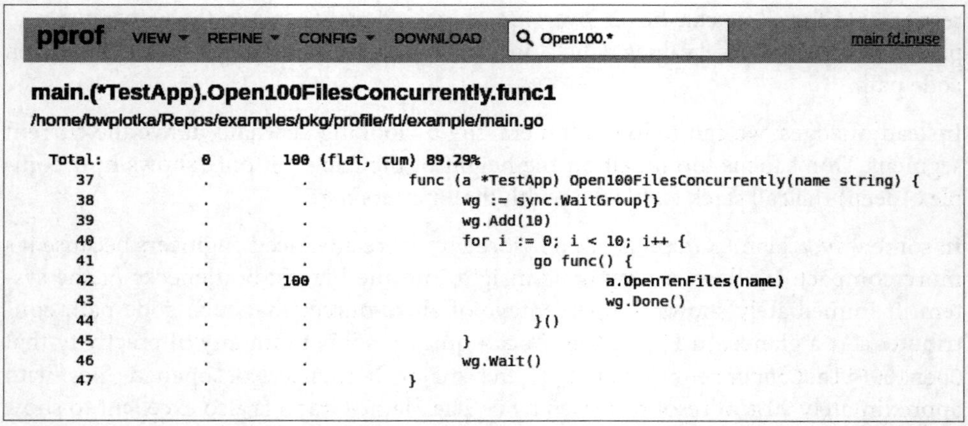

pprof VIEW ▾ REFINE ▾ CONFIG ▾ DOWNLOAD Open100.* main fd.inuse

main.(*TestApp).Open100FilesConcurrently.func1

/home/bwplotka/Repos/examples/pkg/profile/fd/example/main.go

```
Total:      0      100 (flat, cum) 89.29%
            .        .           func (a *TestApp) Open100FilesConcurrently(name string) {
            .        .               wg := sync.WaitGroup{}
            .        .               wg.Add(10)
            .        .               for i := 0; i < 10; i++ {
            .        .                   go func() {
            .      100                       a.OpenTenFiles(name)
            .        .                       wg.Done()
            .        .                   }()
            .        .               }
            .        .               wg.Wait()
            .        .           }
```

Figure 9-7. The Source view of Example 9-2 focused on the `Open100FilesConcurrently` search

For me, there is something special in the Source view. Seeing the open file descriptors, allocation points, CPU time, etc., directly mapped to a code statement in your source code gives a bigger understanding and awareness than seeing lines as a bunch of boxes in Figure 9-4. For the standard library code, or when you provide a binary (as mentioned for the Disassemble view), you can also click on a function to display its assembly code!

The Source view is incredibly useful when attempting to estimate the "Complexity Analysis" on page 240 of the code we profile. I recommend using the Source view if you can't fully wrap your head around the part of the code that uses the resource and why.

Finally, the Disassemble view is useful for advanced profiling. It provides the Source view, but at the assembly level (see "Assembly" on page 115). It allows checking compilation details around the problematic code. This view requires a provided binary built from the same source code as the program you took the profile from. For example, for my case with the *fd.inuse* file, I have to provide a statically built binary via a path using `go tool pprof -http :8080 pkg/profile/fd/example/main fd.pprof`.[13]

13 Note that currently there are some bugs in this view in `pprof`. When you are missing binary, the UI shows `no matches found for regexp:`. Search also does not work, but you can use the built-in browser search to find what you want (e.g., using Ctrl+F).

Currently, no mechanism will check if you are using the correct program binary for the profile you analyze. Therefore, the results might be, by accident, correct or totally wrong. The result in the error case is nondeterministic, so ensure you provide the correct binary!

The `pprof` tool is an amazing way to confirm, in a data-driven way, your initial guesses about the efficiency of your application and what causes the potential problems. The amazing thing about the skills you acquired in this section is that the mentioned text and visual representations of the `pprof` profiles are not only used by the native `pprof` tooling. Similar views and techniques are used among many other profiling tools and paid vendor services, like Polar Signals (*https://oreil.ly/HowVb*), Grafana Phlare (*https://oreil.ly/Ru0Hu*) Google Profiler (*https://oreil.ly/mJu6V*), Datadog's Continuous Profiler (*https://oreil.ly/WF9fG*), Pyroscope project (*https://oreil.ly/eKyK7*), and more!

It is also quite likely that your Go IDE[14] supports rendering and gathering `pprof` profiles out of the box. Using IDE is great as it can integrate directly into your source code and enable smooth navigation through locations. However, I prefer `go tool pprof` and `pprof` tool-based cloud projects like the Parca project (*https://oreil.ly/2PKkx*) since we often have to profile on the macrobenchmarks level (see "Macrobenchmarks" on page 306).

With the format and visualization descriptions complete, let's dive into how to obtain profiles from your Go program.

Capturing the Profiling Signal

Recently we started treating profiling as a fourth observability signal (*https://oreil.ly/zlAis*). This is because profiling, in many ways, is very similar to the previously discussed signals in Chapter 6, like metrics, logging, and tracing. For example, similar to other signals, we need instrumentation and reliable experiments to obtain meaningful data.

We discussed how to write custom instrumentation in "pprof Format" on page 332, and we will go through common existing profilers available in Go runtime. However, it's not enough to be able to fetch profiles about various resource usage in our program—we also need to know how to trigger situations that would give us the information about the efficiency bottleneck we want.

14 For example, the plugin in VSCode (*https://oreil.ly/eaooe*) or GoLand (*https://oreil.ly/YT9cs*).

Fortunately, we already went through "Reliability of Experiments" on page 256 and "Benchmarking Levels" on page 266 that explained reliable experiments. Profiling practices are designed to integrate with our benchmarking process naturally. This enables a pragmatic optimization workflow that fits well in our TFBO loop ("Efficiency-Aware Development Flow" on page 102):

1. We perform a benchmark on the desired level (micro, macro, or production) to assure the efficiency of our program.
2. If we are not happy with the result, we can rerun the same benchmark while also capturing the profile during or at the end of the experiment to find the efficiency bottleneck.

Always-On Profiling

You can design your workflow to not need to rerun the benchmark for profiling capturing. In "Microbenchmarks" on page 275, I recommended always capturing your profiles on most of your Go benchmarks. In "Continuous Profiling" on page 373, you will learn how to profile continuously at macro or production levels!

Having instrumentation and the right experiment (reusing benchmarks) is great. Still, we also need to learn how to trigger and transfer the profile from the instrumentation of your choice to analysis with the tools you learned in "go tool pprof Reports" on page 340.

We need to know the API for the profiler we want to use for that purpose. As we learned in Chapter 6, similar to other signals, we generally have two main types of instrumentation: autoinstrumentation and manual. Regarding the former model, there are many ways to obtain profiles about our Go program without adding a single line of code! With technology like eBPF (*https://oreil.ly/8mqs6*), we can have instrumentation for virtually any resource usage of our Go program. Many open source projects, start-ups, or established vendors are on the mission to make this space accessible and easier to use.

However, everything is a trade-off. The eBPF is still early technology that works only on Linux. It has some portability challenges across Linux kernel versions and nontrivial maintainability costs. It is also usually a generic solution that will never have the same reliability and ability to provide semantic, application-level profiles as we can now with more manual, in-process profilers. Finally, this is a Go programming language book, so I would love to share how to create, capture, and use native inprocess profilers.

The API for using instrumentation depends on the implementation. For example, you can write a profiler that will save a profile on a disk every minute or every time some event occurs (e.g., when a certain Linux signal is captured (*https://oreil.ly/xCW7u*)). However, generally in the Go community, we can outline three main patterns of triggering and saving profiles:

Programmatically triggered
: Most profilers you will see and use in Go can be manually inserted into your code to save profiles when you want. This is what I used in Example 9-2 to capture the *fd.pprof* file we were analyzing in "go tool pprof Reports" on page 340. The typical interface has a signature similar to the `WriteTo(w io.Writer) error` (used in Example 9-1) that captures samples that were recorded from the beginning of the program run. The profile in `pprof` format is then written to a writer of your choice (typically a file).

  Some profilers set an explicit starting point when the profiler starts recording samples. This is true, for example, for the CPU profiler (see "CPU" on page 367) that has a signature like `StartCPUProfile(w io.Writer) error` to start the cycle, and then `StopCPUProfile()` to end the profiling cycle.

  This pattern of using the profiles is great for quick tests in the development environment or when used in the microbenchmarks code (see "Microbenchmarks" on page 275). Usually, however, developers don't use it directly. Instead, they often use it as a building block for two other patterns: Go benchmark integrations and HTTP handlers:

Go benchmark integrations
: As presented in an example command I typically use for Go benchmarks in Example 8-4, you can fetch all standard profiles from a microbenchmark by specifying flags in the `go test` tool. Almost all profiles explained in "Common Profile Instrumentation" on page 360 can be enabled using the `-memprofile`, `-cpuprofile`, `-blockprofile`, and `-mutexprofile` flags. No need to put custom code into your benchmark unless you want to trigger the profile at a certain moment. There's no support for custom profiles at the moment.

HTTP handlers
: Finally, an HTTP server is the most common way to capture profiles for programs at macro and production levels. This pattern is especially useful for backend Go applications, which by default accept HTTP connections for normal use. It's then fairly easy to add special HTTP handlers for profiling and other monitoring functionalities (e.g., the Prometheus `/metrics` endpoint). Let's explore this pattern next.

The standard Go library provides HTTP server handlers for all profilers using the `pprof.Profile` structure, for example, our Example 9-1 profiler or any of the standard profiles explained in "Common Profile Instrumentation" on page 360. You can add these handlers to your `http.Server` in a few code lines in your Go program, as presented in Example 9-5.

Example 9-5. Creating the HTTP server with debug handlers for custom and standard profilers

```
import (
    "net/http"
    "net/http/pprof"

    "github.com/felixge/fgprof"
)

// ...

m := http.NewServeMux() ❶
m.HandleFunc("/debug/pprof/", pprof.Index) ❷
m.HandleFunc("/debug/pprof/profile", pprof.Profile) ❸
m.HandleFunc("/debug/fgprof/profile", fgprof.Handler().ServeHTTP) ❹

srv := http.Server{Handler: m}

// Start server...
```

❶ The `Mux` structure allows registering HTTP server handlers on specific HTTP paths. Importing `_ "net/http/pprof"` will register standard profiles in the default global mux (`http.DefaultServeMux`) by default. However, I always recommend creating a new empty `Mux` instead of using a global one to be explicit for what paths you are registering. That's why I register them manually in my example.

❷ The `pprof.Index` handler exposes a root HTML index page that lists quick statistics and links to profilers registered using `pprof.NewProfile`. An example view is presented in Figure 9-8. Additionally, this handler forwards to each profiler referenced by name; for example, `/debug/pprof/heap` will forward to the heap profiler (see "Heap" on page 360). Finally, this handler adds links to `cmdline` and `trace` handlers, which provides further debugging capabilities, and to the `profile` registered line below.

❸ The standard Go CPU is not using `pprof.Profile`, so we have to register that HTTP path explicitly.

❹ The same profile-capturing method can be used for third-party profilers, e.g., the profiler for "Off-CPU Time" on page 369 called `fgprof`.

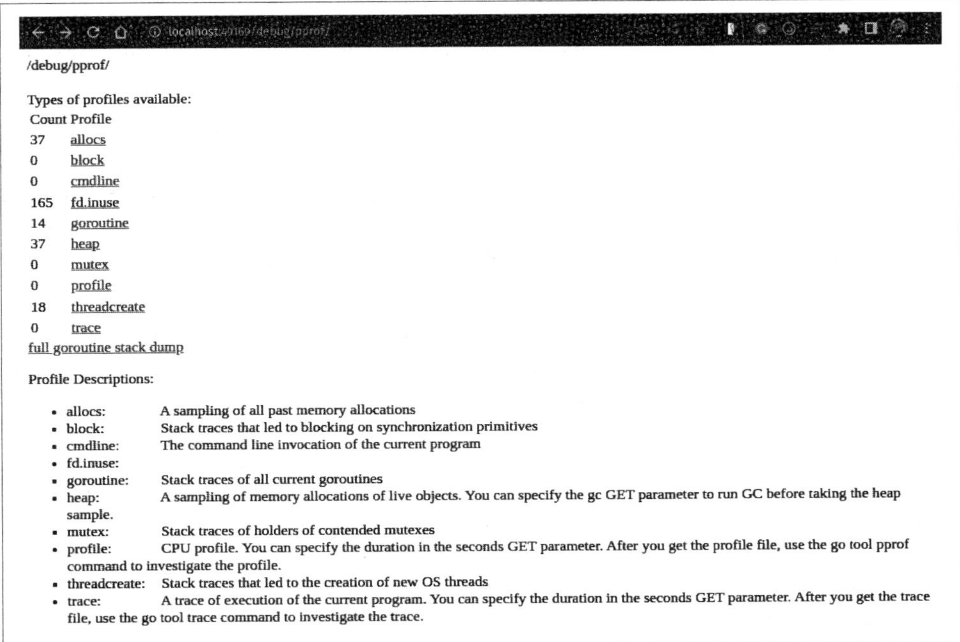

/debug/pprof/
Types of profiles available:
Count Profile
37 allocs
0 block
0 cmdline
165 fd.inuse
14 goroutine
37 heap
0 mutex
0 profile
18 threadcreate
0 trace
full goroutine stack dump
Profile Descriptions:
allocs: A sampling of all past memory allocations
block: Stack traces that led to blocking on synchronization primitives
cmdline: The command line invocation of the current program
fd.inuse:
goroutine: Stack traces of all current goroutines
heap: A sampling of memory allocations of live objects. You can specify the gc GET parameter to run GC before taking the heap sample.
mutex: Stack traces of holders of contended mutexes
profile: CPU profile. You can specify the duration in the seconds GET parameter. After you get the profile file, use the go tool pprof command to investigate the profile.
threadcreate: Stack traces that led to the creation of new OS threads
trace: A trace of execution of the current program. You can specify the duration in the seconds GET parameter. After you get the trace file, use the go tool trace command to investigate the trace.

Figure 9-8. The served HTML page from the debug/pprof/ *path of the server created in Example 9-5*

The index page is nice to have if you forget what name the profiler uses or what profilers you have available in your Go program. Notice that our custom Example 9-1 profiler is also on this list (`fd.inuse` with 165 files[15]), because it was created using `pprof.NewProfile`. For programs that do not import the `fd` package that has the code presented in Example 9-1, this index page would miss the `fd.inuse` line.

A nice debugging page is not the primary purpose of the HTTP handlers. Their fundamental benefit is that a human operator or automation can dynamically capture the profiles from outside, triggering them in the most relevant moments of the macro test, incident, or normal production run. In my experience, I have found four ways of using the profilers via the HTTP protocol:

- You can click on the link for the desired profiler in the HTML page visible in Figure 9-8, for example, `heap`. This will open the `http://<address>/debug/pprof/heap?debug=1` URL that prints the count of samples per stack trace in the current moment—a simplified memory profile in text format.

15 Funny enough, the 165 number is excessive. Making this screenshot gave me the insight that I have a bug in the `labeler` code. I was not closing the temporary file.

- Removing the `debug` parameter will download the desired profile in `pprof` format; e.g., the `http://<address>/debug/pprof/heap` URL in the browser will download the memory profile explained in "Heap" on page 360 to a local file. You can then open this file using `go tool pprof`, as I explained in "go tool pprof Reports" on page 340.
- You can point the `pprof` tool directly to the profiler URL to avoid the manual process of downloading the file. For example, we can open a web profiler viewer for a memory profile if we run in our terminal `go tool pprof -http :8080 http://<address>/debug/pprof/heap`.
- Finally, we can use another server to collect those profiles to a dedicated database periodically, e.g., using the Phlare (*https://oreil.ly/Ru0Hu*) or Parca (*https://oreil.ly/2PKkx*) projects explained in "Continuous Profiling" on page 373.

To sum up, use whatever you find more convenient for the program you are analyzing. Profiling is great for understanding the efficiency of complex production applications in a microservice architecture, so the pattern of the HTTP API for capturing profiles is usually what I use. The Go benchmark profiling is perhaps the most useful for the micro level. The mentioned access patterns are commonly used in the Go community, but it doesn't mean you can't innovate and write the capturing flow that will fit to your workflow better.

To explain the view types in "go tool pprof Reports" on page 340, `pprof` format, and custom profilers, I created the simplest possible file descriptor profiling instrumentation (Example 9-1). Fortunately, we don't need to write our instrumentation to have robust profiling for common machine resources. Go comes with a few standard profilers, well maintained and used by the community and users worldwide. Plus, I will mention a useful bonus profiler from the open source community. Let's unpack those in the next section.

Common Profile Instrumentation

In Chapters 4 and 5, I explained two main resources we have to optimize for—CPU time and memory. I also discussed how those could impact latency. The whole space can be intimidating at first, given the complexity and the concern given in "Reliability of Experiments" on page 256. This is why it's critical to understand what common profiling implementations Go has and how to use them. We will start with heap profiling.

Heap

The `heap` profile, also sometimes referred to as the `alloc` profile, provides a reliable way to find the main contributors of the memory allocated on the heap (explained in

"Go Memory Management" on page 172). However, similar to the `go_memstats_heap` metric mentioned in "Memory Usage" on page 234, it only shows memory blocks allocated on the heap, not memory allocated on stack, or custom `mmap` calls. Still, the heap part of Go program memory usually causes the biggest problem; thus, the heap profile tends to be very useful, in my experience.

You can redirect the heap profile to `io.Writer` using `pprof.Lookup("heap").WriteTo(w, 0)` (*https://oreil.ly/kMjqJ*), with `-memprofile` on Go benchmark, or by calling the `/debug/pprof/heap` URL with handlers, as in Example 9-5.[16]

The memory profiler has to be efficient for it to be feasible for practical purposes. That's why the `heap` profiler is sampled and deeply integrated with the Go runtime allocator flow (*https://oreil.ly/NF1ni*) that is responsible for allocating values, pointers, and memory blocks (see "Values, Pointers, and Memory Blocks" on page 176). The sampling can be controlled by the `runtime.MemProfileRate` variable (*https://oreil.ly/iJaAU*) (or the `GODEBUG=memprofilerate=X` environment variable) and is defined as the average number of bytes that must be allocated to record a profile sample. By default, Go records a sample per every 512 KB of allocated memory on the heap.

What Memory Profile Rate Should You Choose?

I would recommend not changing the default value of 512 KB. It is low enough for practical bottleneck analysis for most Go programs, and cheap enough so we can always have it on.

For more detailed profiling values or to optimize a smaller size of allocations on the critical path, consider changing it to one byte to record all allocations in your program. However, this can impact your application's latency and CPU time (which will be visible on the CPU profile). Still, it might be fine for your memory-focused benchmark.

If you have multiple allocations in a single function, it is often useful to analyze the heap profile in `lines` granularity (add the `&g=lines` URL parameter in the web viewer). An example heap profile of `labeler` in the `e2e` framework (see "Go e2e Framework" on page 310) is presented in Figure 9-9.

16 The same profile is also available via `/debug/pprof/alloc`. The only difference is that the `alloc` profile has `alloc_space` as the default value type.

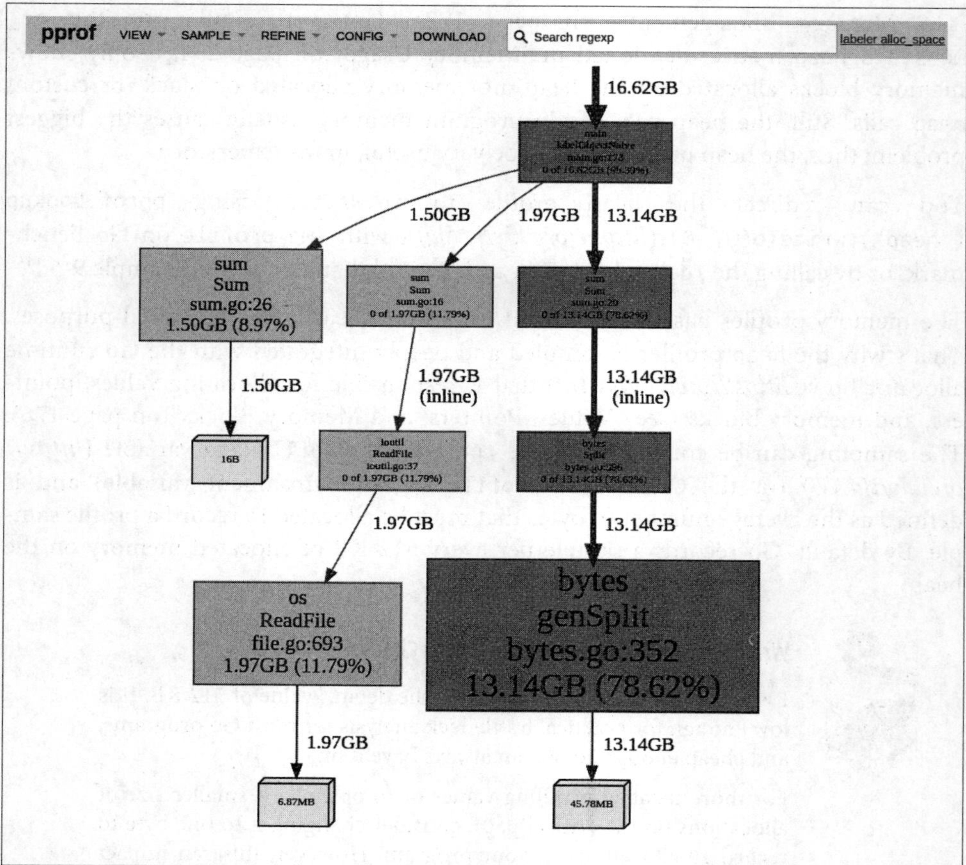

Figure 9-9. The zoomed-in Graph view for the `heap` profile from the `labeler` `Sum` code from Example 4-1 in `alloc_space` dimension and `lines` granularity

The unique aspect of the `heap` profile is that it has four value (sample) types, which you can choose in a new SAMPLE menu item. The currently selected value type is presented in the top right-hand corner. Each type is useful in a different way:

`alloc_space`
: In this mode, the sample value means a total number of allocated bytes by location on the heap since the start of your program. This means that we will see all the memory that was allocated in the past, but most likely is already released by the garbage collection.

Don't be surprised to see huge values here! For example, if the program runs for a longer time and one function allocates 100 KB every minute, it means ~411 GB after 30 days. This looks scary, but the same application might just use a maximum of 10 MB of physical memory during those 30 days.

The total historical allocations are great to see in the code that in total allocated the largest amount of bytes in the past, which can lead to problems with the maximum memory used by that program. Even if the allocations made by certain locations were small but very frequent, it might be caused by the impact of the garbage collection (see "Garbage Collection" on page 185). The `alloc_space` is also very useful for spotting past events that allocated large space.

For example, in Figure 9-9 we see 78.6% of cumulative memory used by the `bytes.Split` function. This knowledge will be extremely valuable in the example in "Optimizing Memory Usage" on page 395. As we already saw in "Go Benchmarks" on page 277, the number of allocations is way larger than the dataset, so there must be a way to find a less expensive memory solution to splitting a string into lines.

Resetting Cumulative Allocations

We can't reset the heap profiler programmatically, for example, to start recording allocations from a certain moment.

However, as you will learn in "Comparing and Aggregating Profiles" on page 378, we can perform operations like subtracting the `pprof` values. So for example, we can capture the heap profile at moment A, then 30 seconds later at moment B, and create a "delta" heap profile that will show what allocation happened during those 30 seconds.

There is also a hidden feature for Go `pprof` HTTP handlers. When capturing the `heap` profile, you can add a `seconds` parameter! For example, with Example 9-5 you can call `http://<address>/debug/pprof/heap?seconds=30s` to remotely capture a delta heap profile!

`alloc_objects`
: Similar to `alloc_space`, the value tells us about the number of allocated memory blocks, not the actual space. This is mainly useful for finding the latency bottlenecks caused by frequent allocations.

`inuse_space`
: This mode shows the currently allocated bytes on the heap—the allocated memory minus the released memory at each location. This value type is great for cases when we want to find the memory bottleneck in a specific moment of the program.[17]

  Finally, this mode is excellent for finding memory leaks. The memory that was constantly allocated and never released will stand out in the profile.

Finding the Source of the Memory Leaks

The `heap` profile shows the code that allocated memory blocks, not the code (e.g., variables) that currently reference those memory blocks. To discover the latter, we could use the `view core` utility (*https://oreil.ly/c4rGl*) that analyzes the currently formed heap. This is, however, not trivial.

Instead, try to statically analyze the code path first to find where the created structures might be referred. But even before that, check the `goroutine` profile in the next section first. We will discuss this problem in "Don't Leak Resources" on page 426.

`inuse_objects`
: The value shows the current number of allocated memory blocks (objects) on the heap. This is useful to reveal the amount of live objects on the heap, which represents well the amount of work for garbage collection (see "Garbage Collection" on page 185). Most of the CPU-bound work of garbage collection is in the mark phase that has to traverse through objects in a heap. So the more we have, the larger the negative impact allocation might be.

Knowing how to use `heap` profiles is a must-have skill for every Go developer interested in the efficiency of their programs. Focus on the code with the biggest contribution of allocations space. Don't worry about the absolute numbers that might not correlate with the memory you use with other observability tools (see "Memory Usage" on page 234). With higher memory profile rates, you see only a portion of the allocations that statically matter.

---

17 Unfortunately, given I took the snapshot when the load test finished, the current amount of spaces contributed by code toward the heap is minimal and does not represent any interesting event that happened in the past. You will see this value type being more useful in "Continuous Profiling" on page 373.

Goroutine

The `goroutine` profiler can show us how many goroutines are running and what code they are executing. This includes all goroutines waiting on I/O, locks, channels, etc. There is no sampling for this profile—all goroutines except system goroutines (*https://oreil.ly/bg2fB*) are always captured.[18]

Similar to the `heap` profile, we can redirect this profile to `io.Writer` using `pprof.Lookup("goroutine").WriteTo(w, 0)`, with `-goroutineprofile` on Go benchmark, or by calling the `/debug/pprof/goroutine` URL with handlers, as in Example 9-5. The overhead of capturing a `goroutine` profile can be significant for Go programs with a larger number of goroutines or when you care about every 10 ms of your program latency.

The key value of the goroutine profile is to give you an awareness of what most of your code goroutines are doing. In some cases, you might be surprised how many goroutines your program requires to fulfill some functionality. Seeing a large (and perhaps increasing) number of goroutines doing the same thing might indicate a memory leak.

Remember that, as mentioned in Figure 9-3, for Go developers, by design, there is no link between the new goroutine and the goroutine that created it.[19] For this reason, the root location we see in the profile is always the first statement or function where the goroutine is called.

The example Graph view for our `labeler` program is presented in Figure 9-10. We can see that `labeler` does not do a lot. In the zoom-out view, we can see there are only 13 goroutines, and none of the locations are the application logic—only profiler goroutine, signal goroutine, and a few HTTP server ones polling connection bytes. This indicates that perhaps the server is waiting on a TCP connection for the incoming request.

18 See the excellent goroutine profiler overview (*https://oreil.ly/U8tCN*).

19 Technically speaking, the Go scheduler records that information (*https://oreil.ly/g3tl2*). It can be exposed to us when stack is retrieved with `GODEBUG=tracebackancestors=X`.

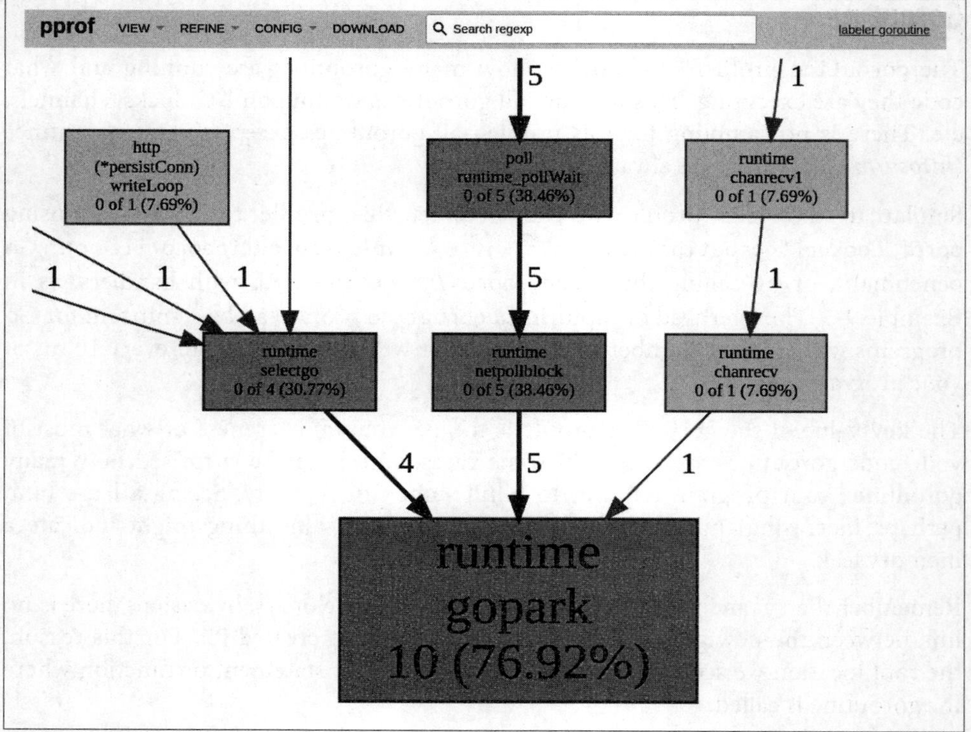

Figure 9-10. The zoomed-in Graph view for the `goroutine` profile from the `labeler` `Sum` code from Example 4-1

Still, Figure 9-10 makes you aware of a few common functions you can typically find in the goroutine view:

`runtime.gopark`
: The `gopark` (*https://oreil.ly/Zqf2K*) is an internal function that keeps the goroutine waiting for the state until an external callback will get it back to work. Essentially it is a way for the runtime scheduler to pause (park) goroutines when they are waiting for things a bit longer—for example, channel communication, network I/O, or sometimes mutex locks.

`runtime.chanrecv` *and* `runtime.chansend`
: As the name suggests, a goroutine in the `chanrecv` function is receiving messages or waiting for something to be sent in the channel. Similarly, it is in `chansend` if it is sending a message or waiting for the channel to have a buffer room.

`runtime.selectgo`
: You will see this if the goroutine is waiting or checking cases in the `select` statement (*https://oreil.ly/T52Kg*).

`runtime.netpollblock`
: The `netpoll` function (*https://oreil.ly/5Iw71*) sets the goroutine to wait until the I/O bytes are received from the network connection.

As you can see, it's fairly easy to track the functions' meaning, even if you are seeing them in your profile for the first time.

CPU

We profile the CPU to find the parts of code that use CPU time the most. Reducing that allows us to reduce the cost of running our program and enable easier system scalability. For the CPU-bound programs, shaving some CPU usage also means reduced latency.

Profiling the CPU usage is proven to be very hard. The first reason for this is that the CPU just does a lot in a single moment—the CPU clocks can perform billions of operations per second. Understanding the full distribution of all the cycles across our program code is hard to track without slowing down significantly. The multi-CPU core programs make this problem even harder.

At the time of writing this book, Go 1.19 provides a CPU profiler integrated into the Go runtime. Any CPU profiler adds some overhead, so it can't just run in the background. We have to start and stop it for the whole process explicitly. Like other profilers, we can do that programmatically through the `pprof.StartCPUProfile(w)` and `pprof.StopCPUProfile()` functions. We can use the `-cpuprofile` flag on Go benchmark or the `/debug/pprof/profile?seconds=<integer>` URL with handlers in Example 9-5.

CPU Profile Has Its Start and End

Don't be surprised if the `profile` HTTP handler does not return the response immediately, as with other profiles! The HTTP handler will start the CPU profiler, run it for the number of seconds provided in the `seconds` parameter (30 seconds if not specified), and only then return the HTTP request.

The current implementation is heavily sampled. When the profiler starts, it schedules the OS-specific timers to interrupt the program execution at the specified rate. On Linux, this means using either `settimer` (*https://oreil.ly/tQNJK*) or `timer_create` (*https://oreil.ly/WdjVW*) to set up timers for each OS thread, and in the Go runtime, listening for the `SIGPROF` (*https://oreil.ly/dcQTf*) signal. The signal interrupts the Go runtime, which then obtains the current stack trace of the goroutine executing on that OS thread. The sample is then queued into a pre-allocated ring buffer, which is then scraped by the `pprof` writer every 100 milliseconds.[20]

The CPU profiling rate is currently hardcoded[21] to 100 Hz, so it will record, in theory, one sample from each OS thread every 10 ms of the CPU time (not real time). There are plans (*https://oreil.ly/VXEPO*) to make this value configurable in the future.

Despite the CPU profile being one of the most popular efficiency workflows, it's a complex problem to solve. It will serve you well for the typical cases, but it's not perfect. For example, there are known problems on some OSes like the BSD[22] and various inaccuracies in some specific cases (*https://oreil.ly/Ar8Up*). In the future, we might see some improvements in this space, with new proposals (*https://oreil.ly/zDSEq*) being currently considered that use hardware-based performance monitor units (PMUs) (*https://oreil.ly/75AHf*).

The example CPU profile showing the distribution of CPU time taken by each function for the `labeler` is presented in Figure 9-11. Given the inaccuracies from the lower sampling rate, the function granularity view might lead to better conclusions.

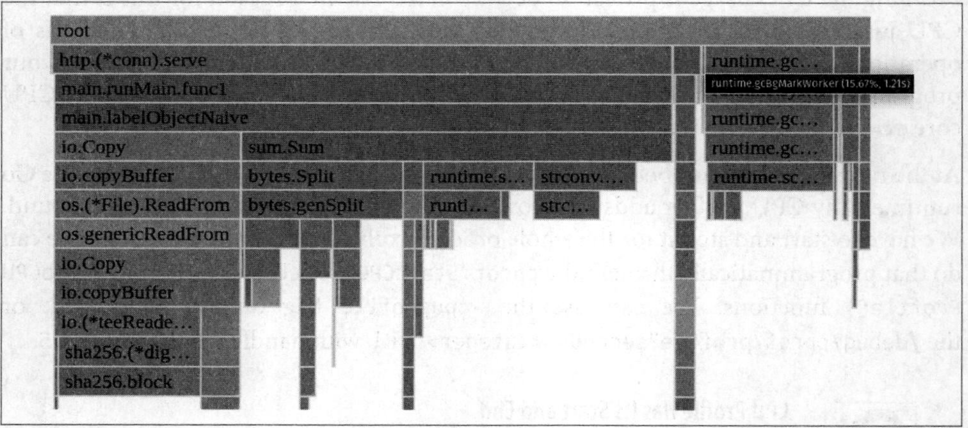

Figure 9-11. The Flame Graph view for the 30-second CPU profile from the `labeler` `Sum` code from Example 4-1 at `functions` granularity

20 See the proposal for the next iteration of the potential CPU profiler (*https://oreil.ly/8vy83*) for a detailed description.

21 Technically speaking, there is one very hacky way of setting different profiling CPU rates. You can call `runtime.SetCPUProfileRate()` (*https://oreil.ly/M8HwB*) with the rate you want right before `pprof.StartCPUProfile(w)`. The `pprof.StartCPUProfile(w)` will try to override the rate, but it will fail due to the bug (*https://oreil.ly/8JBxX*). Change the rate only if you know what you are doing—100 Hz is usually a good default. Values higher than 250–500 Hz are not supported by most of the OS timers anyway.

22 See this issue (*https://oreil.ly/E0W5v*) for a currently known list of OSes with certain problems.

The CPU profile comes with two value types:

Samples
: The sample value indicates the number of samples observed at the location.

CPU
: Each sample value represents the CPU time.

From Figure 9-11, we can see what we have to focus on if we want to optimize CPU time or latency caused by the amount of work by our `labeler` Go program. From the Flame Graph view, we can outline five major parts:

`io.Copy`
: This function used by the code responsible for copying the file from local object storage takes 22.6% of CPU time. Perhaps we could utilize local caching to save that CPU time.

`bytes.Split`
: This splits lines in Example 4-1 and takes 19.69%, so this function might be checked if there is any way we can split it into lines with less work.

`gcBgMarkWorker`
: This function takes 15.6%, which indicates there was a large number of objects alive on the heap. Currently, the GC takes some portion of CPU time for garbage collection.

`runtime.slicebytetostring`
: It indicates a nontrivial amount of CPU time (13.4%) is spent converting bytes to string. Thanks to the Source view, I could track it to `num, err := strconv.ParseInt(string(line), 10, 64)` line. This reveals a straightforward optimization of trying to come up with a function that parses integers directly from the byte slice.

`strconv.ParseInt`
: This function uses 12.4% of CPU. We might want to check if there is any unnecessary work or checks we could remove by writing our parsing function (spoiler: there is).

Turns out, such a CPU profile is valuable even if it is not entirely accurate. We will try the mentioned optimizations in "Optimizing Latency" on page 383.

Off-CPU Time

It is often forgotten, but the typical goroutines mostly wait for work instead of executing on the CPU. This is why when looking to optimize the latency of our program's functionality, we can't just look at CPU time.[23] For all programs, especially the I/O-bound ones, your process might take a lot of time sleeping or waiting. Specifically, we can define four categories that compose the entire program execution, presented in Figure 9-12.

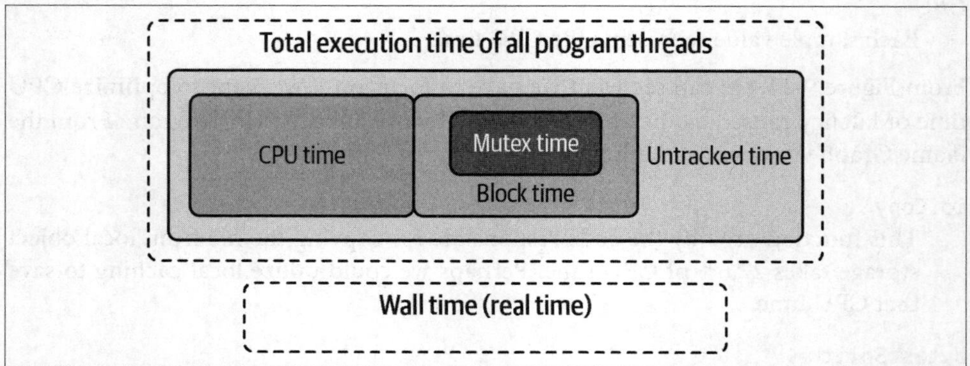

Figure 9-12. The process execution time composition[24]

The first observation is that the total execution time is longer than the wall time, so real time elapsed when executing this program. It's not because computers can slow time somehow; it's because all Go programs are multithreaded (or even multigoroutines in Go), so the total measured execution time will always be longer than real time. We can outline four categories of execution time:

CPU time
: The time our program actively spent using CPU, as explained in "CPU" on page 367.

Block time
: The mutex time, plus the time our process spent waiting for Go channel communication (e.g., `<-ctx.Done()`, as discussed in "Go Runtime Scheduler" on page 138), so all synchronization primitives. We can profile that time using the `block` profiler. It's not enabled by default, so we need to turn it on by setting a nonzero block profiling rate using `runtime.SetBlockProfileRate(int)` (*https://oreil.ly/GwjwY*). This specifies the number of nanoseconds spent blocked for one blocking event sample. Then we can use `pprof.Lookup` in Go, `-blockprofile` in Go benchmark, or the `/debug/pprof/block` HTTP handler to capture `contention` and `delay` value types.

23 In fact, even CPU time includes waiting for a memory fetch, as discussed in "CPU and Memory Wall Problem" on page 126. This is, however, included in the CPU profile.

24 This view is heavily inspired by the Felix's great guide (*https://oreil.ly/nwVwF*).

Mutex time

The time spent on lock contentions (e.g., the time spent in `sync.RWMutex.Lock` (*https://oreil.ly/chnpS*)). Like block profile, it's disabled by default and can be enabled with `runtime.SetMutexProfileFraction(int)` (*https://oreil.ly/olg45*). Fraction specifies that `1/`*`<fraction>`* lock contentions should be tracked. Similarly, we can use `pprof.Lookup` in Go, `-mutexprofile` in Go benchmark, or the `/debug/pprof/mutex` HTTP handler to capture `mutex` and `delay` value types.

Untracked off-CPU time

The goroutines that are sleeping, waiting for CPU time, I/O (e.g., from disk, network, or external device), syscalls, and so on are not tracked by any standard profiling tool. To discover the impact of that latency, we need to use different tools as explained next.

Do We Have to Measure or Find Bottlenecks in Off-CPU Time?

Program threads spend a lot of time off-CPU. This is why the main reason your program is slow might not be its CPU time. For example, suppose the execution of your program takes 20 seconds, but it waits 19 seconds on an answer from the database. In that case, we might want to look at bottlenecks in the database (or mitigate the database slowness in our code) instead of optimizing the CPU time.

Generally, it is recommended to use tracing to find the bottlenecks in the wall time (latency) of our functionality. Especially, distributed tracing allows us to narrow down our optimization focus to what takes the most time in the request of functionality flow. Go has built-in tracing instrumentation (*https://oreil.ly/pKeI1*), but it only instruments Go runtime, not our application code. However, we discussed basic tracing instrumentation compatible with the cloud-native standards like OpenTelemetry (*https://oreil.ly/sPiw9*) to achieve application-level tracing.

There is also an amazing profiler called the Full Go Profiler (`fgprof`) (*https://oreil.ly/4WWHN*) out there focused on tracking both CPU and off-CPU time. While it's not (*https://oreil.ly/ri1Kb*) officially recommended yet and has known limitations (*https://oreil.ly/8Lk9t*), I found it very useful, depending on what kind of Go program I analyze. The `fgprof` profile can be exposed using the HTTP handler mentioned in Example 9-5. The example view of the `fgprof` profile for `labeler` service is presented in Figure 9-13.

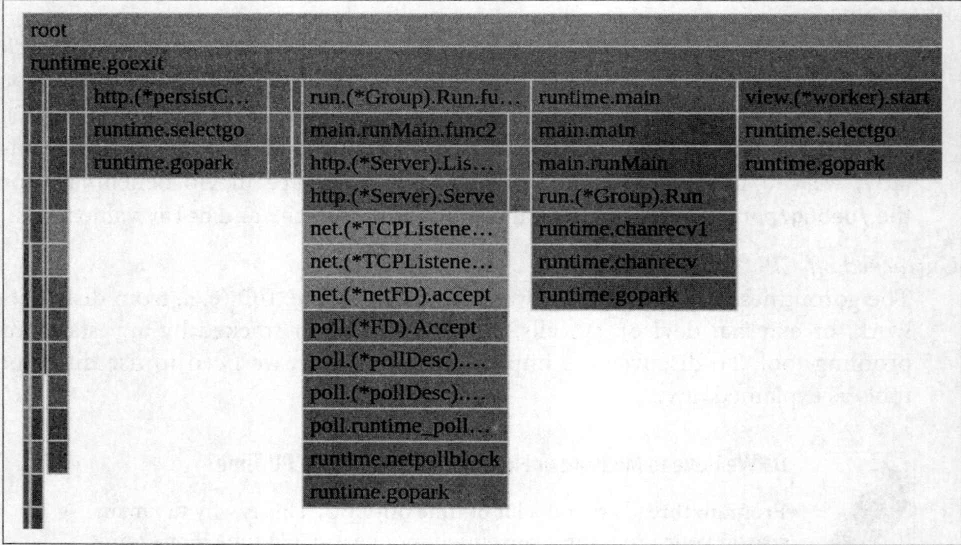

Figure 9-13. The Flame Graph view for the 30-seconds-`fgprof` profile from the `labeler` `Sum` code from Example 4-1 at `functions` granularity

From the profile, we can quickly tell that for most of the wall time, the `labeler` service is simply waiting for the signal interrupt or HTTP requests! If we are interested in improving the maximum rate of the incoming requests that `labeler` can serve, we can quickly find that `labeler` is not the problem, but rather the testing client is not sending requests fast enough.[25]

To sum up, in this section, I presented the most common profiler implementations that are used[26] in the Go community. There are also tons of closed-box monitoring profilers like Linux `perf` and `eBPF`-based profiles, but they are outside the scope of this book. I prefer the ones I mentioned as they are free (open source!), explicit, and relatively easy to use and understand.

Let's now look at some lesser-known tools and practices I found useful when profiling Go programs.

---

25 This can be confirmed in Example 8-19 code, where the `k6s` script has only one user that waits 500 ms between HTTP calls.

26 I skipped the `threadcreate` profile present in the Go `pprof` package as it's known to be broken since 2013 (*https://oreil.ly/b8MpS*) with little priority to be fixed in the future.

Tips and Tricks

There are three more advanced yet incredibly useful tricks for profiling I would love you to know. These helped me analyze software bottlenecks even more effectively. So let's go through them!

Sharing Profiles

Typically, we don't work on software projects alone. Instead, we are in a bigger team, which shares responsibilities and reviews each other's code. Sharing is caring, so similar to "Sharing Benchmarks with the Team (and Your Future Self)" on page 294, we should focus on presenting our bottlenecks results and findings with team members or other interested parties.

We download or check multiple `pprof` profiles in the typical workflow. In theory, we could name them descriptively to avoid confusion and send them to each other using any file-sharing solution like Google Drive or Slack. This, however, tends to be cumbersome because the recipient has to download the `pprof` file and run `go tool pprof` locally to analyze.

Another option is to share a screenshot of the profile, but we have to choose some partial view, which can be cryptic for others. Perhaps others would like to analyze the profile using a different view or value type. Maybe they want to find the sampling rate or narrow the profile down to some code path. With just a screenshot, you are missing all of those interactive capabilities.

Fortunately, some websites allow us to save `pprof` files for others or our future self and analyze them without downloading that profile. For example, the Polar Signals (*https://oreil.ly/HowVb*) company hosts an entirely free *pprof.me* (*https://pprof.me*) website that allows exactly that. You can upload your profile (note that it will be shared publicly!) and share the link with team members, who can analyze it using common `go tools pprof` reports views (see "go tool pprof Reports" on page 340). I use it all the time with my team.

Continuous Profiling

In the open source ecosystem, continuous profiling was perhaps one of the most popular topics in 2022. It means automatically collecting useful profiles from our Go program at every configured interval instead of being manually triggered.

In many cases, the efficiency problem happens somewhere in the remote environment where the program is running. Perhaps it happened in the past in response to some event that is now hard to reproduce. Continuous profiling tools allow us to have our profiling "always on" and retrospectively look at profiles from the past.

> Say you see an increase in resource usage – say, CPU usage. And then you take a one-time profile to try to figure out what's using more resources. Continuous profiling is essentially doing this all the time. (...) When you have all this data over time, you can compare the entire lifetime of a version of a process to a newly rolled-out version. Or you can compare two different points in time. Let's say there's a CPU or memory spike. We can actually understand what was different in our processes down to the line number. It's super powerful, and it's an extension of the other tools already useful in observability, but it shines a different light on our running programs.
>
> —Frederic Branczyk, "Grafana's Big Tent: Continuous Profiling with Frederic Branczyk" (*https://oreil.ly/Jp9gQ*)

Continuous profiling emerged in the cloud-native open source community as the fourth observability signal, but it's not new. The concept was introduced first in 2010 by the "Google-Wide Profiling: A Continuous Profiling Infrastructure For Data Centers" research paper (*https://oreil.ly/FbHY8*) by Gang Ren et al., which proved that profiling can be used against production workloads continuously without the major overhead, and helped in efficiency optimizations at Google.

We have recently seen open source projects that made this technology more accessible. I have personally used the continuous profiling tool for a couple of years already to profile our Go services, and I love it!

You can quickly set up continuous profiling using the open source Parca project (*https://oreil.ly/X8003*). In many ways, it is similar to the Prometheus project (*https://oreil.ly/2Sa3P*). Parca is a single binary Go program that periodically captures profiles using the HTTP handlers we discussed in "Capturing the Profiling Signal" on page 355 and stores them in a local database. Then we can search for profiles, download them, or even use the embedded `tool pprof` like a viewer to analyze them.

You can use it anywhere: set up continuous profiling on your production, remote environment, or macrobenchmarking environment that might run in the cloud or on your laptop. It might not make sense on a microbenchmarks level, as we run tests in the smallest possible scope, which can be profiled for the full duration of the benchmark (see "Microbenchmarks" on page 275).

Adding continuous profiling with Parca to our `labeler` macrobenchmark in Example 8-19 requires only a few lines of code and a simple YAML configuration, as presented in Example 9-6.

Example 9-6. Starting continuous profiling container in Example 8-19 between `labeler` creation and `k6` script execution

```
labeler := ...

parca := e2e.NewInstrumentedRunnable(e, "parca").
    WithPorts(map[string]int{"http": 7070}, "http").
```

```
    Init(e2e.StartOptions{
        Image: "ghcr.io/parca-dev/parca:main-4e20a666", ❶
        Command: e2e.NewCommand("/bin/sh", "-c",
          `cat << EOF > /shared/data/config.yml && \
    /parca --config-path=/shared/data/config.yml
object_storage: ❷
  bucket:
    type: "FILESYSTEM"
    config:
      directory: "./data"
scrape_configs: ❸
- job_name: "%s"
  scrape_interval: "15s"
  static_configs:
    - targets: [ '`+labeler.InternalEndpoint("http")+`' ]
  profiling_config:
    pprof_config: ❹
      fgprof:
        enabled: true
        path: /debug/fgprof/profile
        delta: true
EOF
`),
        User:      strconv.Itoa(os.Getuid()),
        Readiness: e2e.NewTCPReadinessProbe("http"),
    })
testutil.Ok(t, e2e.StartAndWaitReady(parca))
testutil.Ok(t, e2einteractive.OpenInBrowser("http://"+parca.Endpoint("http"))) ❺

k6 := ...
```

❶ The `e2e` framework (*https://oreil.ly/f0IJo*) runs all workloads in the container, so we do that for the Parca server. We use the container image build from the official project page (*https://oreil.ly/ETsNV*).

❷ The basic configuration of the Parca server has two parts. The first is object storage configuration: where we want to store Parca's database internal data files. Parca uses FrostDB columnar storage (*https://oreil.ly/A9y23*) to store debugging information and profiles. To make it easy, we can use the local filesystem as our most basic object storage.

❸ The second important configuration is the scrape configuration that allows us to put certain endpoints as targets to profile capturing. In our case, I only put the `labeler` HTTP endpoint on the local network. I also specified to get the profile every 15 seconds. For always-on production use, I would recommend larger intervals, e.g., one minute.

❹ The common profile—like a heap, CPU, goroutine block, and mutex—are enabled by default (*https://oreil.ly/pcZmg*). However, we have to manually allow other profiles, like the `fgprof` profile discussed in "Off-CPU Time" on page 369.

❺ Once Parca starts, we can use the `e2einteractive` package to open the Parca UI to explore viewer-like presentations of our profiles during or after the `k6` script finishes.

Thanks to continuously profiling, we don't need to wait until our benchmark (using the `k6` load tester) finishes—we can jump to our UI straightaway to see profiles every 15 seconds, live! Another great thing about continuous profiling is that we can extract metrics from the sum of all sample values taken from each profile over time. For example, Parca can give us a graph of heap memory usage for the `labeler` container over time, taken from periodic `heap inuse_alloc` profiles (discussed in Figure 9-9). The result, presented in Figure 9-14, should have values very close to the `go_memstats_heap_total` metric mentioned in "Memory Usage" on page 234.

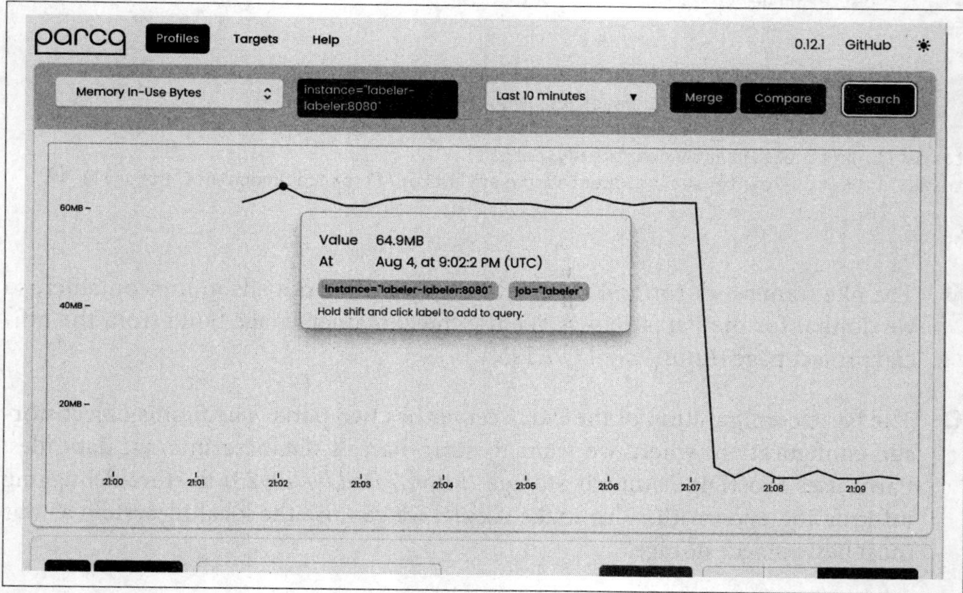

Figure 9-14. Screenshot of Parca UI result showing the `labeler` Figure 9-9 `inuse_alloc` profiles over time

You can now click on samples in the graph, representing the moment of taking the profile snapshot. Thanks to continuous form, you can choose the time that interests you the most, perhaps the moment when the memory usage was the highest! Once clicked, the Flame Graph of that specific profile appears, as presented in Figure 9-15.

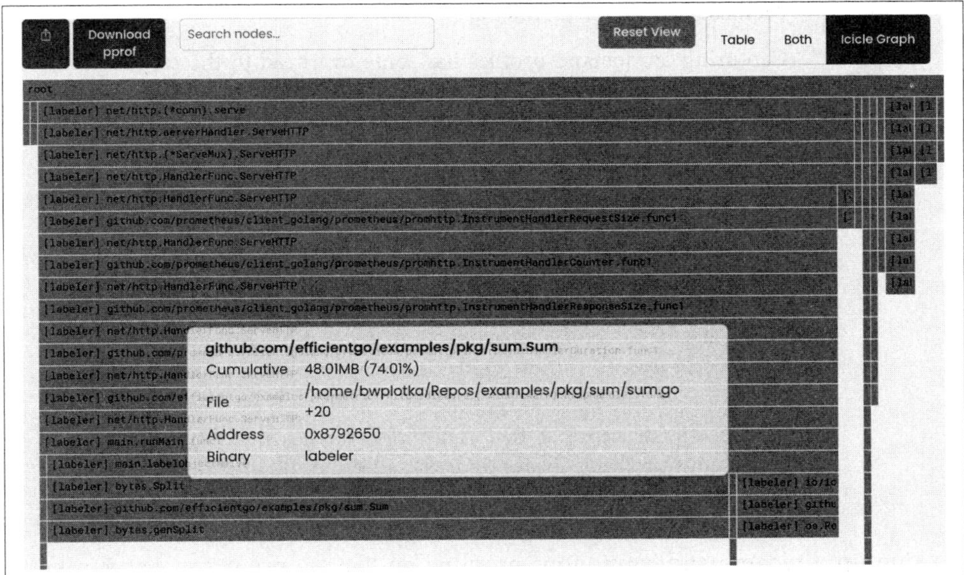

Figure 9-15. Screenshot of Parca UI Flame Graph (called Icicle Graph in Parca) when you click the specific profile from Figure 9-4

The Parca maintainers decided to use a different visual style for the Flame Graphs than the `go tool pprof` tool in "Flame Graph" on page 352. However, as many other tools in the profiling space, it uses the same semantics. This means we can use our analyzing skills from `go tool pprof` specifics with different UIs like Parca.

In the profile view, we can download the `pprof` file we selected. We can share the profile as discussed in "Sharing Profiles" on page 373, filter view, or choose different views. We also see a Flame Graph representing the function's contributions to the live objects in a heap for the selected time. We could not easily capture that manually. In Figure 8-5, I captured the profile after the interesting event happened, so I had to use `alloc_space` that shows the total allocations from when the program started. For a long-living process, this view might be very noisy and show situations that I am not interested in. Even worse, the process might have restarted after certain events, like panics or OOMs. Doing such a heap profile after restart will tell us nothing. A similar problem occurs with every other profile that only shows the current or specific moment, like goroutines, CPUs, or our custom file descriptor profile.

This is where continuous profiling proves to be extremely helpful. It allows us to have profiles captured whenever an interesting event occurs, so we can quickly jump into the UI and analyze for efficiency bottlenecks. For example, in Figure 9-15, we can see the `bytes.Split` as the function that uses the most memory on the heap at the current moment.

Overhead of Continuous Profiling

Capturing on-demand profiles has some overhead to the running Go program. However, capturing multiple profiles periodically makes this overhead continuous throughout the application run, so ensure your profilers do not cause your efficiency to drop below the expected level.

Try to understand the overhead of profiling in your programs. The standard default Go profilers aim to not add more than 5% of the CPU overhead for a single process. You can control that by changing the continuous profiling interval or the sampling of profiles. It is also useful to profile only one of many of the same replicas in large deployments (*https://oreil.ly/yAACa*) to amortize collection cost.

In our infrastructure at Red Hat (*https://oreil.ly/6CSV7*), we run continuous profiling always on with a one-minute interval, and we keep only a few days' worth of profiles.

To sum up, I recommend continuous profiling on live Go programs that you know might need continuous efficiency improvement in the future. Parca is one open source example, but there are other projects or vendors[27] that allow you to do the same. Just be careful, as profiling might be addictive!

Comparing and Aggregating Profiles

The `pprof` format has one more interesting characteristic. By design, it allows certain aggregations or comparison for multiple profiles:

Subtracting profiles

You can subtract one profile from another. This is useful to reduce noise and narrow down to the event or component you care about. For example, you can have a heap profile from one run of your Go program when you load tested simultaneously with some `A` and `B` events. Then, you can subtract the heap second profile you have from the same Go program that was load tested with only the `B` event to check what the impact was purely from the `A` event. The `go tool pprof` allows you to subtract one profile from another using the `-base` flag—for example, `go tool pprof heap-AB.pprof -base heap-B.pprof`.

27 Phlare (*https://oreil.ly/Ru0Hu*), Pyroscope (*https://oreil.ly/eKyK7*), Google Cloud Profiler (*https://oreil.ly/OGoVR*), AWS CodeGuru Profiler (*https://oreil.ly/urVE0*), or Datadog continuous profiler (*https://oreil.ly/El7zq*), to name a few.

Comparing profiles
: The comparison (*https://oreil.ly/NHfZP*) is similar to subtracting; instead of removing matching sample values, it provides negative or positive delta numbers between profiles. This is useful to measure the change of the contribution of a particular function before and after optimization. You can also use `go tool pprof` to compare your profiles using `-diff_base`.

Merging profiles
: It is less known in the community, but you can merge multiple profiles into one! The merging functionality allows us to combine profiles representing the current situation. For example, we could take dozens of short CPU profiles into a single profile of all the CPU work across a longer duration. Or perhaps we could merge multiple heap profiles to the aggregate profile of all heap objects from multiple time points.

  The `go tool pprof` does not support this. However, you can write your own Go program that does it using the `google/pprof/profile.Merge` function (*https://oreil.ly/bvoSL*).

I wasn't using these mechanics very often because I was easily confused with multiple local `pprof` files when working with the `go tool pprof` tool. This changed when I started working with more advanced profiling tools like Parca. As you can see in Figure 9-14, there is a Compare button to compare two particular profiles, and a Merge button to combine all profiles from the focused time range into one profile. With the UI, it is much easier to select what profiles you want to compare or aggregate, and how!

Summary

Profiling space for Go might be nuanced, but it's not that difficult to utilize once you know the basics. In this chapter, we went through all the profiling aspects from the common profilers, through capturing patterns and `pprof` format, to standard visualization techniques. Finally, we touched on advanced techniques like continuous profiling, which I recommend trying.

Profile First, Ask Questions Later

I would suggest using profiling in any shape that fits in your daily optimization workflow. Ask questions like what is causing the slowdown or high resource usage in your code only after you have already captured the profiles from your program.

I believe this is not the end of the innovations in this space. Thanks to common efficient profiling formats like `pprof` that allow interoperability across different tools and profilers, we will see more tools, UI, useful visualizations, or even correlations with different observability signals mentioned in Chapter 6.

Furthermore, more eBPF profiles are emerging in the open source ecosystem, making profiling cheaper and more uniform across programming languages. So be open-minded and try different techniques and tools to find out what works best for you, your team, or your organization.

CHAPTER 10

Optimization Examples

It's finally time to collect all the tools, skills, and knowledge you gathered from the previous chapters and apply some optimizations! In this chapter, we will try to reinforce the pragmatic optimization flow by going through some examples.

We will attempt to optimize the naive implementation of the `Sum` from Example 4-1. I will show you how the TFBO (from "Efficiency-Aware Development Flow" on page 102) can be applied to three different sets of efficiency requirements.

> Optimizations/pessimizations don't generalize very well. It all depends on the code, so measure each time and don't cast absolute judgments.
>
> —Bartosz Adamczewski, Tweet (*https://oreil.ly/oW3ND*) (2022)

We will use our optimization stories as a foundation for some optimization patterns summarized in the next chapter. Learning about thousands of optimization cases that happened in the past is not very useful. Every case is different. The compiler and language change, so any "brute-force" attempt to try those thousands of optimizations one by one is not pragmatic.[1] Instead, I have focused on equipping you with the knowledge, tools, and practices that will let you find a more efficient solution to your problem!

Please don't focus on particular optimizations, e.g., the specific algorithmic or code changes I applied. Instead, try to follow how I came up with those changes, how I found what piece of code to optimize first, and how I assessed the change.

1 For example, I already know about a `strconv.ParseInt` optimization (*https://oreil.ly/IZxm7*) coming to Go 1.20, which would change the memory efficiency of the naive Example 4-1 without any optimization from my side.

We will start in "Sum Examples" by introducing the three problems. Then we will take the `Sum` and perform the optimizations in "Optimizing Latency" on page 383, "Optimizing Memory Usage" on page 395, and "Optimizing Latency Using Concurrency" on page 402. Finally, we will mention some other ways we could solve our goals in "Bonus: Thinking Out of the Box" on page 411. Let's go!

Sum Examples

In Chapter 4, we introduced a simple `Sum` implementation in Example 4-1 that sums large numbers of integers provided in a file.[2] Let's leverage all the learning you have gained and use it to optimize Example 4-1. As we learned in "Resource-Aware Efficiency Requirements" on page 86, we can't "just" optimize—we have to have some goal in mind. In this section, we will repeat the efficiency optimization flow three times, each time with different requirements:

- Lower latency with a maximum of one CPU used
- Minimal amount of memory
- Even lower latency with four CPU cores available for the workload

The terms *lower* or *minimal* are not very professional. Ideally, we have some more specific numbers to aim for, in a written form like a RAER. A quick Big O analysis can tell us that the `Sum` runtime complexity is at least O(*N*)—we have to revisit all lines at least once to compute the sum. Thus, the absolute latency goal, like "`Sum` has to be faster than 100 milliseconds," won't work as its problem space depends on the input. We can always find big enough input that violates any latency goals.

One way to address this is to specify the maximum possible input with some assumptions and latency goals. The second is to define the required runtime complexity as a function that depends on input—so throughput. Let's do the latter and specify the amortized latency function for the `Sum`. We can do the same with memory. So let's be more specific. Imagine that, for my hardware, a system design stakeholder came up with the following required goals for the `Sum` in Example 4-1:

- Maximum latency of 10 nanoseconds per line (10 * *N* nanoseconds) with maximum one CPU used
- Latency as above and a maximum of 10 KB of memory allocated on the heap for any input

2 If you are interested in what input files I used, see the code I used (*https://oreil.ly/0SMxA*) for generating the input.

- Maximum latency of 2.5 nanoseconds per line (2.5 * N nanoseconds) with maximum four CPU used

What If We Can't Match This Goal?

It might be the case that the goals we initially aimed for will be hard to achieve due to underestimation of the problem, new requirements, or new knowledge. This is fine. In many cases, we can try to renegotiate the goals. For example, as we dissected in "Optimization Design Levels" on page 98, every optimization beyond a certain point costs more and more in time, effort, risk, and readability, so it might be cheaper to add more machines, CPUs, or RAM to the problem. The key is to estimate those costs roughly and help stakeholders decide what's best for them.

Following the TFBO flow, before we optimize, we first have to benchmark. Fortunately, we already discussed designs of benchmarks for the `Sum` code in "Go Benchmarks" on page 277, so we can go ahead and use Example 8-13 for our benchmarks. I used the command presented in Example 10-1 to perform 5 10-second benchmarks with a 2 million integer input file and limited to 1 CPU.

Example 10-1. The command to invoke the benchmark

```
export ver=v1 && go test -run '^$' -bench '^BenchmarkSum$' \
    -benchtime 10s -count 5 -cpu 1 -benchmem \
    -cpuprofile=${ver}.cpu.pprof -memprofile=${ver}.mem.pprof | tee ${ver}.txt
```

With Example 4-1, the preceding benchmark yielded the following results: 101 ms, 60.8 MB space allocated, and 1.60 million allocations per operation. Therefore, we will use that as our baseline.

Optimizing Latency

Our requirements are clear. We need to make the `Sum` function in Example 4-1 faster to achieve a throughput of at least 10 * N nanoseconds. The baseline results give us 50 * N nanoseconds. Time to see if there are any quick optimizations!

In “Complexity Analysis” on page 240, I shared a detailed complexity of the `Sum` function that clearly outlines the problems and bottlenecks. However, I used information from this section to define that. For now, let’s forget that we discussed such complexity and try to find all the information from scratch.

The best way is to perform a bottleneck analysis using the profiles explained in Chapter 9. I captured the CPU profile on every benchmark with Example 8-4, so I could quickly bring the Flame Graph of the CPU time, as presented in Figure 10-1.

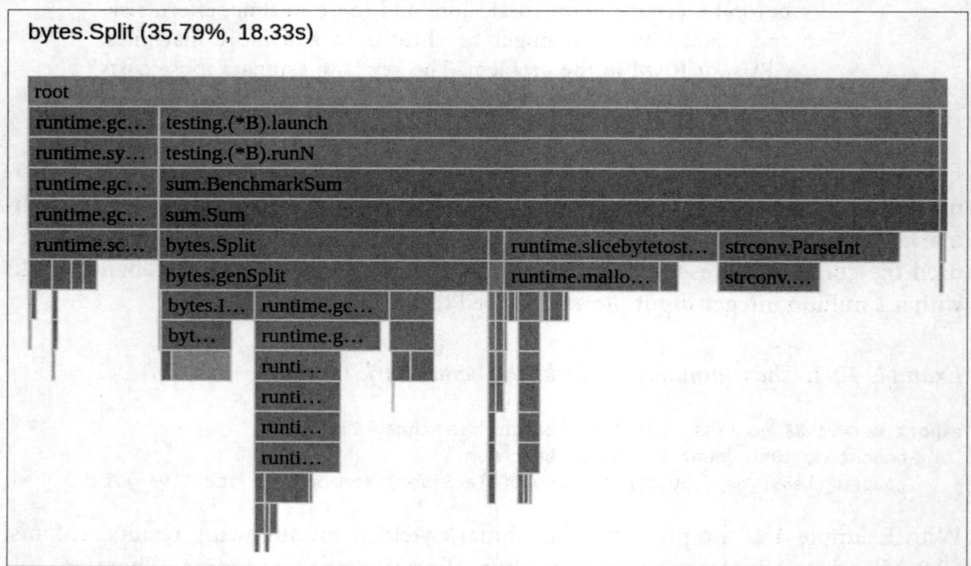

Figure 10-1. Flame Graph view of Example 4-1 CPU time with function granularity

Profiling gives us a great overview of the situation. We see four clear major contributors to the CPU time usage:

- `bytes.Split`
- `strconv.ParseInt`
- Runtime function `runtime.slicebytetostr...`, which ends with `runtime.malloc`, meaning we spent a lot of CPU time allocating memory
- Runtime function `runtime.gcBgMarkWorker`, which indicates GC runs

The CPU profile gives us a list of functions we can go through and potentially cut out some CPU usage. However, as we learned in "Off-CPU Time" on page 369, the CPU time might not be a bottleneck here. Therefore, we must first confirm if our function here is CPU bound, I/O bound, or mixed.

One way of doing this is by manually reading the source code. We can see that the only external medium used in Example 4-1 is a file, which we use to read bytes from. The rest of the code should only perform computations using the memory and CPU.

This makes this code a mixed-bound job, but how mixed? Should we start with file reads optimization or CPU time?

The best way to find this out is the data-driven way. Let's check both CPU and off-CPU latency thanks to the full goroutine profile (`fgprof`) discussed in "Off-CPU Time" on page 369. To collect it in the Go benchmark, I quickly wrapped our benchmark from Example 8-13 with the `fgprof` profile in Example 10-2.

Example 10-2. Go benchmark with fgprof profiling

```
// BenchmarkSum_fgprof recommended run options:
// $ export ver=v1fg && go test -run '^$' -bench '^BenchmarkSum_fgprof' \
//    -benchtime 60s  -cpu 1 | tee ${ver}.txt ❶
func BenchmarkSum_fgprof(b *testing.B) {
    f, err := os.Create("fgprof.pprof")
    testutil.Ok(b, err)

    defer func() { testutil.Ok(b, f.Close()) }()

    closeFn := fgprof.Start(f, fgprof.FormatPprof)
    BenchmarkSum(b) ❷
    testutil.Ok(b, closeFn())
}
```

❶ To get more reliable results, we have to measure for longer than five seconds. Let's measure for 60 seconds to be sure.

❷ To reuse code and have better reliability, we can execute the same Example 8-13 benchmark, just wrapped with the `fgprof` profile.

The resulting `fgprof.pprof` profile after 60 seconds is presented in Figure 10-2.

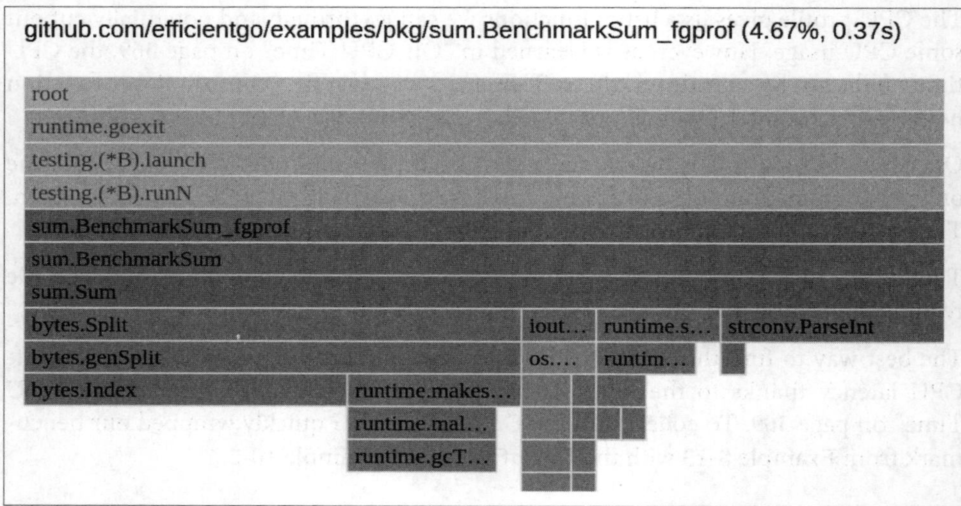

Figure 10-2. Flame Graph view of Example 4-1 CPU and off-CPU time with function granularity

The full goroutine profile confirms that our workload is a mix of I/O (5%[3]) and CPU time (majority). So while we have to worry about latency introduced by file I/O at some point, we can optimize CPU time first. So let's go ahead and focus on the biggest bottleneck first: the `bytes.Split` function that takes almost 36% of the `Sum` CPU time, as seen in Figure 10-1.

Optimize One Thing at a Time

Thanks to Figure 10-1, we found four main bottlenecks. However, I have chosen to focus on the biggest one in our first optimization in Example 10-3.

It is important to iterate one optimization at a time. It feels slower than if we would try to optimize all we know about now, but in practice, it is more effective. Each optimization might affect the other and introduce more unknowns. We can draw more reliable conclusions, e.g., compare the contributions percentage between profiles. Furthermore, why eliminate four bottlenecks if optimizing first might be enough to match our requirements?

3 There is a small segment in Figure 10-2 that shows `ioutil.ReadFile` latency with 0.38% of all samples. When we unfold the `ReadFile`, the `syscall.Read` (which we could assume is an I/O latency) takes 0.25%, given the `sum.BenchmarkSum_fgprof` contributes to 4.67% of overall wall time (the rest is taken by benchmarking and CPU profiling). The (0.25 * 100%)/4.67 is equal to 5.4%.

Optimizing bytes.Split

To figure out where the CPU time is spent in `bytes.Split`, we have to try to understand what this function does and how. By definition (*https://oreil.ly/UqAg8*), it splits a large byte slice into smaller slices based on the potentially multicharacter separator `sep`. Let's quickly look at the Figure 10-1 profile and focus on that function using the `Refine` options. This would show `bytes.Index` (*https://oreil.ly/DQrCS*), and impact allocations and garbage collections with functions like `makeslice` and `runtime.gcWriteBarrierDX`. Furthermore, we could quickly look into the Go source code for the `genSplit` (*https://oreil.ly/pCMH1*) used by `bytes.Split` to check how it's implemented. This should give us a few warning signals. There might be things that `bytes.Split` does but might not be necessary for our case:

- `genSplit` goes through the slices first to count how many slices we expect (*https://oreil.ly/Wq6F4*) to have.
- `genSplit` allocates a two-dimensional byte slice (*https://oreil.ly/YzXdr*) to put the results in. This is scary because for a large 7.2 MB byte slice with 2 million lines, it will allocate a slice with 2 million elements. A memory profile confirms that a lot of memory is allocated by this line.[4]
- Then it will iterate two million times using the `bytes.Index` (*https://oreil.ly/8diMw*) function we saw in the profile. That is two million times we will go and gather bytes until the next separator.
- The separator in `bytes.Split` is a multicharacter, which requires a more complicated algorithm. Yet we need a simple, single-line newline separator.

Unfortunately, such an analysis of the mature standard library functions might be difficult for more beginner Go developers. What parts of this CPU time or memory usage are excessive, and what aren't?

What always helps me to answer this question is to go back to the algorithm design phase and try to design my own simplest splitting-lines algorithm tailored for the `Sum` problem. When we understand what a simple, efficient algorithm could look like and we are happy with it, we can then start challenging existing implementations. It turns out there is a very simple flow that might work for Example 4-1. Let's go through it in Example 10-3.

4 We can further inspect that using "Heap" on page 360 profile, which would in my tests show us that 78.6% of the total 60.8 MB of allocation per operation is taken by `bytes.Split`!

Example 10-3. Sum2 is Example 4-1 with optimized CPU bottleneck of `bytes.Split`

```
func Sum2(fileName string) (ret int64, _ error) {
    b, err := os.ReadFile(fileName)
    if err != nil {
        return 0, err
    }

    var last int ❶
    for i := 0; i < len(b); i++ {
        if b[i] != '\n' { ❷
            continue
        }
        num, err := strconv.ParseInt(string(b[last:i]), 10, 64)
        if err != nil {
            return 0, err
        }

        ret += num
        last = i + 1
    }
    return ret, nil
}
```

❶ We record the index of the last seen newline, plus one, to tell where the next line starts.

❷ Compared to `bytes.Split`, we can hardcode a new line as our separator. In one loop iteration, while reusing the `b` byte slice, we can find the full line, parse the integer, and perform the sum. This algorithm is also often called "in place."

Before we come to any conclusion, we have to first check if our new algorithm works functionally. After successfully verifying it using the unit test, I ran Example 8-13 with the `Sum2` function instead of `Sum` to assess its efficiency. The results are optimistic, with 50 ms and 12.8 MB worth of allocations. Compared to `bytes.Split`, we could perform 50% less work while using 78% less memory. Knowing that `bytes.Split` was responsible for ~36% of CPU time and 78.6% of memory allocations, such an improvement tells us we completely removed this bottleneck from our code!

Standard Functions Might Not Be Perfect for All Cases

The preceding example of working optimization asks why the `bytes.Split` function wasn't optimal for us. Can't the Go community optimize it?

The answer is that `bytes.Split` and other standard or custom functions you might import on the internet could be not as efficient as the tailored algorithm for your requirements. Such a popular function has to be, first of all, reliable for many edge cases that you might not have (e.g., multicharacter separator). Those are often optimized for cases that might be more involved and complex than our own.

It doesn't mean we have to rewrite all imported functions now. No, we should just be aware of the possibility of easy efficiency gains by providing a tailored implementation for critical paths. Still, we should use known and battle-tested code like a standard library. In most cases, it's good enough!

Is our Example 10-3 optimization our final one? Not quite—while we improved the throughput, we are at the 25 * *N* nanoseconds mark, still far from our goal.

Optimizing runtime.slicebytetostring

The CPU profile from the Example 10-3 benchmark should give us a clue about the next bottleneck, shown in Figure 10-3.

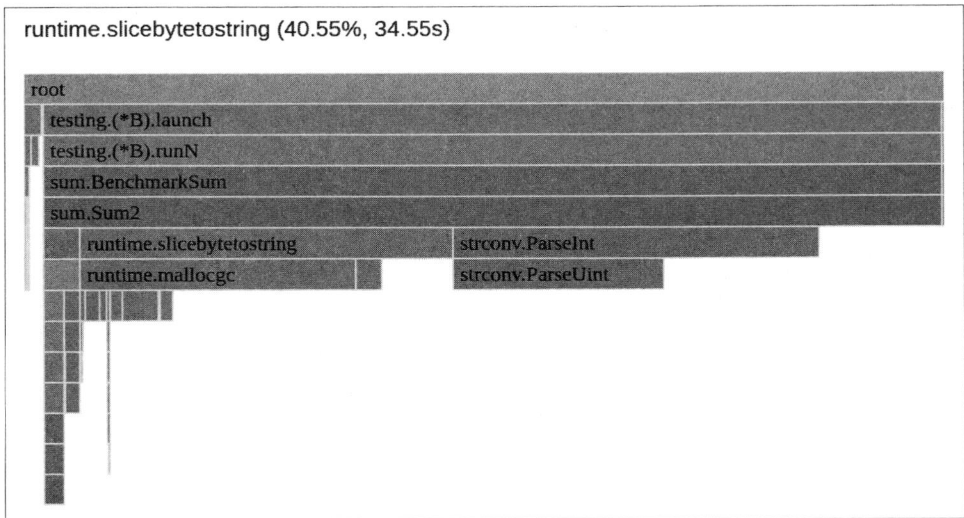

Figure 10-3. Flame Graph view of Example 10-3 CPU time with function granularity

As the next bottleneck, let's take this odd runtime.slicebytetostring function that spends most of its CPU time allocating memory. If we look for it in the Source or Peek view, it points us to the num, err := strconv.ParseInt(string(b[last:i]), 10, 64) line in Example 10-3. Since this CPU time contribution is not accounted for to strconv.ParseInt (a separate segment), it tells us that it has to be executed before we invoke strconv ParseInt, yet in the same code line. The only dynamically executed things are the b byte slice subslicing and conversion to string. On further inspection, we can tell that the string conversion is expensive here.[5]

What's interesting is that string (*https://oreil.ly/7dv5w*) is essentially a special byte slice (*https://oreil.ly/fYwwq*) with no Cap field (capacity in string is always equal to length). As a result, at first it might be surprising that the Go compiler spends so much time and memory on this. The reason is that string(<byte slice>) is equivalent to creating a new byte slice with the same number of elements, copying all bytes to a new byte, and then returning the string from it. The main reason for copying is that, by design, string type is immutable (*https://oreil.ly/I4fER*), so every function can use it without worrying about potential races. There is, however, a relatively safe way to convert []byte to string. Let's do that in Example 10-4.

Example 10-4. Sum3 is Example 10-3 with optimized CPU bottleneck of string conversion

```
// import "unsafe"

func zeroCopyToString(b []byte) string {
    return *((*string)(unsafe.Pointer(&b))) ❶
}

func Sum3(fileName string) (ret int64, _ error) {
    b, err := os.ReadFile(fileName)
    if err != nil {
        return 0, err
    }

    var last int
    for i := 0; i < len(b); i++ {
        if b[i] != '\n' {
            continue
        }
        num, err := strconv.ParseInt(zeroCopyToString(b[last:i]), 10, 64)
        if err != nil {
```

5 We can deduce that from the runtime.slicebytetostring function name in the profile. We can also split this line into three lines (string conversion in one, subslicing in the second, and invoking the parsing function in the third) and profile again to be sure.

```
            return 0, err
        }

        ret += num
        last = i + 1
    }
    return ret, nil
}
```

❶ We can use the `unsafe` package to remove the type information from `b` and form an `unsafe.Pointer`. Then we can dynamically cast this to different types, e.g., `string`. It is unsafe because if the structures do not share the same layout, we might have memory safety problems or nondeterministic values. Yet the layout is shared between `[]byte` and `string`, so it's safe for us. It is used in production in many projects, including Prometheus, known as `yoloString` (*https://oreil.ly/QmqCn*).

The `zeroCopyToString` allows us to convert file bytes to string required by `ParseInt` with almost no overhead. After functional tests, we can confirm this by using the same benchmark with the `Sum3` function again. The benefit is clear—`Sum3` takes 25.5 ms for 2 million integers and 7.2 MB of allocated space. This means it is 49.2% faster than Example 10-3 when it comes to CPU time. The memory usage is also better, with our program allocating almost precisely the size of the input file—no more, no less.

Deliberate Trade-offs

With unsafe, no-copy bytes to string conversion, we enter a deliberate optimization area. We introduced potentially unsafe code and added more nontrivial complexity to our code. While we clearly named our function `zeroCopyToString`, we have to justify and use such optimization only if necessary. In our case, it helps us reach our efficiency goals, so we can accept these drawbacks.

Are we fast enough? Not yet. We are almost there with 12.7 * *N* nanoseconds throughput. Let's see if we can optimize something more.

Optimizing strconv.Parse

Again, let's look at the newest CPU profile from the Example 10-4 benchmark to see the latest bottleneck we could try to check, as shown in Figure 10-4.

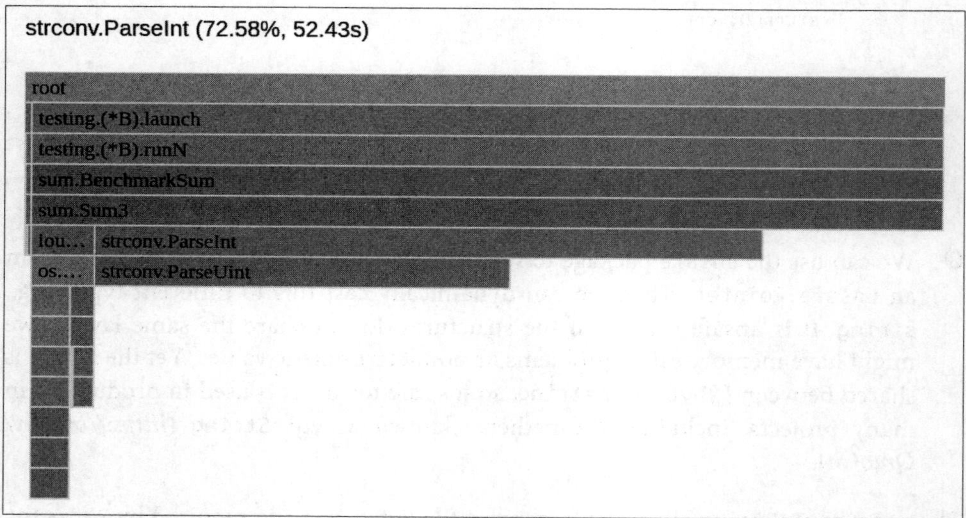

Figure 10-4. Flame Graph view of Example 10-4 CPU time with function granularity

With `strconv.Parse` using 72.6%, we can gain a lot if we can improve its CPU time. Similar to `bytes.Split`, we should check its profile and implementation (*https://oreil.ly/owR53*). Following both paths, we can immediately outline a couple of elements that feel like excessive work:

- We check for an empty string twice, in `ParseInt` (*https://oreil.ly/gqJpb*) and `ParseUint` (*https://oreil.ly/BB9Ie*). Both are visible as nontrivial CPU time used in our profile.
- `ParseInt` allows us to parse to integers with different bases and bit sizes. We don't need this generic functionality or extra input to check our `Sum3` code. We only care about 64-bit integers of base 10.

One solution here is similar to `bytes.Split`: finding or implementing our own `ParseInt` function that focuses on efficiency—does what we need and nothing more. The standard library offers the `strconv.Atoi` function (*https://oreil.ly/CpZeF*), which looks promising. However, it still requires strings as input, which forces us to use unsafe package code. Instead, let's try to come up with our own quick implementation. After a few iterations of testing and microbenchmarking my new `ParseInt` function,[6] we can come up with the fourth iteration of our sum functionality, presented in Example 10-5.

6 In benchmarks, I also found that my `ParseInt` is also faster by 10% to `strconv.Atoi` for the `Sum` test data.

Example 10-5. `Sum4` is Example 10-4 with optimized CPU bottleneck of string conversion

```
func ParseInt(input []byte) (n int64, _ error) {
    factor := int64(1)
    k := 0

    if input[0] == '-' {
        factor *= -1
        k++
    }

    for i := len(input) - 1; i >= k; i-- {
        if input[i] < '0' || input[i] > '9' {
            return 0, errors.Newf("not a valid integer: %v", input)
        }

        n += factor * int64(input[i]-'0')
        factor *= 10
    }
    return n, nil
}

func Sum4(fileName string) (ret int64, err error) {
    b, err := os.ReadFile(fileName)
    if err != nil {
        return 0, err
    }

    var last int
    for i := 0; i < len(b); i++ {
        if b[i] != '\n' {
            continue
        }
        num, err := ParseInt(b[last:i])
        if err != nil {
            return 0, err
        }

        ret += num
        last = i + 1
    }
    return ret, nil
}
```

The side effect of our integer parsing optimization is that we can tailor our `ParseInt` to parse from a byte slice, not a string. As a result, we can simplify our code and avoid unsafe `zeroCopyToString` conversion. After tests and benchmarks, we see that `Sum4` achieves 13.6 ms, 46.66% less than Example 10-4, with the same memory allocations.

The full comparison of our sum functions is presented in Example 10-6 using our beloved `benchstat` tool.

Example 10-6. Running `benchstat` on the results from all four iterations with a two million line file

```
$ benchstat v1.txt v2.txt v3.txt v4.txt
name \ (time/op)   v1.txt       v2.txt        v3.txt        v4.txt
Sum                 101ms ± 0%    50ms ± 2%    25ms ± 0%    14ms ± 0% ❶

name \ (alloc/op) v1.txt        v2.txt        v3.txt        v4.txt
Sum               60.8MB ± 0%  12.8MB ± 0%  7.2MB ± 0%  7.2MB ± 0%

name \ (allocs/op) v1.txt        v2.txt        v3.txt        v4.txt
Sum                 1.60M ± 0%   1.60M ± 0%  0.00M ± 0%  0.00M ± 0%
```

❶ Notice that `benchstat` can round some numbers for easier comparison with the large number from *v1.txt*. The *v4.txt* result is 13.6 ms, not 14 ms, which can make a difference in throughput calculations.

It seems like our hard work paid off. With the current results, we achieved 6.9 * *N* nanoseconds throughput, which is more than enough to fulfill our first goal. However, we only checked it with two million integers. Are we sure the same throughput can be maintained with larger or smaller input sizes? Our Big O runtime complexity O(*N*) would suggest so, but I ran the same benchmark with 10 million integers just in case. The 67.8 ms result gives the 6.78 * *N* nanoseconds throughput. This more or less confirms our throughput number.

The code in Example 10-5 is not the fastest or most memory-efficient solution possible. There might be more optimizations to the algorithm or code to improve things further. For example, if we profile Example 10-5, we would see a relatively new segment, indicating 14% of total CPU time used. It's `os.ReadFile` code that wasn't so visible on past profiles, given other bottlenecks and something we didn't touch with our optimizations. We will mention its potential optimization in "Pre-Allocate If You Can" on page 441. We could also try concurrency (which we will do in "Optimizing Latency Using Concurrency" on page 402). However, with one CPU, we cannot expect a lot of gains here.

What's important is that there is no need to improve anything else in this iteration, as we achieved our goal. We can stop the work and claim success! Fortunately, we did not need to add magic or dangerous nonportable tricks to our optimization flow. Only readable and easier deliberate optimizations were required.

Optimizing Memory Usage

In the second scenario, our goal is focused on memory consumption while maintaining the same throughput. Imagine we have a new business customer for our software with `Sum` functionality that needs to run on an IoT device with little RAM available for this program. As a result, the requirement is to have a streaming algorithm: no matter the input size, it can only use 10 KB of heap memory in a single moment.

Such a requirement might look extreme at first glance, given the naive code in Example 4-1 has a quite large space complexity. If a 10 million line, 36 MB file requires 304 MB of heap memory for Example 4-1, how can we ensure the same file (or bigger!) can take a maximum of 10 KB of memory? Before we start to worry, let's analyze what we can do on this subject.

Fortunately, we already did some optimization work that improved memory allocations as a side effect. Since the latency goal still applies, let's start with `Sum4` in Example 10-5, which fulfills that. The space complexity of `Sum4` seems to be around O(*N*). It still depends on the input size and is far from our 10 KB goal.

Moving to Streaming Algorithm

Let's pull up the heap profile from the `Sum4` benchmark in Figure 10-5 to figure out what we can improve.

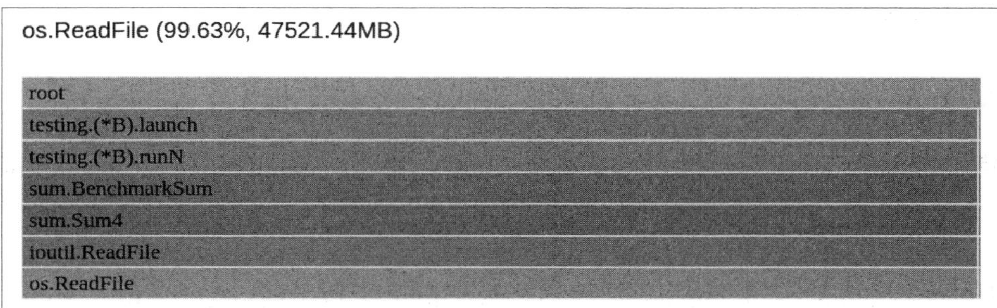

Figure 10-5. Flame Graph view of Example 10-5 heap allocations with function granularity (`alloc_space`)

The memory profile is very boring. The first line allocates 99.6% of memory in Example 10-5. We essentially read the whole file into memory so we can iterate over the bytes in memory. Even if we waste some allocation elsewhere, we can't see it because of excessive allocation from `os.ReadFile`. Is there anything we can do about that?

During our algorithm, we must go through all the bytes in the file; thus, we have to read all bytes eventually. However, we don't need to read all of them to memory at the same time. Technically, we only need a byte slice big enough to hold all digits for an integer to be parsed. This means we can try to design the external memory algorithm (*https://oreil.ly/Dr3MB*) to stream bytes in chunks. We can try using the existing bytes scanner from the standard library—the `bufio.Scanner` (*https://oreil.ly/CqiG7*). For example, `Sum5` in the Example 10-7 implementation uses it to scan enough memory to read and parse a line.

Example 10-7. `Sum5` is Example 10-5 with `bufio.Scanner`

```
func Sum5(fileName string) (ret int64, err error) {
    f, err := os.Open(fileName) ❶
    if err != nil {
        return 0, err
    }
    defer errcapture.Do(&err, f.Close, "close file") ❷

    scanner := bufio.NewScanner(f)
    for scanner.Scan() { ❸
        num, err := ParseInt(scanner.Bytes())
        if err != nil {
            return 0, err
        }

        ret += num
    }
    return ret, scanner.Err() ❹
}
```

❶ Instead of reading the whole file into memory, we open the file descriptor here.

❷ We have to make sure the file is closed after the computation so as not to leak resources. We use `errcapture` to get notified about potential errors in the deferred file `Close`.

❸ The scanner `.Scan()` method tells us if we hit the end of the file. It returns true if we still have bytes to result in splitting. The split is based on the provided function in the `.Split` method. By default, `ScanLines` (*https://oreil.ly/YUpLU*) is what we want.

❹ Don't forget to check the scanner error! With such iterator interfaces, it's very easy to forget to check its error.

To assess efficiency, now focusing more on memory, we can use the same Example 8-13 with `Sum5`. However, given our past optimizations, we've moved dangerously close to what can be reasonably measured within the accuracy and overhead of our tools for input files on the order of a million lines. If we got into microsecond latencies, our measurements might be skewed, given limits in the instrumentation accuracy and benchmarking tool overheads. So let's increase the file to 10 million lines. The benchmarked `Sum4` in Example 10-5 for that input results in 67.8 ms and 36 MB of memory allocated per operation. The `Sum5` with the scanner outputs 157.1 ms and 4.33 KB per operation.

In terms of memory usage, this is great. If we look at the implementation, the scanner allocates an initial 4 KB (*https://oreil.ly/jbpJc*) and uses it for reading the line. It increases this if needed when the line is longer, but our file doesn't have numbers longer than 10 digits, so it stays at 4 KB. Unfortunately, the scanner isn't fast enough for our latency requirement. With a 131% slowdown to `Sum4`, we hit 15.6 * *N* nanoseconds latency, which is too slow. We have to optimize latency again, knowing we still have around 6 KB to allocate to stay within the 10 KB memory goal.

Optimizing bufio.Scanner

What can we improve? As usual, it's time to check the source code and profile of Example 10-7 in Figure 10-6.

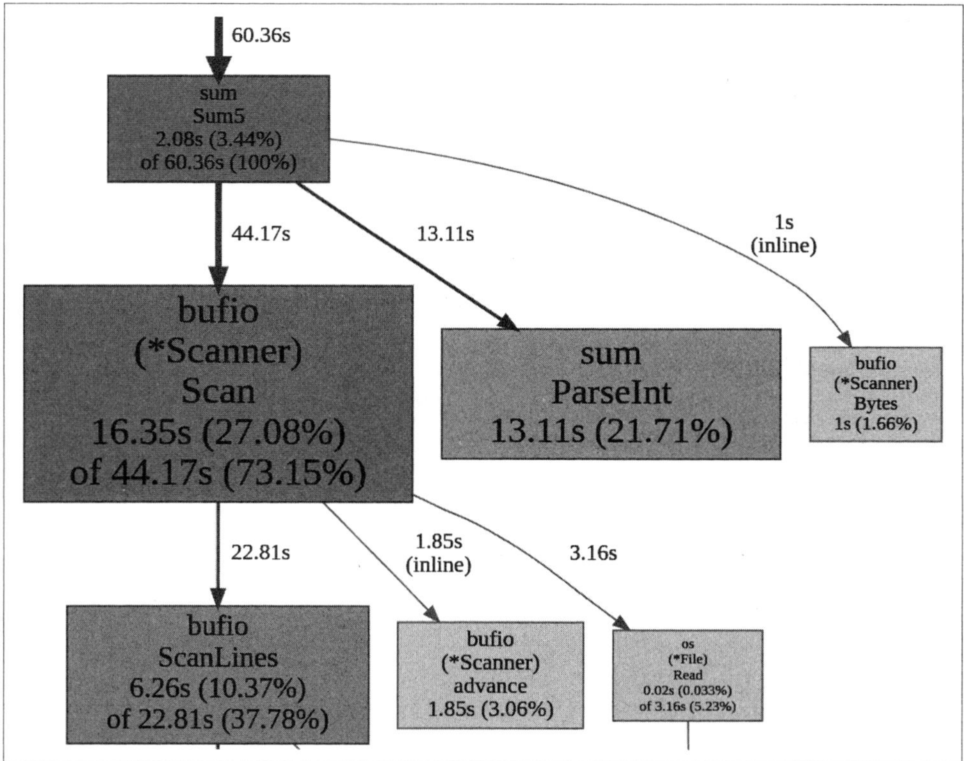

Figure 10-6. Graph view of Example 10-7 CPU time with function granularity

The commentary on the Scanner structure in the standard library gives us a hint. It tells us that "Scanner is for safe, simple jobs" (*https://oreil.ly/6eXZE*). The ScanLines is the main bottleneck here, and we can swap the implementation with a more efficient one. For example, the original function removes carriage return (CR) control characters (*https://oreil.ly/wwUbC*), which wastes cycles for us as our input does not have them. I managed to provide optimized ScanLines, which improves the latency by 20.5% to 125 ms, which is still too slow.

Similar to previous optimizations, it might be worth writing a custom streamed scanning implementation instead of bufio.Scanner. The Sum6 in Example 10-8 presents a potential solution.

Example 10-8. Sum6 is Example 10-5 with buffered read

```
func Sum6(fileName string) (ret int64, err error) {
    f, err := os.Open(fileName)
    if err != nil {
        return 0, err
    }
    defer errcapture.Do(&err, f.Close, "close file")

    buf := make([]byte, 8*1024) ❶
    return Sum6Reader(f, buf)
}

func Sum6Reader(r io.Reader, buf []byte) (ret int64, err error) { ❷
    var offset, n int
    for err != io.EOF {
        n, err = r.Read(buf[offset:]) ❸
        if err != nil && err != io.EOF { ❹
            return 0, err
        }
        n += offset ❺

        var last int
        for i := range buf[:n] { ❻
            if buf[i] != '\n' {
                continue
            }
            num, err := ParseInt(buf[last:i])
            if err != nil {
                return 0, err
            }

            ret += num
            last = i + 1
        }

        offset = n - last
```

```
        if offset > 0 {
            _ = copy(buf, buf[last:n]) ❼
        }
    }
    return ret, nil
}
```

❶ We create a single 8 KB buffer of bytes we will use for reading. I chose 8 KB and not 10 KB to leave some headroom within our 10 KB limit. The 8 KB also feels like a great number given the OS page is 4 KB, so we know it will need only 2 pages.

This buffer assumes that no integer is larger than ~8,000 digits. We can make it much smaller, even down to 10, as we know our input file does not have numbers with more than 9 digits (plus the newline). However, this would make the algorithm much slower due to the certain waste explained in the next steps. Additionally, even without waste reading, 8 KB is faster than reading 8 bytes 1,024 times due to overhead.

❷ This time, let's separate functionality behind the convenient `io.Reader` interface. This will allow us to reuse `Sum6Reader` in the future.[7]

❸ In each iteration, we read the next 8 KB, minus `offset` bytes from a file. We start reading more file bytes after `offset` bytes to leave potential room for digits we didn't parse yet. This can happen if we read bytes that split some numbers into parts, e.g., we read `...\n12` and `34/n...` in two different chunks.

❹ In the error handling, we excluded the `io.EOF` sentinel error, which indicated we hit the end of the file. That's not an error for us—we still want to process the remaining bytes.

❺ The number of bytes we have to process from the buffer is exactly `n + offset`, where `n` is the number of bytes read from a file. The end of file `n` can be smaller than what we asked for (length of the `buf`).

❻ We iterate over `n` bytes in the `buf` buffer.[8] Notice that we don't iterate over the whole slice because in an `err == io.EOF` situation, we might read less than 10

---

7 Interestingly enough, just adding a new function call and interface slows down the program by 7% per operation on my machine, proving that we are on a very high efficiency level already. However, given reusability, perhaps we can afford that slowdown.

8 As an interesting fact, if we replace this line with a technically simpler loop like `for i := 0; i < n; i++ {`, the code is 5% slower! Don't take it as a rule (always measure!), as it probably depends on your workload, but it's interesting to see the `range` loop (without a second argument) be more efficient here.

KB of bytes, so we need to process only `n` of them. We process all lines found in our 10 KB buffer in each loop iteration.

7. We calculate `offset`, and if there is a need for one, we shift the remaining bytes to the front. This creates a small waste in CPU, but we don't allocate anything additional. Benchmarks will tell us if this is fine or not.

Our `Sum6` code got a bit bigger and more complex, so hopefully, it gives good efficiency results to justify the complexity. Indeed, after the benchmark, we see it takes 69 ms and 8.34 KB. Just in case, let's put Example 10-8 to the extra test by computing an even larger file—100 million lines. With bigger input, `Sum6` yields 693 ms and around 8 KB. This gives us a 6.9 * *N* nanoseconds latency (runtime complexity) and space (heap) complexity of ~8 KB, which satisfies our goal.

Careful readers might still be wondering if I didn't miss anything. Why is space complexity 8 KB, not 8 + *x* KB? There are some additional bytes allocated for 10 million line files and even more bytes for larger ones. How do we know that at some point for a hundred-times larger file, the memory allocation would not exceed 10 KB?

If we are very strict and tight on that 10 KB allocation goal, we can try to figure out what happens. The most important thing is to validate that there is nothing that grows allocation with the file size. This time the memory profile is also invaluable, but to understand things fully, let's ensure we record all allocations by adding `runtime.MemProfileRate = 1` in our `BenchmarkSum` benchmark. The resulting profile is presented in Figure 10-7.

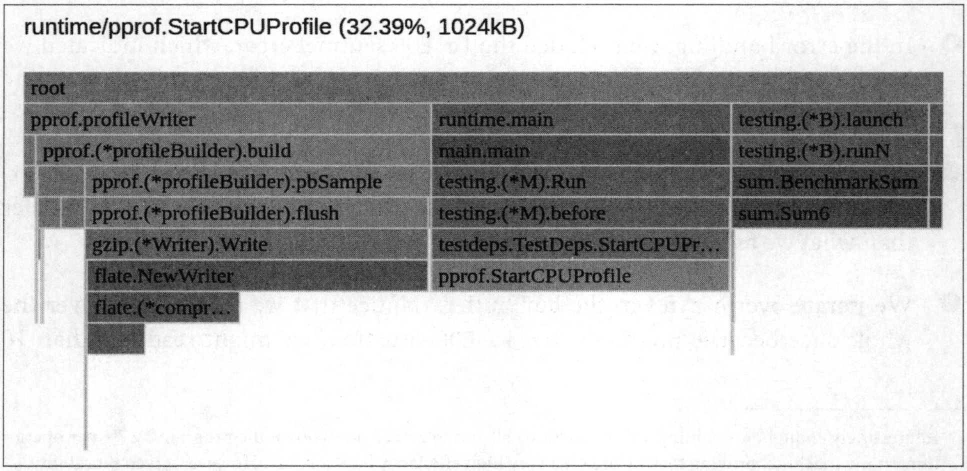

Figure 10-7. Flame Graph view of Example 10-8 memory with function granularity and profile rate 1

We can see more allocations from the `pprof` package than our function. This indicates a relatively large allocation overhead by the profiling itself! Still, it does not prove that `Sum` does not allocate anything else on the heap than our 8 KB buffer. The Source view turns out to be helpful, presented in Figure 10-8.

```
github.com/efficientgo/examples/pkg/sum.Sum6
/home/bwplotka/Repos/examples/pkg/sum/sum.go

 Total:      7.82MB      7.82MB (flat, cum) 98.17%
    240           .           .                 return 0, nil, nil
    241           .           .         }
    242           .           .
    243           .           .         // Sum6 is like Sum4, but trying to us
    244           .           .         // Assuming no integer is larger than
    245           .           .         func Sum6(fileName string) (ret int64,
    246           .           .                 f, err := os.Open(fileName)
    247           .           .                 if err != nil {
    248           .           .                         return 0, err
    249           .           .                 }
    250           .           .                 defer errcapture.Do(&err, f.Clos
    251           .           .
    252      7.82MB      7.82MB                 buf := make([]byte, 8*1024)
    253           .           .                 return Sum6Reader(f, buf)
    254           .           .         }
```

Figure 10-8. Source view of Example 10-8 memory with profile rate 1 after benchmark with 1,000 iterations and 10 MB input file

It shows that `Sum6` has only one heap allocation point. We can also benchmark without CPU profiling, which now gives stable 8,328 heap allocated bytes for any input size.

Success! Our goal is met, and we can move to the last task. The overview of each iteration's achieved result is shown in Example 10-9.

Example 10-9. Running `benchstat` on the results from all 3 iterations with a 10 million line file

```
$ benchstat v1.txt v2.txt v3.txt v4.txt
name \ (time/op)    v4-10M.txt   v5-10M.txt    v6-10M.txt
Sum                 67.8ms ± 3%  157.1ms ± 2%  69.4ms ± 1%

name \ (alloc/op) v4-10M.txt   v5-10M.txt    v6-10M.txt
Sum               36.0MB ± 0%    0.0MB ± 3%   0.0MB ± 0%

name \ (allocs/op)  v4-10M.txt   v5-10M.txt    v6-10M.txt
Sum                   5.00 ± 0%    4.00 ± 0%    4.00 ± 0%
```

Optimizing Latency Using Concurrency

Hopefully, you are ready for the last challenge: getting our latency down even more to the 2.5 nanoseconds per line level. This time we have four CPU cores available, so we can try introducing some concurrency patterns to achieve it.

In "When to Use Concurrency" on page 145, we mentioned the clear need for concurrency to employ asynchronous programming or event handling in our code. We talked about relatively easy gains where our Go program does a lot of I/O operations. However, in this section, I would love to show you how to improve the speed of our `Sum` in the Example 4-1 code using concurrency with two typical pitfalls. Because of the tight latency requirement, let's take an already optimized version of `Sum`. Given we don't have any memory requirements, and `Sum4` in Example 10-5 is only a little slower than `Sum6`, yet has a smaller amount of lines, let's take that as a start.

A Naive Concurrency

As usual, let's pull out the Example 10-5 CPU profile, shown in Figure 10-9.

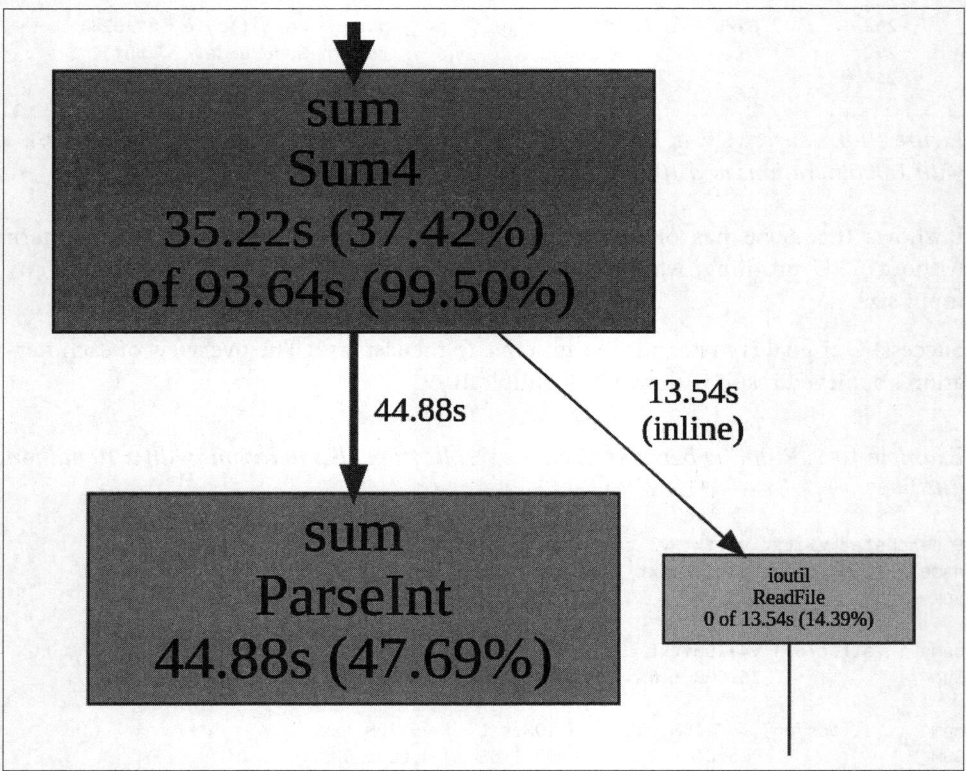

Figure 10-9. Graph view of Example 10-5 CPU time with function granularity

As you might have noticed, most of Example 10-5 CPU time comes from `ParseInt` (47.7%). Since we're back to reading the whole file at the beginning of the program, the rest of the program is strictly CPU bound. As a result, with only one CPU we couldn't expect better latency with the concurrency (*https://oreil.ly/rsLff*). However, given that within this task we have four CPU cores available, our task now is to find a way to evenly split the work of parsing the file's contents with as little coordination[9] between goroutines as possible. Let's explore three example approaches to optimize Example 10-5 with concurrency.

The first thing we have to do is find computations we can do independently at the same time—computations that do not affect each other. Because the sum is commutative, it does not matter in what order numbers are added. The naive, concurrent implementation could parse the integer from the string and add the result atomically to the shared variable. Let's explore this rather simple solution in Example 10-10.

Example 10-10. Naive concurrent optimization to Example 10-5 that spins a new goroutine for each line to compute

```
func ConcurrentSum1(fileName string) (ret int64, _ error) {
    b, err := os.ReadFile(fileName)
    if err != nil {
        return 0, err
    }

    var wg sync.WaitGroup
    var last int
    for i := 0; i < len(b); i++ {
        if b[i] != '\n' {
            continue
        }

        wg.Add(1)
        go func(line []byte) {
            defer wg.Done()
            num, err := ParseInt(line)
            if err != nil {
                // TODO(bwplotka): Return err using other channel.
                return
            }
            atomic.AddInt64(&ret, num)
        }(b[last:i])
        last = i + 1
    }
    wg.Wait()
    return ret, nil
```

9 We discussed synchronization primitives in "Go Runtime Scheduler" on page 138.

After the successful functional test, it's time for benchmarking. Similar to previous steps, we can reuse the same Example 8-13 by simply replacing `Sum` with `ConcurrentSum1`. I also changed the `-cpu` flag to 4 to unlock the four CPU cores. Unfortunately, the results are not very promising—for a 2 million line input, it takes about 540 ms and 151 MB of allocated space per operation! Almost 40 times more time than the simpler, noncurrent Example 10-5.

A Worker Approach with Distribution

Let's check the CPU profile in Figure 10-10 to learn why.

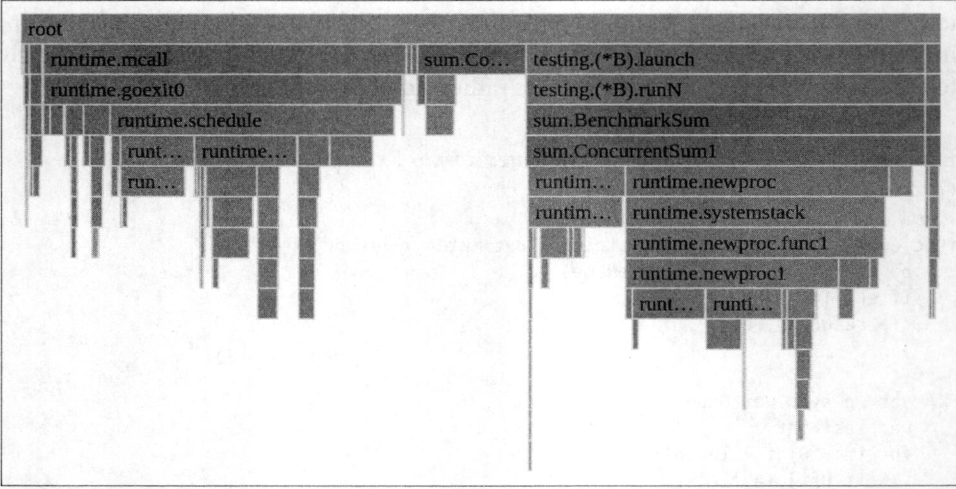

Figure 10-10. Flame Graph view of Example 10-10 CPU time with function granularity

The Flame Graph clearly shows the goroutine creation and scheduling overhead indicated by blocks called `runtime.schedule` and `runtime.newproc`. There are three main reasons why Example 10-10 is too naive and not recommended for our case:

- The concurrent work (parsing and adding) is too fast to justify the goroutine overhead (both in memory and CPU usage).
- For larger datasets, we create potentially millions of goroutines. While goroutines are relatively cheap and we can have hundreds of them, there is always a limit, given only four CPU cores to execute. So you can imagine the delay of the scheduler that tries to fairly schedule millions of goroutines on four CPU cores.
- Our program will have a nondeterministic performance depending on the number of lines in the file. We can potentially hit a problem of unbounded concurrency since we will spam as many goroutines as the external file has lines (something outside our program control).

That is not what we want, so let's improve our concurrent implementation. There are many ways we could go from here, but let's try to address all three problems we notice. We can solve problem number one by assigning more work to each goroutine. We can do that thanks to the fact that addition is also associative and cumulative. We can essentially group work into multiple lines, parse and add numbers in each goroutine, and add partial results to the total sum. Doing that automatically helps with problem number two. Grouping work means we will schedule fewer goroutines. The question is, what is the best number of lines in a group? Two? Four? A hundred?

The answer most likely depends on the number of goroutines we want in our process and the number of CPUs available. There is also problem number three—unbounded concurrency. The typical solution here is to use a worker pattern (sometimes called goroutine pooling). In this pattern, we agree on a number of goroutines up front, and we schedule all of them at once. Then we can create another goroutine that will distribute the work evenly. Let's see an example implementation of that algorithm in Example 10-11. Can you predict if this implementation will be faster?

Example 10-11. Concurrent optimization of Example 10-5 that maintains a finite set of goroutines that computes a group of lines. Lines are distributed using another goroutine.

```
func ConcurrentSum2(fileName string, workers int) (ret int64, _ error) {
    b, err := os.ReadFile(fileName)
    if err != nil {
        return 0, err
    }

    var (
        wg     = sync.WaitGroup{}
        workCh = make(chan []byte, 10)
    )

    wg.Add(workers + 1)
    go func() {
        var last int
        for i := 0; i < len(b); i++ {
            if b[i] != '\n' {
                continue
            }
            workCh <- b[last:i]
            last = i + 1
      }
        close(workCh) ❶
        wg.Done()
    }()

    for i := 0; i < workers; i++ {
        go func() {
```

```
            var sum int64
            for line := range workCh { ❷
                num, err := ParseInt(line)
                if err != nil {
                    // TODO(bwplotka): Return err using other channel.
                    continue
                }
                sum += num
            }
            atomic.AddInt64(&ret, sum)
            wg.Done()
        }()
    }
    wg.Wait()
    return ret, nil
}
```

❶ Remember, the sender is usually responsible for the closing channel. Even if our flow does not depend on it, it's a good practice to always close channels after use.

❷ Beware of common mistakes. The `for _, line := range <-workCh` would sometimes compile as well, and it looks logical, but it's wrong. It will wait for the first message from the `workCh` channel and iterate over single bytes from the received byte slice. Instead, we want to iterate over messages.

Tests pass, so we can start benchmarking. Unfortunately, on average, this implementation with 4 goroutines takes 207 ms to complete a single operation (using 7 MB of space). Still, this is 15 times slower than simpler, sequential Example 10-5.

A Worker Approach Without Coordination (Sharding)

What's wrong this time? Let's investigate the CPU profile presented in Figure 10-11.

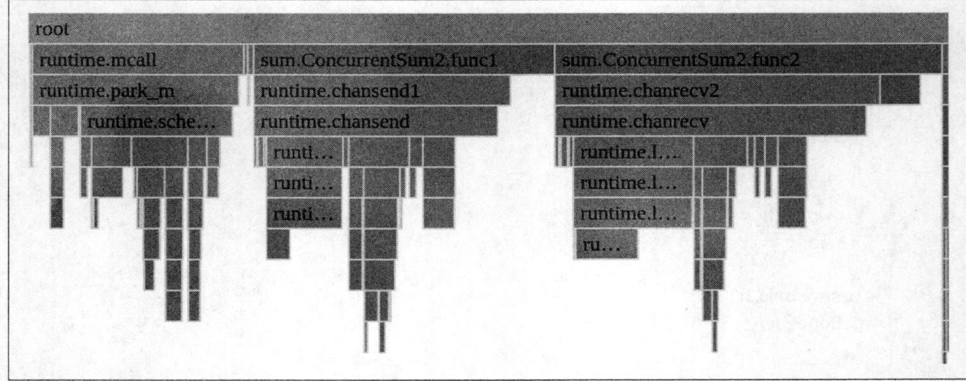

Figure 10-11. Flame Graph view of Example 10-11 CPU time with function granularity

If you see a profile like this, it should immediately tell you that the concurrency overhead is again too large. We still don't see the actual work, like parsing integers, since this work has outnumbered the overhead. This time the overhead is caused by three elements:

`runtime.schedule`
: The runtime code responsible for scheduling goroutines.

`runtime.chansend`
: In our case, waiting on the lock to send to our single channel.

`runtime.chanrecv`
: The same as `chansend` but waiting on a read from the receive channel.

As a result, parsing and additions are faster than the communication overhead. Essentially, coordination and distribution of the work take more CPU resources than the work itself.

We have multiple options for improvement here. In our case, we can try to remove the effort of distributing the work. We can accomplish this via a coordination-free algorithm that will shard (split) the workload evenly across all goroutines. It's coordination free because there is no communication to agree on which part of the work is assigned to each goroutine. We can do that thanks to the fact that the file size is known up front, so we can use some sort of heuristic to assign each part of the file with multiple lines to each goroutine worker. Let's see how this could be implemented in Example 10-12.

Example 10-12. Concurrent optimization of Example 10-5 that maintains a finite set of goroutines that computes groups of lines. Lines are sharded without coordination.

```
func ConcurrentSum3(fileName string, workers int) (ret int64, _ error) {
    b, err := os.ReadFile(fileName)
    if err != nil {
        return 0, err
    }

    var (
        bytesPerWorker = len(b) / workers
        resultCh       = make(chan int64)
    )

    for i := 0; i < workers; i++ {
        go func(i int) {
            // Coordination-free algorithm, which shards
            // buffered file deterministically.
            begin, end := shardedRange(i, bytesPerWorker, b) ❶

            var sum int64
```

```
            for last := begin; begin < end; begin++ {
                if b[begin] != '\n' {
                    continue
                }
                num, err := ParseInt(b[last:begin])
                if err != nil {
                    // TODO(bwplotka): Return err using other channel.
                    continue
                }
                sum += num
                last = begin + 1
            }
            resultCh <- sum
        }(i)
    }

    for i := 0; i < workers; i++ {
        ret += <-resultCh
    }
    close(resultCh)
    return ret, nil
}
```

❶ `shardedRange` is not supplied for clarity. This function takes the size of the input file and splits into `bytesPerWorker` shards (four in our case). Then it gives each worker the `i`-th shard. You can see the full code here (*https://oreil.ly/By9wO*).

Tests pass too, so we confirmed that Example 10-12 is functionally correct. But is it faster? Yes! The benchmark shows 7 ms and 7 MB per operation, which is almost twice as fast as sequential Example 10-5. Unfortunately, this puts us in 3.4 * *N* nanoseconds throughput, which is failing our goal of 2.5 * *N*.

A Streamed, Sharded Worker Approach

Let's profile in Figure 10-12 one more time to check if we can improve anything easily.

The CPU profile shows that the work done by our goroutines takes the most CPU time. However, ~10% of CPU time is spent reading all bytes, which we can also try to do concurrently. This effort does not look promising at first glance. However, even if we would remove all 10% of the CPU time, 10% better throughput gives us only the 3.1 * *N* nanoseconds number, so not enough.

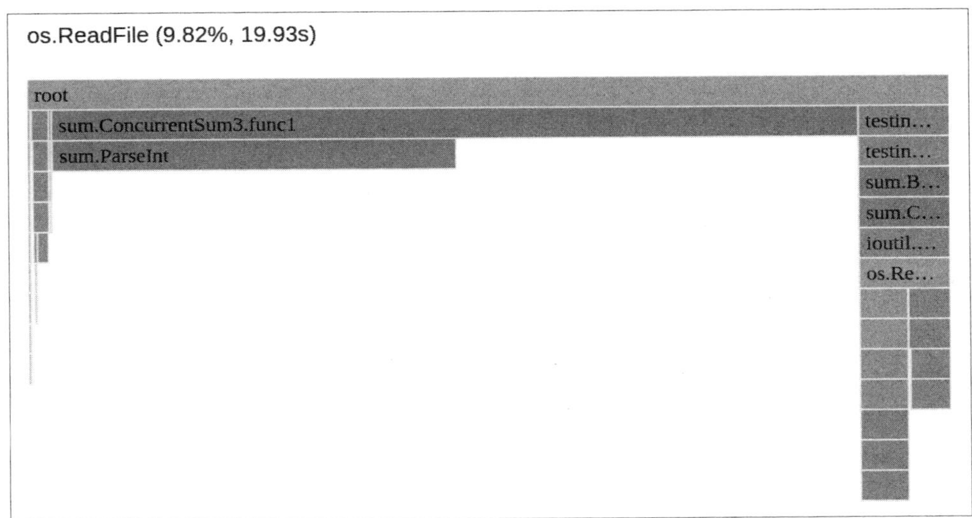

Figure 10-12. Flame Graph view of Example 10-12 CPU time with function granularity

This is where we have to be vigilant, though. As you can imagine, reading files is not a CPU-bound job, so perhaps the actual real time spend on that 10% of CPU time makes `os.ReadFile` a bigger bottleneck, thus a better option for us to optimize. As in "Optimizing Latency" on page 383, let's perform a benchmark wrapped with the `fgprof` profile! The resulting full goroutine profile is presented in Figure 10-13.

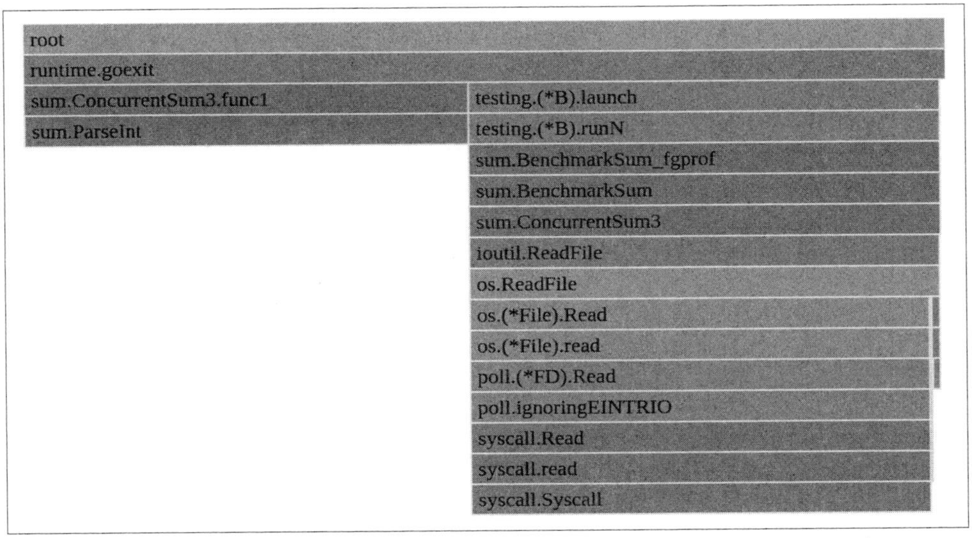

Figure 10-13. Flame Graph view of Example 10-12 full goroutine profile with function granularity

The fgprof profile shows that a lot can be gained in latency if we try to read files concurrently, as it currently takes around 50% of the real time! This is way more promising, so let's try to move file reads to worker goroutines. The example implementation is shown in Example 10-13.

Example 10-13. Concurrent optimization of Example 10-12 that also reads from a file concurrently using separate buffers

```
func ConcurrentSum4(fileName string, workers int) (ret int64, _ error) {
    f, err := os.Open(fileName)
    if err != nil {
        return 0, err
    }
    defer errcapture.Do(&err, f.Close, "close file")

    s, err := f.Stat()
    if err != nil {
        return 0, err
    }

    var (
        size           = int(s.Size())
        bytesPerWorker = size / workers
        resultCh       = make(chan int64)
    )

    if bytesPerWorker < 10 {
        return 0, errors.New("can't have less bytes per goroutine than 10")
    }

    for i := 0; i < workers; i++ {
        go func(i int) {
            begin, end := shardedRangeFromReaderAt(i, bytesPerWorker, size, f)
            r := io.NewSectionReader(f, int64(begin), int64(end-begin)) ❶

            b := make([]byte, 8*1024)
            sum, err := Sum6Reader(r, b) ❷
            if err != nil {
                // TODO(bwplotka): Return err using other channel.
            }
            resultCh <- sum
        }(i)
    }

    for i := 0; i < workers; i++ {
        ret += <-resultCh
    }
    close(resultCh)
    return ret, nil
}
```

❶ Instead of splitting the bytes from the input file in memory, we tell each goroutine what bytes from the file it can read. We can do this thanks to the `SectionReader` (*https://oreil.ly/j4cQd*), which returns a reader that only allows reading from a particular section. There is a small complexity in `shardedRangeFromReaderAt` (*https://oreil.ly/PwNty*) to make sure we read all lines (we don't know where the newlines in a file are), but it can be done in the relatively easy algorithm presented here.

❷ We can reuse Example 10-8 for this job as it knows how to use any `io.Reader` implementation, so in our example, both `*os.File` and `*io.SectionReader`.

Let's assess the efficiency of that code. Finally, after all this work, Example 10-13 yields an astonishing 4.5 ms per operation for 2 million lines, and 23 ms for 10 million lines. This takes us into ~2.3 * *N* nanosecond throughput, which satisfies our goal! A full comparison of latencies and memory allocations for successful iterations is presented in Example 10-14.

Example 10-14. Running `benchstat` on the results from all four iterations with a two million line file

```
name \ (time/op)   v4-4core.txt  vc3.txt       vc4.txt
Sum-4              13.3ms ± 1%   6.9ms ± 6%    4.5ms ± 3%

name \ (alloc/op)  v4-4core.txt  vc3.txt       vc4.txt
Sum-4              7.20MB ± 0%   7.20MB ± 0%  0.03MB ± 0%
```

To summarize, we went through three exercises showcasing the optimization flow focused on different goals. I also have some possible concurrency patterns that allow utilizing our multicore machines. Generally, I hope you saw how critical benchmarking and profiling were throughout this journey! Sometimes the results might surprise you, so always seek confirmation of your ideas.

There is, however, another way to solve those exercises in an innovative way that might work for certain use cases. Sometimes it allows us to avoid the huge optimization effort we did in the past three sections. Let's take a look!

Bonus: Thinking Out of the Box

Given the challenging goals we set in this chapter, I spent a lot of time optimizing and explaining optimization for the naive `Sum` implementation in Example 4-1. This showed you some optimization ideas, practices, and generally a mental model I use during optimization efforts. But hard optimization work is not always an answer—there are numerous ways to reach our goals.

For example, what if I told you there is a way to get amortized runtime complexity of a few nanoseconds and zero allocations (and just four more code lines)? Let's see Example 10-15.

Example 10-15. Adding simplest caching to Example 4-1

```
var sumByFile = map[string]int64{} ❶

func Sum7(fileName string) (int64, error) {
    if s, ok := sumByFile[fileName]; ok {
        return s, nil
    }

    ret, err := Sum(fileName)
    if err != nil {
        return 0, err
    }

    sumByFile[fileName] = ret
    return ret, nil
}
```

❶ `sumByFile` represents the simplest storage for cache. There are tons of more production read-caching implementations you can consider as well. We can write our own that will be goroutine safe. If we need more involved eviction policies, I would recommend HashiCorp's golang-lru (*https://oreil.ly/nnYoM*) and the even more optimized Dgraph's ristretto (*https://oreil.ly/QNshi*). For distributed systems, you should use distributed caching services like Memcached (*https://oreil.ly/fudbQ*), Redis (*https://oreil.ly/1ovP1*), or peer-to-peer caching solutions like groupcache (*https://oreil.ly/vJONo*).

The functional test passes, and the benchmarks show amazing results—for 100 million line files, we see 228 ns and 0 bytes allocated! This example is, of course, a very trivial one. It's unlikely our optimization journey is always as easy as that. Simple caching is limited and can't be used if the file input constantly changes. But what if we can?

Think smart, not hard. It might be the case that we don't need to optimize Example 4-1 because the same input files are constantly used. Caching a single sum value for each file is cheap—even if we would have a million of those files, we can cache all using a few megabytes. If that's not the case, perhaps the file content often repeats, but the filename is unique. In that case, we could calculate the checksum of the file and cache based on that. It would be faster than parsing all lines into integers.

Focus on the goal and be smart and innovative. For example, a hard, week-long, deep optimization effort might not be worth it if there is some smart solution that avoids that work!

Summary

We did it! We optimized the initial naive implementation of Example 4-1 using the TFBO flow from "Efficiency-Aware Development Flow" on page 102. Guided by the requirements, we managed to improve the `Sum` code significantly:

- We improved the runtime complexity from around 50.5 * *N* nanoseconds (where N is a number of lines) to 2.25 * *N*. This means around 22 times faster latency, even though both naive and most optimized algorithms are linear (we optimized O(*N*) constants).
- We improved the space complexity from around 30.4 * *N* bytes to 8 KB, which means our code had O(*N*) asymptotic complexity but now has constant space complexity. This means the new `Sum` code will be much more predictable for the users and more friendly for the garbage collector.

To sum up, sometimes efficiency problems require a long and careful optimization process, as we did for `Sum`. On the other hand, sometimes, you can find quick and pragmatic optimization ideas that fulfill your goals quickly. Nevertheless, we all learned a lot from the exercises in this chapter (including me!).

Let's move to the last chapter of this book, where we will summarize some learning and patterns we saw during our exercises in this chapter, and what I have seen in the community from my experience.

CHAPTER 11
Optimization Patterns

With all we've learned from the past 10 chapters, it's time to go through various patterns and common pitfalls I found when developing efficient code in Go. As I mentioned in Chapter 10, the optimization suggestion doesn't generalize well. However, given you should know at this point how to assess code changes effectively, there is no harm in stating some common patterns that improve efficiency in certain cases.

Be a Mindful Go Developer

Remember that most optimization ideas you will see here are highly deliberate. This means we have to have a good reason to add them as they take the developer's time to get right and maintain in the future. Even if you learn about some common optimization, ensure it improves efficiency for your specific workload.

Don't use this chapter as a strict manual but as a list of potential options you did not think about. Nevertheless, always stick to the observability, benchmarking, and profiling tools we learned in previous chapters to ensure the optimizations you do are pragmatic, follow YAGNI (*https://oreil.ly/G9OLQ*), and are needed.

We will start with "Common Patterns" on page 416, where I describe some high-level optimization patterns we could see from optimization examples in Chapter 10. Then I will introduce you to the "The Three Rs Optimization Method" on page 421, an excellent memory optimization framework from the Go (and Prometheus) community.

Finally, in "Don't Leak Resources" on page 426, "Pre-Allocate If You Can" on page 441, "Overusing Memory with Arrays" on page 446, and "Memory Reuse and Pooling" on page 449, we will go through a set of specific optimizations, tips, and gotchas I wish I'd known when I started my journey with making Go code more efficient. I have chosen the most common ones that are worth being aware of!

Let's start with common optimization patterns. Some of them I used in previous chapters.

Common Patterns

How can you find optimizations? After benchmarking, profiling, and studying the code, the process requires us to figure out a better algorithm, data structure, or code that will be more efficient. Of course, this is easier said than done.

Some practice and experience help, but we can outline a few patterns that repeat in our optimization journeys. Let's now walk through four generic patterns we see in the programming community and literature: doing less work, and trading functionality for efficiency, trading space for time, and trading time for space.

Do Less Work

The first thing we should focus on is avoiding unnecessary work. Especially in "Optimizing Latency" on page 383, we improved the CPU time multiple times by removing a lot of unnecessary code. It might feel simplistic, but it's a powerful pattern we often forget. If some portion of the code is critical and requires optimization, we can go through bottlenecks (e.g., lines of code with large contributions we see in Source view as we discussed in "go tool pprof Reports" on page 340) and check if we can:

Skip unnecessary logic

Can we remove this line? For example, in "Optimizing Latency" on page 383, `strconv.ParseInt` had a lot of checks that weren't needed in our implementation. We can use the assumptions and requirements we have to our advantage and trim down the functionality that isn't strictly needed. This also includes potential resources we can clean early or any resource leaks (see "Don't Leak Resources" on page 426).

Generic Implementations

It's very tempting to approach programming problems with a generic solution. We are trained to see patterns, and programming languages offer many abstractions and object-oriented paradigms to reuse more code.

As we could see in "Optimizing Latency" on page 383, while the `bytes.Split` and `strconv.ParseInt` functions are well designed, safe to use, and richer in features, they might not always be suitable for critical paths. Being "generic" has many drawbacks, and efficiency is usually the first victim.

Do things once

Was it done already? Perhaps we already loop over the same array somewhere else, so we could do more things "in place," as we did in Example 10-3.

There might be cases where we validate some invariant even though it was validated before. Or we sort again "just in case," but when we double-check the code, it was sorted already. For example, in the Thanos project, we can do a k-way merge (*https://oreil.ly/LxjZq*) instead of a naive merge and sort again when merging different metric streams because of the invariant that each stream gives metrics in lexicographic order.

Another common example is reusing memory. For instance, we can create a small buffer once and reuse it, as in Example 10-8, instead of creating a new one every time we need it. We can also use caching or "Memory Reuse and Pooling" on page 449.

Leverage math to do less

Using math is an amazing way to reduce the work we have to do. For example, to calculate the number of samples retrieved through the Prometheus API, we don't decode chunks and iterate over all samples to count them. Instead, we estimate the number of samples by dividing the size of the chunk by the average sample size.

Use the knowledge or precomputed information

Many APIs and functions are designed to be smart and automate certain work, even if it means doing more work. One example is pre-allocation possibilities, discussed in "Pre-Allocate If You Can" on page 441.

In another, more complex example, the `minio-go` (*https://oreil.ly/YqDZ6*) object storage client we use in objstore (*https://oreil.ly/l8xHu*) can upload an arbitrary `io.Reader` implementation. However, the implementation requires calculating the checksum before upload. Thus, if we don't give the total expected size of the bytes available in a reader, `minio-go` will use additional CPU cycles and memory to buffer the whole, potentially gigabytes-large object. All this just to calculate a checksum that has to be sometimes sent up front. On the other hand, if we notice this and have the total size handy, providing this information through the API can dramatically improve upload efficiency.

These elements seem like they focus on CPU time and latency, but we can use the same toward memory or any other resource usage. For example, consider a small example in Example 11-1 that shows what it means to do "less work" focused on lower memory usage.

Example 11-1. The function finding if the slice has a duplicated element optimized with an empty struct. Uses "Generics" on page 63.

```
func HasDuplicates[T comparable](slice ...T) bool {
    dup := make(map[T]any, len(slice))
    for _, s := range slice {
        if _, ok := dup[s]; ok {
            return true
        }
        dup[s] = "whatever, I don't use this value"
    }
    return false
}

func HasDuplicates2[T comparable](slice ...T) bool {
    dup := make(map[T]struct{}, len(slice))
    for _, s := range slice {
        if _, ok := dup[s]; ok {
            return true
        }
        dup[s] = struct{}{} ❶
    }
    return false
}
```

❶ Since we don't use the `map` value, we can use the `struct{}` statement, which uses no memory. Thanks to this, the `HasDuplicates2` on my machine is 22% faster and allocates 5 times less memory for a `float64` slice with 1 million elements. The same pattern can be used in places where we don't care about value. For example, for channels we use to synchronize goroutines, we can use `make(chan struct{})` to avoid unnecessary space we don't need.

Usually, there is always room to reduce some effort in our programs. We can use profiling to our advantage to check all expensive parts and their relevance to our problem. Often we can remove or transform those into cheaper forms, gaining efficiency.

Be Strategic!

Sometimes, doing less work now means more work or resource usage later. We can be strategic about this and ensure that our local benchmark doesn't miss the important trade-off elsewhere. This problem is highlighted in "Memory Reuse and Pooling" on page 449, where the macrobenchmark results give opposite conclusions to the microbenchmark.

Trading Functionality for Efficiency

In some cases, we have to negotiate or remove certain functionality to improve efficiency. In "Optimizing Latency" on page 383, we can improve the CPU time by removing support for negative integers in the file. Without this requirement, we can remove the check for negative sign in the Example 10-5 `ParseInt` function! Perhaps this feature is not well used, and it can be traded for cheaper execution!

This is also why accepting all the possible features in the project is often not very sustainable. In many cases, an extra API, extra parameter, or functionality might add a significant efficiency penalty for critical paths, which could be avoided if we just limit the functionality to a minimum.[1]

Trading Space for Time

What else can we do if we limit our program's work to a minimum by reducing unnecessary logic, features, and leaks? Generally, we can shift to systems, algorithms, or code that use less time but cost us more in terms of storage, like memory, disk, and so on. Let's walk through some possible changes like this:[2]

Precomputing result

Instead of computing the same expensive function, we could try to precompute it and store the result in some table lookup or variable.

These days, it's very common to see a compiler adapting optimization like this. The compiler trades compiler latency and program code space for faster execution. For example, statements like `10*1024*1024` or `20 * time.Seconds` can be precomputed by a compiler, so they don't have to be computed at runtime.

But there might be cases of more complex function statements that the compiler can't precompute for us. For example, we could use `regexp.MustCompile("...").MatchString(` in some condition, which is on a critical path. Perhaps it will be efficient to create a variable `pattern := regexp.MustCompile("...")` and operate on `pattern.MatchString(` in that heavily used code instead. On top of that, some cryptographic encryption offer precompute methods (*https://oreil.ly/2VBL4*) that speed up execution.

1 I spoke about this problem at the GitHub Global Maintainers Summit (*https://oreil.ly/z6YHe*).

2 This list was inspired by Chapter 4 in *Writing Efficient Programs* by Jon Louis Bentley.

Caching
: When the computed results heavily depend on the input, precomputing it for one input that is only used from time to time is not very helpful. Instead, we can introduce caching as we did in Example 4-1. Writing our caching solution is a nontrivial effort and should be done with care.[3] There are many caching policies (*https://oreil.ly/UAhqT*), with the Least Recently Used (LRU) being the most popular in my experience. In "Bonus: Thinking Out of the Box" on page 411, I mentioned a few off-the-shelf solutions in open source that we can use.

Augmenting data structure
: We can often change the data structure so certain information can be accessed more easily, or by adding more information to the structure. For example, we can store the size next to a file descriptor to know the file size instead of asking for it every time.

  In addition, we can maintain a map of elements next to the slice we already have in our structure, so we deduplicate or find elements easier (similar to the deduplication map I did in Example 11-1).

Decompressing
: Compression algorithms are great for saving disk or memory space. However, any compression—e.g., string interning, gzip, zstd (*https://oreil.ly/OEx9B*), etc.—have some CPU (thus, time) overhead, so when time is money, we might want to get rid of compression. Be careful, though, as enabled compression can improve program latency, e.g., when used for messages across slow networks. Therefore, spending more CPU time to reduce message size so that we can send more with a smaller number of network packets can potentially be faster.

Ideally, the decision is deliberate. For example, perhaps we know that based on the RAERs, our program can still use more memory, but we are not meeting the latency goal. In such a case, we could check if there is anything we can add, cache, or store that would allow you to spend less time in our program.

Trading Time for Space

If we can spare some latency or extra CPU time but are low on memory during the execution, we can try the opposite rule to the previous one, trading space for time. The methods are usually exactly the opposite of those in "Trading Space for Time" on page 419: compressing and encoding more, removing extra fields from the struct, recomputing results, removing caches, etc.

3 There's a reason some people call caches "a memory leak you don't know about yet" (*https://oreil.ly/KNQP3*).

Trading Space for Time or Time for Space Optimizations Is Not Always Intuitive

Sometimes to save memory resource usage, we have to allocate more first!

For example, in "Overusing Memory with Arrays" on page 446 and "Memory Reuse and Pooling" on page 449, I mention situations where allocating more memory or explicitly copying memory is better, despite looking like more work. So it can save us more memory space in the long run.

To sum up, consider the four general rules as higher-level patterns of possible optimizations. Let me now introduce you to the "three Rs," which helped me a lot to guide some of the optimizations in my efficiency development tasks.

The Three Rs Optimization Method

The three Rs technique is an excellent method to reduce waste. It is generally applicable for all computer resources, but it is often used for ecology purposes (*https://oreil.ly/p6elc*) to reduce literal waste. Thanks to those three ingredients—reduce, reuse, and recycle—we can reduce the impact we have on the Earth's environment and ensure sustainable living.

At FOSDEM (*https://fosdem.org*) 2018, I saw Bryan Boreham's amazing talk (*https://oreil.ly/BLIiT*), where he described using this method to mitigate memory issues. Indeed, the three Rs method is especially effective against memory allocations, which is the most common source of memory efficiency and GC overhead problems. So, let's explore each "R" component and how each can help.

Reduce Allocations

> Attempting to directly affect the pace [e.g., using `GOGC` or `GOMEMLIMIT`] of [garbage] collection has nothing to do with being sympathetic with the collector. It's really about getting more work done between each collection or during the collection. You affect that by reducing the amount or the number of allocations any piece of work adds to heap memory.
>
> —William Kennedy, "Garbage Collection in Go: Part I—Semantics" (*https://oreil.ly/DVdNm*)

There is almost always room to reduce allocations—look for the waste! Some ways to reduce the number of objects our code puts on the heap are obvious (reasonable optimizations like the pre-allocations of slices we saw in Example 1-4).

However, other optimizations require certain trade-offs—typically more CPU time or less readable code, for example:

- String interning (*https://oreil.ly/qJu7u*), where we avoid operating on the `string` type by providing a dictionary and using a much smaller, pointer-free dictionary of integers representing the ID of the string.
- Unsafe conversion from `[]byte` to `string` (and vice versa) without copying memory (*https://oreil.ly/Y10YT*), which potentially saves allocations, but if done wrongly can keep more memory in a heap (discussed in Example 11-15).
- Ensuring that a variable does not escape to the heap can also be considered an effort that reduces allocations.

There are unlimited different ways we could reduce allocations. We already mentioned some earlier. For example, when doing less work, we typically can allocate less! Another tip is to look for reducing allocations on all optimization design levels ("Optimization Design Levels" on page 98), not only code. In most cases, the algorithm must change first so we can have big improvements in the space complexity before we move to the code level.

Reuse Memory

Reusing is also an effective technique. As we learned in "Garbage Collection" on page 185, the Go runtime already reuses memory somehow. Still, there are ways to explicitly reuse objects like variables, slices, or maps for repeated operations instead of recreating them in every loop. We will discuss some techniques in "Memory Reuse and Pooling" on page 449.

Again, utilize all optimization design levels (see "Optimization Design Levels" on page 98). We can choose the designs of systems or algorithms that reuse memory; for example, see "Moving to Streaming Algorithm" on page 395. Another example of a "reuse" optimization on the system level is the TCP protocol. It offers to keep connections alive for reuse, which also helps with the network latency required to establish a new connection.

Be Careful When Reusing

Treating this tip literally is tempting—many try to go as far as reusing every little thing, including variables. As we learned in "Values, Pointers, and Memory Blocks" on page 176, variables are boxes that require some memory, but usually it's on the stack, so we should not be afraid to create more of them if needed. On the contrary, overusing variables can lead to hard-to-find bugs when we shadow variables (*https://oreil.ly/9Dfvb*).

Reusing complex structures can also be very dangerous for two reasons:[4]

- It is often not easy to reset the state of a complex structure before using it a second time (instead of allocating a new one, which creates a deterministic, empty structure).
- We cannot concurrently use those structures, which can limit further optimizations or surprise us and cause data races.

Recycle

Recycling is a minimum of what we must have in our programs if we use any memory. Fortunately, we don't need anything extra in our Go code, as it's the built-in GC's responsibility to recycle unused memory to the OS, unless we utilize advanced utilities like "mmap Syscall" on page 162 or other off-heap memory techniques.

However, if we can't "reduce" or "reuse" more memory, we can sometimes optimize our code or GC configuration, so the recycling is more efficient for the garbage collection. Let's go through some ways to improve recycling:

Optimize the structure of the allocated object

If we can't reduce the number of allocations, maybe we can reduce the number of pointers in our objects! However, avoiding pointers is not always possible, given popular structures like `time` (*https://oreil.ly/3ZmWi*), `string` (*https://oreil.ly/CIoPc*), or slices (*https://oreil.ly/Ow484*), which contain pointers. Especially `string` doesn't look like it, but it is just a special `[]byte`, which means it has a pointer to a byte array. In extreme cases, in certain conditions, it might be worth changing `[]string` into `offsets []int` and `bytes []byte` (*https://oreil.ly/0zi89*) to make it a pointer-free structure!

Another widespread example where it's easy to get very pointer-rich structures is when implementing data structures that are supposed to be marshaled and unmarshaled to different byte formats like JSON, YAML, or

4 See a nice blog post about those here (*https://oreil.ly/KrVnG*).

protobuf (*https://oreil.ly/yZVuB*). It is tempting to use pointers for nested structures to allow optionality of the field (the ability to differentiate if the field was set or not). Some code generation engines like Go protobuf generator (*https://oreil.ly/SeNub*) put all fields as pointers by default. This is fine for smaller Go programs, but if we use a lot of objects (which is common, especially if we use them for messages over the network), we might consider trying to remove pointers from those data structures (many generators and marshalers offer that option).

Reducing the number of pointers in our structures is better for GC and can make our data structure more L-cache friendly, decreasing the program latency. It also increases the chances that the compiler will put the data structure on the stack instead of the heap!

The main downside, however, is more overhead when you pass that struct by value (copy overhead mentioned in "Values, Pointers, and Memory Blocks" on page 176).

GC tuning

I mentioned in "Garbage Collection" on page 185 about two tuning options for Go GC: `GOGC` and `GOMEMLIMIT`.

Adjusting the `GOGC` option from the default 100% value might sometimes positively affect your program efficiency. Moving the next GC collection to happen sooner or later (depending on need) might be beneficial. Unfortunately, it requires lots of benchmarking to find the right number. It also does not guarantee that this tuning will work well for all possible states of your applications. On top of that, this technique has poor sustainability if you change the critical path in your code a lot. Every change requires another tuning session. This is why some bigger companies like Google and Uber (*https://oreil.ly/8YMRi*) invest in automated tools that adjust `GOGC` automatically in runtime!

The `GOMEMLIMIT` is another option you can adjust on top of the `GOGC`. It's a relatively new option for GC to run more frequently when the heap is close to or above the desired soft memory limit.

Using Kubernetes? Use GOMEMLIMIT Together with Pod Memory Limits

Some orchestration systems like Kubernetes allow setting hard resource limits (*https://oreil.ly/4zpkg*) on the workloads. For incompressible resources like memory, when the workload requires more memory as a limit, the system will typically OOM the process.

> The `GOMEMLIMIT` option is designed to help if the GC memory overhead is causing those OOMs (GC reacted to memory spikes). The official guide (*https://oreil.ly/zq6bb*) also suggests that we should leave an additional 5–10% of headroom to account for memory sources the Go runtime is unaware of. Setting the `GOMEMLIMIT` option to 90–95% of the workload memory limit might be quite effective.
>
> If we don't want to oversubscribe memory on our machines (*https://oreil.ly/GYTB9*), we can also set `GOGC=off` to trigger GC only if close to the memory limit, which can save some CPU time.

See a more detailed guide on GC tuning (*https://oreil.ly/3nGzV*) with the interactive visualizations.

Triggering GC and freeing OS memory manually
: In extreme cases, we might want to experiment with manually triggered GC collections using `runtime.GC()`. For example, we might want to trigger GC manually after an operation that allocated a lot of memory and no longer reference it. Note that a manual GC trigger is usually a strong anti-pattern, especially in libraries as it has global effects.[5]

Allocating objects off-heap
: We mentioned trying to allocate objects on the stack first instead of the heap. But the stack and heap are not our only options. There are ways to allocate memory off-heap, so that it's outside of the Go runtime's responsibility to manage.

We can achieve that with the explicit `mmap` syscall (*https://oreil.ly/yko2o*) we learned in "mmap Syscall" on page 162. Some have even tried calling C functions like `jemalloc` through the CGO (*https://oreil.ly/6se5i*).

While possible, we need to acknowledge that doing this can be compared to reimplementing parts of the Go Allocator from scratch, not to mention dealing with the manual allocations and lack of memory safety. It is the last thing we might want to try for the ultimate high-performance Go implementation!

On the bright side, this space is continuously improving. At the time of writing this book, the Go team approved and implemented an exciting proposal (*https://oreil.ly/jXgHY*) behind the `GOEXPERIMENT=arena` environment variable. It allows allocating a set of objects from the contiguous region of memory (`arena`) that lives outside of heap regions managed by GC. As a result, we will be able to iso-

5 For example, in the Prometheus project we removed (*https://oreil.ly/WFbrk*) the manual GC trigger when code conditions changed a little. That decision was based on micro- and macrobenchmarks discussed in Chapter 7.

late, track, and quickly release that memory explicitly when we need it (e.g., when an HTTP request is handled) without waiting or paying for garbage collection cycles. What's special about `arenas` is that it's meant to panic your program when you accidentally use the memory that was unused before assuring a certain level of memory safety. I can't wait to start playing with it once it is released—it might mean safe and easier-to-use off-heap optimizations.

Benchmarking and measuring all the effects of these optimizations is essential before trying any recycle improvements on our production code. Some of these can be considered tricky to maintain and unsafe if used without extensive tests.

To sum up, keep the three Rs method in mind, ideally in the same order: reduce, reuse, and recycle. Let's now dive into some common Go optimizations I have seen in my experience. Some of them might surprise you!

Don't Leak Resources

Resource leak is a common problem that reduces the efficiency of our Go programs. The leak occurs when we create some resource or background goroutine, and after using it, we want it to get released or stopped, but it is accidentally left behind. This might not be noticeable on a smaller scale, but sooner or later this can become a large and hard-to-debug issue. I suggest always clearing something you created, even if you expect to exit the program in the next cycle![6]

"This Program Has a Memory Leak!"

Not every higher memory utilization behavior can be considered a leak. For example, we could generally "waste" more memory for some operations, resulting in a spike in heap usage, but it gets cleared at some point.

Technically a leak is only when, for the same amount of load on the program (e.g., the same amount of HTTP traffic for a long-living service), we use an unbounded amount of resources (e.g., disk space, memory, rows in the database), which eventually run out.

There are cases of unexpected nondeterministic memory usage on the edge of the leak and waste. These are sometimes called pseudo-memory leaks, and we will discuss some of them in "Overusing Memory with Arrays" on page 446.

6 The reason is that we might reuse the same code in a more long-living scenario, where a leak might have much bigger consequences.

Perhaps we might think that memory should be an exception to this rule. The stack memory is automatically removed, and the garbage collection in Go dynamically removes the memory allocated on the heap.[7] There is no way to trigger the cleanup of a memory block other than stop referencing it and waiting (or triggering) a full GC cycle. However, don't let that fool you. There are many cases when the Go developer writes code that leaks memory, despite eventual garbage collection!

There are a few reasons our program leaks memory:

- Our program constantly creates custom `mmap` syscalls and never closes them (or closes them slower than creating them). This will typically end with a process or machine OOM.
- Our program calls too many nested functions, typically infinite or large recursion. Our process will then exit with a stack overflow error.
- We are referencing a slice with a tiny length, but we forgot that its capacity is very large, as explained in "Overusing Memory with Arrays" on page 446.
- Our program constantly creates memory blocks on the heap, which are always referenced by some variables in the execution scope. This typically means we have leaked goroutines or infinitely growing slices or maps.

It's easy to fix memory leaks when we know where they are, but it's not easy to spot them. We often learn about leaks after the fact, when our application has already crashed. Without advanced tools like those in "Continuous Profiling" on page 373, we have to hope to reproduce the problem with local tests, which is not always possible.

Even with the past heap profile, during the leak, we only see memory in the code that allocated memory blocks, not the code that currently references it.[8] Some of the memory leaks, especially those caused by leaked goroutines, can be narrowed down thanks to the goroutine, but not always.

Fortunately, a few best practices can proactively prevent us from leaking any incompressible resource (e.g., disk space, memory, etc.) and avoid that painful leak analysis. Consider the suggestions in this section as something we always care for and use as reasonable optimizations.

---

7 Unless we disabled it using the `GOGC=off` environment variable.

8 For that, we could use tools that analyze the dumped core (*https://oreil.ly/iTXhz*), but they aren't very accessible at the moment, so I would not recommend them.

Control the Lifecycle of Your Goroutines

> Every time you use the go keyword in your program to launch a goroutine, you must know how, and when, that goroutine will exit. If you don't know the answer, that's a potential memory leak.
>
> —Dave Cheney, "Never Start a goroutine Without Knowing How It Will Stop" (*https://oreil.ly/eZKzr*)

Goroutines are an elegant and clean framework for concurrent programming but have some downsides. One is that each goroutine is fully isolated from other goroutines (unless we use an explicit synchronization paradigm). There is no central dispatch in the Go runtime that we could call and, for example, ask to close the goroutines created by the current goroutine (or even check which one it created). This is not a lack of maturity of the framework, but rather a design choice allowing goroutines to be very efficient. As a trade-off, we have to implement potential code that will stop them when the job is done—or, to be specific, the code inside the goroutine to stop itself (the only way!).

The solution is never to create a goroutine and leave it on its own without strict control, even if we think the computation is fast. Instead, when scheduling goroutines, think about two aspects:

How to stop them
: We should always ask ourselves when the goroutine will finish. Will it finish on its own, or do I have to trigger the finish using context, channels, and so on (as in the examples that follow)? Should I be able to abort the goroutine long execution if, e.g., the request was cancelled?

Should my function wait for the goroutine to finish?
: Do I want my code to continue the execution without waiting for my goroutines to finish? Usually, the answer is no, and you should wait for the goroutine to stop, for example, using channels `sync.WaitGroup` (*https://oreil.ly/PQHom*) (e.g., in Example 10-10), `errgroup` (*https://oreil.ly/G1Aqx*), or the excellent `run.Group` (*https://oreil.ly/B1ABL*) abstraction.

There are many cases where it feels safe just to let the goroutines "eventually" stop, but in practice, not waiting for them has dangerous consequences. For example, consider the HTTP server handler that computes some number asynchronously in Example 11-2.

Example 11-2. Showcase of a common leak in a concurrent function

```
func ComplexComputation() int { ❶
    // Some computation...

    // Some cleanup...
```

```
    return 4
}

func Handle_VeryWrong(w http.ResponseWriter, r *http.Request) {
    respCh := make(chan int)

    go func() { ❷
        defer close(respCh) ❸
        respCh <- ComplexComputation()
    }()

    select { ❹
    case <-r.Context().Done():
        return ❺
    case resp := <-respCh:
        _, _ = w.Write([]byte(strconv.Itoa(resp)))
        return
    }
}
```

❶ Small function simulating longer computation. Imagine it takes around two seconds to complete all.

❷ Imagine a handler that schedules asynchronous computation.

❸ Our code does not depend on someone closing the channel, but as a good practice, the sender closes it.

❹ If cancellation happens, we return immediately. Otherwise, we wait for the result. At first glance, the above code does not look too bad. It feels like we control the lifecycle of the scheduled goroutine.

❺ Unfortunately, the detail is hidden in more information. We control the lifecycle only in a good case (when no cancellation occurs). If our code hits this line, we are doing something bad here. We return without caring about the goroutine lifecycle. We don't stop it. We don't wait for it. Even worse, this is a permanent leak, i.e., the goroutine with `ComplexCalculation` will be starved—as no one reads from the `respCh` channel.

While the goroutine looks like it's controlled, it isn't in all cases. This leaky code is commonly seen in the Go codebase because it requires a lot of detailed focus to not forget about every little edge case. As a result of these mistakes, we tend to delay using goroutines in our Go, as it's easy to create leaks like this.

The worst part about leaks is that our Go program might survive long before someone notices the adverse effects of such leaks. For example, running `Handle_VeryWrong` and cancelling it periodically will eventually OOM this Go program, but if we cancel only from time to time and restart our application periodically, without good observability we might never notice it!

Fortunately, an amazing tool allows us to discover those leaks at the unit test level. Therefore, I suggest using a leak test in every unit (or test file) that uses concurrent code. One of them is called `goleak` (*https://oreil.ly/4N4bb*) from Uber, and its basic use is presented in Example 11-3.

Example 11-3. Testing for leaks in Example 11-2 code

```
func TestHandleCancel(t *testing.T) { ❶
    defer goleak.VerifyNone(t) ❷

    w := httptest.NewRecorder()
    r := httptest.NewRequest("", "https://efficientgo.com", nil)

    wg := sync.WaitGroup{}
    wg.Add(1)

    ctx, cancel := context.WithCancel(context.Background())
    go func() {
        Handle_VeryWrong(w, r.WithContext(ctx))
        wg.Done()
    }()
    cancel()

    wg.Wait()
}
```

❶ Let's create tests that verify cancel behavior. This is where the leak is suspected to be triggered.

❷ To verify goroutine leaks, just defer `goleak.VerifyNone` (*https://oreil.ly/bgcwF*) at the top of our test. It runs at the end of our test and fails if any unexpected goroutine is still running. We can also verify whole package tests using the `goloak.VerifyTestMain` method (*https://oreil.ly/zyPjr*).

Running such a test causes the test to fail with the output in Example 11-4.

Example 11-4. Output of two failed runs of Example 11-3

```
=== RUN   TestHandleCancel
    leaks.go:78: found unexpected goroutines:
        [Goroutine 8 in state sleep, with time.Sleep on top of the stack:
        goroutine 8 [sleep]: ❶
        time.Sleep(0x3b9aca00)
            /go1.18.3/src/runtime/time.go:194 +0x12e
        github.com/efficientgo/examples/pkg/leak.ComplexComputation()
```

```
            /examples/pkg/leak/leak_test.go:107 +0x1e
        github.com/efficientgo/examples/pkg/leak.Handle_VeryWrong.func1()
            /examples/pkg/leak/leak_test.go:117 +0x5d
        created by github.com/efficientgo/examples/pkg/leak.Handle_VeryWrong
            /examples/pkg/leak/leak_test.go:115 +0x7d
        ]
--- FAIL: TestHandleCancel (0.44s)
=== RUN   TestHandleCancel
    leaks.go:78: found unexpected goroutines:
        [Goroutine 21 in state chan send, with Handle_VeryWrong.func1 (...):
        goroutine 21 [chan send]: ❷
        github.com/efficientgo/examples/pkg/leak.Handle_VeryWrong.func1()
            /examples/pkg/leak/leak_test.go:117 +0x71
        created by github.com/efficientgo/examples/pkg/leak.Handle_VeryWrong
            /examples/pkg/leak/leak_test.go:115 +0x7d
        ]
--- FAIL: TestHandleCancel (3.44s)
```

❶ We see the goroutines still running at the end of the test and what they were executing.

❷ If we waited a few seconds after cancelling, we could see that the goroutine was still running. However, this time it was waiting on a read from `respCh`, which would never happen.

The solution to such an edge case leak is to fix the Example 11-2 code. So let's go through two potential solutions in Example 11-5 that seem to fix the problem, but still leak in some way!

Example 11-5. (Still) leaking handlers. This time the goroutines left behind eventually stop.

```
func Handle_Wrong(w http.ResponseWriter, r *http.Request) {
    respCh := make(chan int, 1) ❶

    go func() {
        defer close(respCh)
        respCh <- ComplexComputation()
    }()

    select {
    case <-r.Context().Done():
        return
    case resp := <-respCh:
        _, _ = w.Write([]byte(strconv.Itoa(resp)))
        return
    }
}
```

```
func Handle_AlsoWrong(w http.ResponseWriter, r *http.Request) {
    respCh := make(chan int, 1)

    go func() {
        defer close(respCh)
        respCh <- ComplexComputationWithCtx(r.Context()) ❷
    }()

    select {
    case <-r.Context().Done():
        return
    case resp := <-respCh:
        _, _ = w.Write([]byte(strconv.Itoa(resp)))
        return
    }
}

func ComplexComputationWithCtx(ctx context.Context) (ret int) {
    var done bool
    for !done && ctx.Err == nil {
        // Some partial computation...
    }

    // Some cleanup... ❸
    return ret
}
```

❶ The only difference between this code and `HandleVeryWrong` in Example 11-2 is that we create a channel with a buffer for one message. This allows the computation goroutine to push one message to this channel without waiting for someone to read it. If we cancel and wait some time, the "left behind" goroutine will eventually finish.

❷ To make things more efficient, we could even implement a `ComplexComputationWithCtx` that accepts context, which cancels computation and is no longer needed.

❸ Many context-cancelled functions do not finish immediately when the context is cancelled. Perhaps context is checked periodically, or some cleanup might be needed to revert cancelled changes. In our case, we simulate cleanup wait time with sleep.

The examples in Example 11-5 provide some progress, but unfortunately, they still technically leak. In some ways, the leak is only temporary, but it can still cause problems for the following reasons:

Unaccounted resource usage.
: If we used the `Handle_AlsoWrong` function for request A, then A would cancel. As a result, the `ComplexComputation` would accidentally allocate a lot of memory after `Handle_AlsoWrong` finished—it would create a confusing situation. Furthermore, all observability tools would indicate that a spike of memory happened after request A finished, so it would be a false perception that request A is not correlated to the memory problem.

  Accounting problems can have big consequences on the future scalability of our program. For example, imagine that a cancelled request usually takes 200 ms to finish. That's not true—if we accounted for all computations, we would see it's 200 ms with, e.g., 1 second for `ComplexComputation` cleanup latency. This calculation is very important when predicting resource usage for certain traffic given certain machine resources.

We can run out of resources sooner.
: Such "left behind" goroutines can still cause OOM as the usage is nondeterministic. Continuous runs and cancels can still give the impression that the server is ready to schedule another request, and keep adding leaked asynchronous jobs, which can eventually starve the program. This situation fits in the leak definition.

Are we sure they finished?
: Furthermore, leaving behind goroutines gives us no visibility on how long they run and if they finished in all edge cases. Perhaps there is a bug that gets them stuck at some point.

As a result, I would highly suggest never leaving behind goroutines in your code. Fortunately, Example 11-3 marks all three functions (`Handle_VeryWrong`, `Handle_Wrong`, and `Handle_AlsoWrong`) as leaking, which is usually what we want. To fix the leak completely, we can, in our case, always wait for the result channel, as presented in Example 11-6.

Example 11-6. Version of Example 11-2 that is not leaking

```
func Handle_Better(w http.ResponseWriter, r *http.Request) {
    respCh := make(chan int)

    go func() {
        defer close(respCh)
        respCh <- ComplexComputationWithCtx(r.Context())
    }()

    resp := <-respCh ❶
    if r.Context().Err() != nil {
        return
```

```
    }
    _, _ = w.Write([]byte(strconv.Itoa(resp)))
}
```

❶ Always reading from the channel allows us to wait for the goroutine stop. We also respond to cancel as quickly as possible, thanks to propagating proper context to `ComplexComputationWithCtx`.

Last but not least, be careful when you benchmark concurrent code. Always wait in each `b.N` iteration for what you want to define as "an operation." A common leak in benchmarking code with the solution is presented in Example 11-7.

Example 11-7. Showcase of a common leak in benchmarking concurrent code

```
func BenchmarkComplexComputation_Wrong(b *testing.B) { ❶
    for i := 0; i < b.N; i++ {
        go func() { ComplexComputation() }()
        go func() { ComplexComputation() }()
    }
}

func BenchmarkComplexComputation_Better(b *testing.B) { ❷
    defer goleak.VerifyNone(
        b,
        goleak.IgnoreTopFunction("testing.(*B).run1"),
        goleak.IgnoreTopFunction("testing.(*B).doBench"),
    ) ❸

    for i := 0; i < b.N; i++ {
        wg := sync.WaitGroup{}
        wg.Add(2)

        go func() {
            defer wg.Done()
            ComplexComputation()
        }()
        go func() {
            defer wg.Done()
            ComplexComputation()
        }()
        wg.Wait()
    }
}
```

❶ Let's say we want to benchmark concurrent `ComplexComputation`. Scheduling two goroutines might find some interesting slowdowns if any resources are shared between those functions. However, these benchmark results are completely wrong. My machine shows `1860 ns/op`, but if we look carefully, we will see we don't wait for any of those goroutines to complete. As a result, we only measure the latency needed to schedule two goroutines per operation.

❷ To measure the latency of two concurrent computations, we have to wait for their completion, perhaps with `sync.WaitGroup`. This benchmark shows a much more realistic `2000339135 ns/op` (two seconds per operation) result.

❸ We can also use `goleak` on our benchmarks to verify against leaks! However, we need to have a benchmark-specific filter due to this issue (*https://oreil.ly/VTE9t*).

To sum up, control your goroutine lifecycle for reliable efficiency now and in the future! Ensure the goroutine lifecycle as a reasonable optimization.

Reliably Close Things

This might be obvious, but if we create some object that is supposed to be closed after use, we should ensure we don't forget or ignore this. We have to be extra careful if we create an instance of some `struct` or use a function, and we see some kind of "closer," for example:

- It returns `cancel` or `close` closure, e.g., `context.WithTimeout` (*https://oreil.ly/lmvQd*) or `context.WithCancel` (*https://oreil.ly/aVkMY*).[9]
- The returned object has a method with closing, cancelling, or stopping-like semantics, e.g., `io.ReaderCloser.Close()` (*https://oreil.ly/7Lyfs*), `time.Timer.Stop()` (*https://oreil.ly/V7ba8*), or TearDown.
- Some functions do not have a closer method but have a dedicated closing or deleting package-level function, e.g., the corresponding "releasing" function for `os.Create` (*https://oreil.ly/a2nt4*) or `os.Mkdir` (*https://oreil.ly/klgKo*) is `os.Remove` (*https://oreil.ly/DPNIA*).

If we have such a situation, assume the worst: if we don't call that function at the end of using that object, bad things will happen. Some goroutine will not finish, some memory will be kept referenced, or worse, our data will not bet saved (e.g., in case of `os.File.Close()`). We should try to be vigilant. When we use a new abstraction, we should check if it has any closers. Unfortunately, there are no linters that would point out if we forgot to call them.[10]

---

9 Yes! If we don't invoke the returned `context.CancelContext` function, it will keep a goroutine running forever (when `WithContext` was used) or until the timeout (`WithTimeout`).

10 I have only seen linters that check some basic things like if the code closed request body (*https://oreil.ly/DpSLY*), or sql statements (*https://oreil.ly/EVB8M*). There is room to contribute more of those, e.g., in the `semgrep-go` project (*https://oreil.ly/WfmyC*).

Unfortunately, that isn't everything. We can't just defer a call to `Close`. Typically, it also returns the error, which might mean the close could not happen, and this situation has to be handled. For example, `os.Remove` failed because of permission issues and the file was not removed. If we cannot exit the application, retry, or handle the error, we should at least be aware of this potential leak.

Does it mean that `defer` statements are less useful, and we have to have that `if err != nil` boilerplate for all closers? Not really. This is when I would suggest using the `errcapture` (*https://oreil.ly/ucTUB*) and `logerrcapture` (*https://oreil.ly/vb2vn*) packages. See Example 11-8.

Example 11-8. Examples of closing files with `defer`

```
// import "github.com/efficientgo/core/logerrcapture"
// import "github.com/efficientgo/core/errcapture"

func doWithFile_Wrong(fileName string) error {
    f, err := os.Open(fileName)
    if err != nil {
        return err
    }
    defer f.Close() // Wrong! ❶

    // Use file...

    return nil
}

func doWithFile_CaptureCloseErr(fileName string) (err error) { ❷
    f, err := os.Open(fileName)
    if err != nil {
        return err
    }
    defer errcapture.Do(&err, f.Close, "close file") ❷

    // Use file...

    return nil
}

func doWithFile_LogCloseErr(logger log.Logger, fileName string) {
    f, err := os.Open(fileName)
    if err != nil {
        level.Error(logger).Log("err", err)
        return
    }
    defer logerrcapture.Do(logger, f.Close, "close file") ❸

    // Use file...
}
```

❶ Never ignore errors. Especially on a file close, which often flushes some of our writes to disk only on `Close`, we lose data on an error.

❷ Fortunately, we don't need to give up on the amazing Go `defer` logic. Using `errcapture`, we can return an error if `f.Close` returns an error. If `doWithFile_CaptureCloseErr` returns an error and we do `Close`, the potential close error will be appended to the returned one. This is possible thanks to the return argument `(err error)` of this function. This pattern will not work without it!

❸ We can also log the close error if we can't handle it.

If we see any project I was involved in (and influenced to impact patterns like this), I use `errcapture` in all functions that return errors, and I can defer them—a clean and reliable way to avoid some leaks.

Another common example of when we forget to close things is error cases. Suppose we have to open a set of files for later use. Making sure we close them is not always trivial, as shown in Example 11-9.

Example 11-9. Closing files in error cases

```
// import "github.com/efficientgo/core/merrors"

func openMultiple_Wrong(fileNames ...string) ([]io.ReadCloser, error) {
    files := make([]io.ReadCloser, 0, len(fileNames))
    for _, fn := range fileNames {
        f, err := os.Open(fn)
        if err != nil {
            return nil, err // Leaked files! ❶
        }
        files = append(files, f)
    }
    return files, nil
}

func openMultiple_Correct(fileNames ...string) ([]io.ReadCloser, error) {
    files := make([]io.ReadCloser, 0, len(fileNames))
    for _, fn := range fileNames {
        f, err := os.Open(fn)
        if err != nil {
            return nil, merrors.New(err, closeAll(files)).Err() ❷
        }
        files = append(files, f)
    }
    return files, nil
}
```

```
func closeAll(closers []io.ReadCloser) error {
    errs := merrors.New()
    for _, c := range closers {
        errs.Add(c.Close())
    }
    return errs.Err()
}
```

❶ This is often difficult to notice, but if we create more resources that have to be closed, or we want to close them in a different function, `defer` can't be used. This is normally fine, but if we want to create three files and we have an error when opening the second one, we are leaking resources for the first nonclosed file! We cannot just return the files opened so far from `openMultiple_Wrong` and an error because the consistent flow is to ignore anything returned if there was an error. We typically have to close the already opened file to avoid leaks and confusion.

❷ The solution is typically creating a short helper that will iterate over appended closers and close them. For example, we use the `merrors` (*https://oreil.ly/icRMt*) package for convenient error append, because we want to know if any new error happened in any `Close` call.

To sum up, closing things is very important and considered a good optimization. Of course, no single pattern or linter would prevent us from all mistakes, but we can do a lot to reduce that risk.

Exhaust Things

To make things more complex, certain implementations require us to do more work to release all resources fully. For example, an `io.Reader` (*https://oreil.ly/HR89x*) implementation might not give the `Close` method, but it might assume that all bytes will be read fully. On the other hand, some implementations might have a `Close` method, yet still expect us to "exhaust" the reader for efficient use.

One of the most popular implementations that have such behavior are the `http.Request` (*https://oreil.ly/3Gq9j*) and `http.Response` (*https://oreil.ly/3L02L*) body `io.ReadCloser` from the standard library. The problem is shown in Example 11-10.

Example 11-10. An example of the inefficiency of `http/net` Client caused by a wrongly handled HTTP response

```
func handleResp_Wrong(resp *http.Response) error { ❶
    if resp.StatusCode != http.StatusOK {
        return errors.Newf("got non-200 response; code: %v", resp.StatusCode)
    }
    return nil
```

```
}

func handleResp_StillWrong(resp *http.Response) error {
    defer func() {
        _ = resp.Body.Close() ❷
    }()

    if resp.StatusCode != http.StatusOK {
        return errors.Newf("got non-200 response; code: %v", resp.StatusCode)
    }
    return nil
}

func handleResp_Better(resp *http.Response) (err error) {
    defer errcapture.ExhaustClose(&err, resp.Body, "close") ❸

    if resp.StatusCode != http.StatusOK {
        return errors.Newf("got non-200 response; code: %v", resp.StatusCode)
    }
    return nil
}

func BenchmarkClient(b *testing.B) {
    defer goleak.VerifyNone(
        b,
        goleak.IgnoreTopFunction("testing.(*B).run1"),
        goleak.IgnoreTopFunction("testing.(*B).doBench"),
    )

    c := &http.Client{}
    defer c.CloseIdleConnections() ❹

    b.ResetTimer()
    for i := 0; i < b.N; i++ {
        resp, err := c.Get("http://google.com")
        testutil.Ok(b, err)
        testutil.Ok(b, handleResp_Wrong(resp))
    }
}
```

❶ Imagine we are designing a function that handles an HTTP response from a `http.Client.Get` (*https://oreil.ly/uB0Vd*) request. `Get` clearly mentions that the "caller should close resp.Body when done reading from it." This `handleResp_Wrong` is wrong because it leaks two goroutines:

- One doing `net/http.(*persistConn).writeLoop`
- The second doing `net/http.(*persistConn).readLoop`, which is visible when we run `BenchmarkClient` with the `goleak`

❷ The `handleResp_StillWrong` is better, as we stop the main leak. However, we still don't read bytes from the body. We might not need them, but the `net/http` implementations can block the TCP connection if we don't fully exhaust the body. Unfortunately, this is not well-known information. It is briefly mentioned in the `http.Client.Do` (*https://oreil.ly/RegPv*) method description: "If the Body is not both read to EOF and closed, the Client's underlying RoundTripper (typically Transport) may not be able to re-use a persistent TCP connection to the server for a subsequent 'keep-alive' request."

❸ Ideally, we read until the EOF (end of file), representing the end of whatever we are reading. For this reason we created convenient helpers like `ExhaustClose` from `errcapture` (*https://oreil.ly/4LhOs*) or `logerrcapture` (*https://oreil.ly/XRxyA*) that do exactly this.

❹ Client runs some goroutines for each TCP connection we want to keep alive and reuse. We can close them using `CloseIdleConnection` to detect any leaks our code might introduce.

I wish structures like `http.Response.Body` were easier to use. The close and exhaust need for the body are important and should be used as a reasonable optimization. `handleResp_Wrong` fails the `BenchmarkClient` with a leak error. The `handleResp_StillWrong` does not leak any goroutine, so the leak test passes. The "leak" is on a different level, the TCP level, with the TCP connection being unable to reuse, which can cost us extra latency and insufficient file descriptors.

We can see its impact with the results of the `BenchmarkClient` benchmark in Example 11-10. On my machine, it takes 265 ms to call `http://google.com` with `handleResp_StillWrong`. For the version that cleans all resources in `handleResp_Better`, it takes only 188 ms, which is 29% faster![11]

The need for exhaust is also visible in `http.HandlerFunc` code. We should always ensure our server implementation exhausts and closes the `http.Request` body. Otherwise, we will have the same problem as in Example 11-10. Similarly, this can be true for all sorts of iterators; for example, a Prometheus storage can have a `ChunkSeriesSet` iterator (*https://oreil.ly/voRFc*). Some implementations can leak or overuse resources if we forget to iterate through all items until `Next()` equals false.

To sum up, always check the implementation for those nontrivial edge cases. Ideally, we should design our implementations to have obvious efficiency guarantees.

11 Which is quite interesting, considering we do more work in our code. We read through all bytes of the HTML returned by Google. Yet, it's faster as we create fewer TCP connections.

Let's now dive into the pre-allocation technique I mentioned in previous chapters.

Pre-Allocate If You Can

I mentioned pre-allocation in "Optimized Code Is Not Readable" on page 7 as a reasonable optimization. I showed how easy it is to pre-allocate a slice with `make` in Example 1-4 as an optimization to `append`. Generally, we want to reduce the amount of work that code has to do to resize or allocate new items if we know the code has to do it eventually.

The `append` example is important, but there are more examples. It turns out that almost every container implementation that cares about efficiency has some easier pre-allocation methods. See the ones in Example 11-11 with explanations.

Example 11-11. Examples of pre-allocation for some common types

```
const size = 1e6 ❶

slice := make([]string, 0, size) ❷
for i := 0; i < size; i++ {
    slice = append(slice, "something")
}

slice2 := make([]string, size) ❸
for i := 0; i < size; i++ {
    slice2[i] = "something"
}

m := make(map[int]string, size) ❹
for i := 0; i < size; i++ {
    m[i] = "something"
}

buf := bytes.Buffer{} ❺
buf.Grow(size)
for i := 0; i < size; i++ {
    _ = buf.WriteByte('a')
}

builder := strings.Builder{}
builder.Grow(size)
for i := 0; i < size; i++ {
    builder.WriteByte('a')
}
```

❶ Let's assume we know the size we want to grow the containers up front.

❷ `make` with slices allows us to grow the capacity of the underlying arrays to the given size. Thanks to the proactive growth of the array with `make`, the loop with `append` is much cheaper in CPU time and memory allocation. This is because `append` does not need to resize the array when it's too small.

Resizing is quite naive. It simply creates a new, bigger array and copies all elements. A certain heuristic also tells how many new slices are grown. This heuristic was recently changed (*https://oreil.ly/6uIHH*), but it will still allocate and copy a few times until it extends to our expected one million elements. In our case, the same logic is 8 times faster with pre-allocation and allocates 16 MB instead of 88 MB of memory.

❸ We can also pre-allocate the slice's capacity and length. Both `slice` and `slice2` will have the same elements. Both ways are almost equally efficient, so we use one that fits more functionally to what we need to do. However, with `slice2`, we are using all array elements, whereas in `slice`, we can grow it to be bigger but end up using a smaller number if needed.[12]

❹ Map can be created using `make` with an optional number representing its capacity. If we know the size up front, it's more efficient for Go to create the required internal data structure with up-front sizes. The efficiency results show the difference—on my machine, with pre-allocation, such map initialization takes 87 ms, without 179 ms! The total allocated space with pre-allocation is 57 MB, without 123 MB. However, map insertion can still allocate some memory, just much smaller than pre-allocation.

❺ Various buffers and builders offer the `Grow` function that also pre-allocates.

The preceding example is actually something I use very often during almost every coding session. Pre-allocation usually takes the extra line of code, but it is a fantastic, more readable pattern. If you are still not convinced that you won't have a lot of situations when you know the size up front for the slice, let's talk about `io.ReadAll`. We use `io.ReadAll` (*https://oreil.ly/TN7bt*) (previously `ioutil.ReadAll` (*https://oreil.ly/nt1oT*)) functions in the Go community a lot. Did you know you can optimize it significantly by pre-allocating the internal byte slice if you know the size up front? Unfortunately, `io.ReadAll` does not have a `size` or `capacity` argument, but there is a simple way to optimize it, as presented in Example 11-12.

12 This is often used when we know only the worst-case `size`. Sometimes it's worth growing it to the worst case, even if we use less in the end. See "Overusing Memory with Arrays" on page 446.

Example 11-12. Examples of ReadAll optimizations with the benchmark

```
func ReadAll1(r io.Reader, size int) ([]byte, error) {
   buf := bytes.Buffer{}
   buf.Grow(size)
   n, err := io.Copy(&buf, r) ❶
   return buf.Bytes()[:n], err
}

func ReadAll2(r io.Reader, size int) ([]byte, error) {
   buf := make([]byte, size)
   n, err := io.ReadFull(r, buf) ❷
   if err == io.EOF {
      err = nil
   }
   return buf[:n], err
}

func BenchmarkReadAlls(b *testing.B) {
   const size = int(1e6)
   inner := make([]byte, size)

   b.Run("io.ReadAll", func(b *testing.B) {
      b.ReportAllocs()
      for i := 0; i < b.N; i++ {
         buf, err := io.ReadAll(bytes.NewReader(inner))
         testutil.Ok(b, err)
         testutil.Equals(b, size, len(buf))
      }
   })

   b.Run("ReadAll1", func(b *testing.B) {
      b.ReportAllocs()
      for i := 0; i < b.N; i++ {
         buf, err := ReadAll1(bytes.NewReader(inner), size)
         testutil.Ok(b, err)
         testutil.Equals(b, size, len(buf))
      }
   })

   b.Run("ReadAll2", func(b *testing.B) {
      b.ReportAllocs()
      for i := 0; i < b.N; i++ {
         buf, err := ReadAll2(bytes.NewReader(inner), size)
         testutil.Ok(b, err)
         testutil.Equals(b, size, len(buf))
      }
   })
}
```

❶ One way of simulating `ReadAll` is by creating a pre-allocated buffer and using `io.Copy` to copy all bytes.

❷ Even more efficient is pre-allocating a byte slice and using `ReadFull`, which is similar. `ReadAll` does not use the `io.EOF` error sentinel if everything is read, so we need special handling for it.

The results, presented in Example 11-13, speak for themselves. The `ReadAll2` using `io.ReadFull` is over eight times faster and allocates five times less memory for our one million byte slice.

Example 11-13. Results of the benchmark in Example 11-12

```
BenchmarkReadAlls
BenchmarkReadAlls/io.ReadAll
BenchmarkReadAlls/io.ReadAll-12  1210   872388 ns/op  5241169 B/op  29 allocs/op
BenchmarkReadAlls/ReadAll1
BenchmarkReadAlls/ReadAll1-12    8486   165519 ns/op  1007723 B/op  4 allocs/op
BenchmarkReadAlls/ReadAll2
BenchmarkReadAlls/ReadAll2-12    10000  102414 ns/op  1007676 B/op  3 allocs/op
PASS
```

The `io.ReadAll` optimization is very often possible in our Go code. Especially when dealing with HTTP code, the request or response headers often offer a `Content-Length` header that allows pre-allocations.[13] The preceding examples represent only a small subset of types and abstractions that allow pre-allocation. Check the documentation and code of the type we use if we can average eager allocations for better efficiency.

However, there is one more amazing pre-allocation pattern I would like you to know. Consider a simple, singly linked list. If we implement it using pointers, and if we know we will insert millions of new elements on that list, is there a way to pre-allocate things for efficiency? Turns out there might be, as shown in Example 11-14.

Example 11-14. Basic pre-allocation of linked list elements

```
type Node struct {
    next *Node
    value int
}

type SinglyLinkedList struct {
    head *Node

    pool      []Node ❶
    poolIndex int
```

13 For example, this is what we did in Thanos (*https://oreil.ly/8nWCH*) some time ago.

```
}

func (l *SinglyLinkedList) Grow(len int) { ❷
    l.pool = make([]Node, len)
    l.poolIndex = 0
}

func (l *SinglyLinkedList) Insert(value int) {
    var newNode *Node
    if len(l.pool) > l.poolIndex { ❸
        newNode = &l.pool[l.poolIndex]
        l.poolIndex++
    } else {
        newNode = &Node{}
    }

    newNode.next = l.head
    newNode.value = value
    l.head = newNode
}
```

❶ This line makes this linked list a bit special. We maintain a pool of objects in the form of one slice.

❷ Thanks to the pool, we can implement our own `Grow` method, which will allocate a pool of many `Node` objects within one allocation. Generally, it's way faster to allocate one large `[]Node` than millions of `*Node`.

❸ During the insert, we can check if we have room in our pool and take one element from it instead of allocating an individual `Node`. This implementation can be expanded to be more robust, e.g., for subsequent growth, if we hit the capacity limit.

If we benchmarked the insertion of one million elements using the preceding linked list, we would see that the insertion takes four times less time with one eager allocation and the same space with just one allocation instead of one million.

The simple pre-allocation with slices and maps presented in Example 11-11 have almost no downsides, so they can be treated as reasonable optimizations. The pre-allocation presented in Example 11-14, on the other hand, should be done with care, deliberately, and with benchmarks as it's not without trade-offs.

First, the problem is that potential deletion logic or allowing the `Grow` call multiple times is not trivial to implement. The second issue is that a single `Node` element is now connected to a very large single memory block. Let's dive into this problem in the next section.

Overusing Memory with Arrays

As you probably know, slices are very powerful in Go. They offer robust flexibility for using arrays (*https://oreil.ly/YhOdH*) that is used daily in the Go community. But with power and flexibility comes responsibility. There are many cases where we might end up overusing memory, which some might call a "memory leak." The main problem is that those cases will never appear in "Go Benchmarks" on page 277, because it's related to garbage collection and will not release memory we thought could be released. Let's explore this problem in Example 11-15, which tests potential deletion in `SinglyLinkedList` introduced in Example 11-14.

Example 11-15. Reproducing memory overuse for a linked list that used pre-allocation in Example 11-14

```
func (l *SinglyLinkedList) Delete(n *Node) { /* ... */ } ❶

func TestSinglyLinkedList_Delete(t *testing.T) { ❷
    l := &SinglyLinkedList{}
    l.Grow(size)
    for k := 0; k < size; k++ {
        l.Insert(k)
    }
    l.pool = nil // Dispose pool. ❸
    _printHeapUsage() ❹

    // Remove all but last.
    for curr := l.head; curr.next != nil; curr = curr.next { ❺
        l.Delete(curr)
    }
    _printHeapUsage() ❻

    l.Delete(l.head)
    _printHeapUsage() ❼
}

func _printHeapUsage() {
    m := runtime.MemStats{}

    runtime.GC()
    runtime.ReadMemStats(&m)
    fmt.Println(float64(m.HeapAlloc)/1024.0, "KB")
}
```

❶ Let's add deletion logic to the linked list, which removes the given element.

❷ Using a microbenchmark to assess the efficiency of `Delete` would show us that when `Grow` was used, the deletion was only marginally faster. However, to show-case the memory overuse problem, we would need the macrobenchmarks test (see "Macrobenchmarks" on page 306). Alternatively, we can write a brittle interactive test as we did here.[14]

❸ Notice we are trying our best for the GC to remove the deleted node. However, we `nil` the `pool` variable, so the slice we used to create all nodes in the list is not referenced anywhere.

❹ We use a manual trigger for the GC and print of the heap, which is not very reliable generally as it contains allocations from background runtime work. However, it's good enough here to show us the problem. The pre-allocated list showed 15,818.5 KB in one of the runs, and 15,813.0 KB for the run without `Grow`. Don't look at the difference between those, but how this value changed for pre-allocated.

❺ Let's remove all but one element.

❻ In a perfect world, we would expect to hold only memory for one `Node`, right? This is the case for the non-pre-allocated list—189.85 KB on the heap. On the other hand, for the pre-allocated list, we can observe a certain problem: the heap is still big, with 15,831.2 KB on it!

❼ Only after all the elements do we see a small heap size for both cases (around 190 KB for both).

This problem is important to understand, and we have it every time we work with structs with arrays. The representation of what happens when all but one element is deleted in both cases is shown in Figure 11-1.

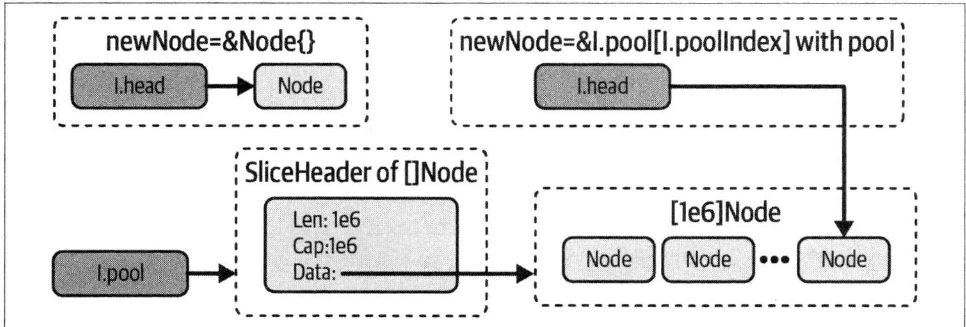

Figure 11-1. The heap's state with references with one node in the list. On the left, created without a pool, on the right with it.

14 This is great as a quick showcase, but does not work well as a reliable efficiency assessment.

When we allocate an individual object, we see that it receives its own memory block that can be managed in isolation. If we use pooling or subslicing (e.g., `buf[1:2]`) from a bigger slice, the GC will see that the big memory block for continuous memory used by the array is referenced. It's not smart enough to see that only 1% of it is used and could be "clipped."

The solution is to avoid pooling or come up with a more advanced pool that can be grown or shrunk (maybe even automatically). For example, if half of the objects are deleted, we can "clip" the array behind our linked list nodes. Alternatively, we can add the on-demand `ClipMemory` method, as presented in Example 11-16.

Example 11-16. Example implementation of clipping too-big memory block

```
func (l *SinglyLinkedList) ClipMemory() {
    var objs int
    for curr := l.head; curr != nil; curr = curr.next {
        objs++
    }

    l.pool = make([]Node, objs) ❶
    l.poolIndex = 0
    for curr := l.head; curr != nil; curr = curr.next {
        oldCurr := curr
        curr = &l.pool[l.poolIndex]
        l.poolIndex++

        curr.next = oldCurr.next ❷
        curr.value = oldCurr.value

        if oldCurr == l.head {
            l.head = curr ❸
        }
    }
}
```

❶ At this moment, we get rid of the reference to the old `[]Node` slice and create a smaller one.

❷ As we saw in Figure 11-1, there are still other references to bigger memory blocks from each element in the list. So we need to perform a copy using a new pool of objects to ensure the GC can remove that old bigger pool.

❸ Let's not forget about the last pointer, `l.head`, which would otherwise still point to the old memory block.

We can now use the `ClipMemory` when we delete some items to resize the underlying memory block.

As presented in Example 11-15, the overuse of memory is more common than we might think. However, we don't need such specific pooling to experience it. Subslicing and using clever zero copy functions like in Example 10-4 (`zeroCopyToString`) are very much prone to this problem.[15]

This section is not to demotivate you from pre-allocating things, subslicing, or experimenting with reusing byte slices. Rather it's a reminder to always keep in mind how Go manages memory (as discussed in "Go Memory Management" on page 172) when we attempt to do more advanced things with slices and underlying arrays.

Remember that Go benchmarking does not cover memory usage characteristics, as mentioned in "Microbenchmarks Versus Memory Management" on page 299. Move to the "Macrobenchmarks" on page 306 level to verify all efficiency aspects if you suspect you are affected by this problem.

Since we mentioned pooling, let's dive into the last section. What are the other ways to reuse and pool memory in Go? It turns out that sometimes not pooling anything might be better!

Memory Reuse and Pooling

Memory reuse allows using the same memory blocks for subsequent operations. If the operation we perform requires a bigger `struct` or `slice` and we perform a lot of them in a quick sequence, it's wasteful to allocate a new memory block every time because:

- Allocation of memory with guaranteed zero-ing of the memory block takes CPU time.
- We put more work into the GC, so more CPU cycles are used.
- The GC is eventual, so our maximum heap size can grow uncontrollably.

15 In the Prometheus project ecosystem, we experienced such a problem many times. For example, chunk pooling caused us to keep arrays that were way bigger than required, so we introduced the `Compact` method (*https://oreil.ly/ORx1C*). In Thanos, I introduced a (probably too) clever `ZLabel` construct (*https://oreil.ly/Z3Q8n*) that avoided expensive copy of strings for metric labels. It turned out to be beneficial for cases when we were not keeping the label strings for longer. For example, it was better to perform when we did a lazy copy (*https://oreil.ly/5o6sH*).

I already presented some memory reuse techniques in Example 10-8, using a small buffer to process files chunk by chunk. Then, in Example 11-14, I showed how we could allocate one bigger memory block at once and use that as our pool of objects.

The logic of reusing objects, especially byte slices, is often enabled by many popular implementations, such as `io.CopyBuffer` or `io.ReadFull`. Even our `Sum6Reader (r io.Reader, buf []byte)` from Example 10-8 allows further reuse of the buffer. However, memory reuse is not always so easy. Consider the following example of byte slice reuse in Example 11-17.

Example 11-17. Simple buffering or byte slice

```
func processUsingBuffer(buf []byte) {
	buf = buf[:0] ❶

	for i := 0; i < 1e6; i++ {
		buf = append(buf, 'a')
	}

	// Use buffer...
}

func BenchmarkProcess(b *testing.B) {
	b.Run("alloc", func(b *testing.B) {
		for i := 0; i < b.N; i++ {
			processUsingBuffer(nil) ❷
		}
	})
	b.Run("buffer", func(b *testing.B) {
		buf := make([]byte, 1e6)
		b.ResetTimer()
		for i := 0; i < b.N; i++ {
			processUsingBuffer(buf) ❸
		}
	})
}
```

❶ Because our logic uses `append`, we need to zero the length of the slice while reusing the same underlying array for efficiency.

❷ We can simulate no buffer by simply passing `nil`. Fortunately, Go handles nil slices in the operations like `buf[:0]` or `append([]byte(nil), 'a')`.

❸ Reusing the buffer is better in this case. On my machine, benchmarks show that each operation with reused buffer is almost two times faster and allocates zero bytes.

The preceding example looks excellent, but the real code contains complications and edge cases. Two main problems sometimes block us from implementing such naive memory reuse, as in Example 11-17:

- We know the buffer size will be similar for most operations, but we don't know the exact number. This can be easily fixed by passing an empty buffer and reusing the grown underlying array from the first operation.
- We might run the `processUsingBuffer` code concurrently at some point. Sometimes with four workers, sometimes with one thousand, sometimes with one. In this case, we could implement this by maintaining a static number of buffers. The number could be the maximum goroutines we want to run concurrently or less with some locking. This obviously can have a lot of waste if the number of goroutines is dynamically changing and is sometimes zero.

For those reasons, the Go team came up with the `sync.Pool` (*https://oreil.ly/BAQwU*) structure that performs a particular form of memory pooling. It's important to understand that memory pooling is not the same as typical caching.

> The type that Brad Fitzpatrick requested [`sync.Pool`] is actually a pool: A set of interchangeable values where it doesn't matter which concrete value you get out, because they're all identical. You wouldn't even notice when, instead of getting a value from the pool, you get a newly created one. Caches, on the other hand, map keys to concrete values.
>
> —Dominik Honnef, "What's Happening in Go Tip" (*https://oreil.ly/z6AUf*)

The `sync.Pool` from the standard library is implemented purely as a very short, temporary cache for the same type of free memory blocks that last until more or less the next GC invocation. It uses quite smart logic that makes it thread-safe yet avoids locking as much as possible for efficient access. The main idea behind `sync.Pool` is to reuse memory that the GC did not yet release. Since we keep those memory blocks around until eventual GC, why not make them accessible and useful? The example of using `sync.Pool` in Example 11-17 is presented in Example 11-18.

Example 11-18. Simple buffering using `sync.Pool`

```
func processUsingPool(p *sync.Pool) {
    buf := p.Get().([]byte) ❶
    buf = buf[:0]

    for i := 0; i < 1e6; i++ {
        buf = append(buf, 'a')
    }
    defer p.Put(buf) ❷

    // Use buffer...
```

```
}

func BenchmarkProcess(b *testing.B) {
    b.ReportAllocs()

    p := sync.Pool{
        New: func() any { return []byte{} }, ❸
    }
    b.ResetTimer()
    for i := 0; i < b.N; i++ {
        processUsingPool(&p) ❹
    }
}
```

❶ `sync.Pool` pools an object of the given type, so we must cast it to the type we put or create. When `Get` is involved, we either allocate a new object or use one of the pooled ones.

❷ To use the pool effectively, we need to put back the object to reuse. Remember to never put back the object you are still using to avoid races!

❸ The `New` closure specifies how a new object will be created.

❹ For our example, the implementation with `sync.Pool` is very efficient. It's over 2 times faster than without reuse, with an average of 2 KB of space allocated versus 5 MB allocated per operation from code that does not reuse the buffer.

While results look very promising, pooling using `sync.Pool` is a more advanced optimization that can bring more efficiency bottlenecks than optimizations if wrongly used. The first problem is that, as with any other complex structure that works with slices, using it is prone to errors. Consider the code with benchmark in Example 11-19.

Example 11-19. Common, hard-to-spot bug while using `sync.Pool` and defer

```
func processUsingPool_Wrong(p *sync.Pool) {
    buf := p.Get().([]byte)
    buf = buf[:0]

    defer p.Put(buf) ❶

    for i := 0; i < 1e6; i++ {
        buf = append(buf, 'a')
    }

    // Use buffer...
}
```

```
func BenchmarkProcess(b *testing.B) {
    p := sync.Pool{
        New: func() any { return []byte{} },
    }
    b.ResetTimer()
    for i := 0; i < b.N; i++ {
        processUsingPool_Wrong(&p) ❷
    }
}
```

❶ There is a bug in this function that defies the point of using `sync.Pool`—`Get` will always allocate an object in our case. Can you spot it?

The problem is that the `Put` might be deferred to the correct time, but its argument is evaluated at the moment of the `defer` schedule. As a result, the `buf` variable we are putting might point to a different slice if `append` will have to grow it.

❷ As a result, the benchmark will show that this `processUsingPool_Wrong` operation is twice as slow as the `alloc` case in Example 11-17 that always allocates. Using `sync.Pool` to only `Get` and never `Put` is slower than straight allocation (`make([]byte)` in our case).

However, the real difficulty comes from the specific `sync.Pool` characteristic: it only pools objects for a short duration, which is not reflected by our typical microbenchmark like in Example 11-18. We can see the difference if we trigger GC manually in our benchmark, done for demonstration in Example 11-20.

Example 11-20. Common, hard-to-spot bug while using `sync.Pool` and `defer`, triggering GC manually

```
func BenchmarkProcess(b *testing.B) {
    b.Run("buffer-GC", func(b *testing.B) {
        buf := make([]byte, 1e6)
        b.ResetTimer()
      for i := 0; i < b.N; i++ {
            processUsingBuffer(buf) ❶
            runtime.GC()
            runtime.GC()
        }
    })

    b.Run("pool-GC", func(b *testing.B) {
        p := sync.Pool{
            New: func() any { return []byte{} },
        }
        b.ResetTimer()
        for i := 0; i < b.N; i++ {
            processUsingPool(&p) ❷
```

```go
            runtime.GC()
            runtime.GC()
        }
    })
}
```

❶ The second surprise comes from the fact that in our initial benchmarks, the `process*` operations are performed quickly, one after another. However, on a macro level that might not be true. This is fine for `processUsingBuffer`. If the GC runs once or twice in the meantime for our simple buffered solution, the allocation and latency (adjusted with GC latency) stay the same because we keep the memory references in our `buf` variable. The next `processUsingBuffer` will be as fast as always.

❷ This is not the case for the standard pool. After two GC runs, the `sync.Pool` is, by design, fully cleaned from all objects,[16] which results in performance worse than `alloc` in Example 11-17.

As you can see, it's fairly easy to make mistakes using `sync.Pool`. The fact that it does not preserve the pool after garbage collection might be beneficial in cases where we don't want to keep pooled objects for a longer duration. However, in my experience, it makes it very hard to work with due to nondeterministic behavior caused by the combination of nontrivial `sync.Pool` implementation with an even more complex GC schedule.

To show the potential damage when `sync.Pool` is applied to the wrong workloads, let's try to optimize the memory use of the `labeler` service from "Go e2e Framework" on page 310 using optimized buffered code from Example 10-8 and four different buffering techniques:

`no-buffering`
: `Sum6Reader` without buffering—always allocates a new buffer.

`sync-pool`
: With `sync.Pool`.

16 If you are interested in the specific implementation details, check out this amazing blog post (*https://oreil.ly/oMh6I*).

`gobwas-pool`
: With `gobwas/pool` (*https://oreil.ly/VZjYW*) that maintains multiple buckets of `sync.Pool`. In theory, it should work well for byte slices that might require different buffer sizes.

`static-buffers`
: With four static buffers that offer a buffer for a maximum of four goroutines.

The main problem is that the Example 10-8 workload might not look immediately like a wrong fit. The small allocation of `make([]byte, 8*1024)` per operation is the only one we make during the computation, so pooling to save the total memory usage might feel like a valid choice. The microbenchmark also shows amazing results. The benchmarks perform sequential `Sum6` operations on two different files (50% of the time, we use files with 10 million numbers, 50% with 100 million). The results are shown in Example 11-21.

Example 11-21. The microbenchmark results with one hundred iterations that compare labeler `labelObject` logic using Example 10-8 and four different buffering versions

```
name                    time/op
Labeler/no-buffering     430ms ± 0%
Labeler/sync-pool        435ms ± 0%
Labeler/gobwas-pool      438ms ± 0%
Labeler/static-buffers 434ms ± 0%

name                    alloc/op
Labeler/no-buffering     3.10MB ± 0%
Labeler/sync-pool        62.0kB ± 0%
Labeler/gobwas-pool      94.5kB ± 0% ❶
Labeler/static-buffers 62.0kB ± 0%

name                    allocs/op
Labeler/no-buffering      3.00 ± 0%
Labeler/sync-pool         3.00 ± 0%
Labeler/gobwas-pool       3.00 ± 0%
Labeler/static-buffers  2.00 ± 0%
```

❶ The bucketed pool is slightly more memory intensive, but this is expected, as two separate pools are maintained. However, ideally, we expect to see larger benefits from that split on a larger scale.

We see that the `sync.Pool` version and static buffer are winning in terms of memory allocations. The latency is more or less similar, given most of Example 10-8 is spent on integer parsing, not allocating the buffer.

Unfortunately, on the macro level, for a 5-minute test per version with 2 virtual users in `k6s` performing a sum on 10 million lines and then 100 million line files, we see that the reality is different than what Example 11-21 showed. What's good is that the `labeler` without buffering allocates significantly more (3.3 GB in total) during that load than other versions (500 MB on average), as visible in Figure 11-2.

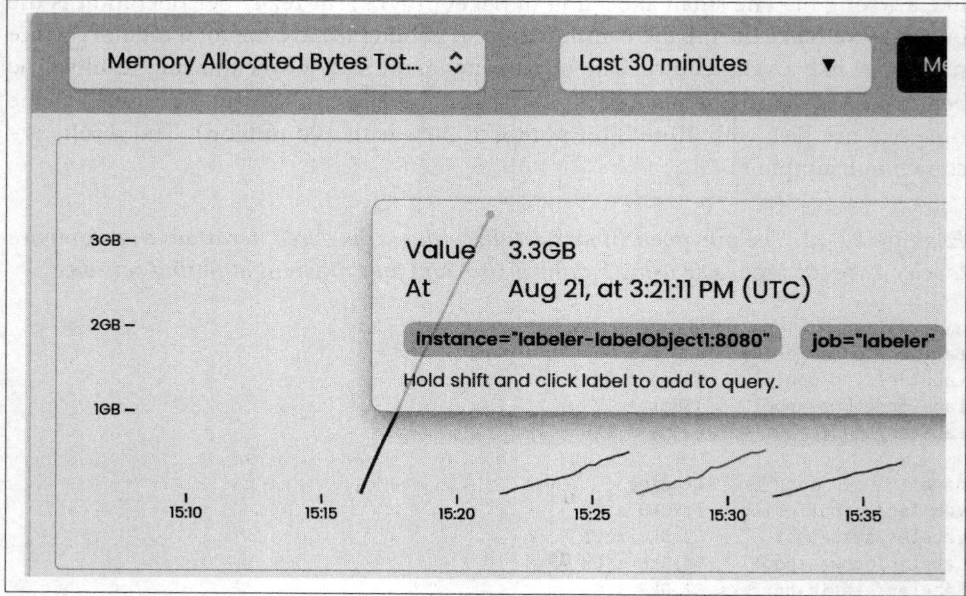

Figure 11-2. The Parca Graph for the total memory allocated during macrobenchmark from heap profiles. Four lines indicate runs of four different versions in order: `no-buffering`, `sync-pool`, `gobwas-pool`, and `static-buffers`.

However, it seems that such allocations are not a huge problem for the GC, as the simplest, no buffering solution `labelObject1` has similar average latency to others (same CPU usage as well), but also the lowest maximum heap usage, as visible in Figure 11-3.

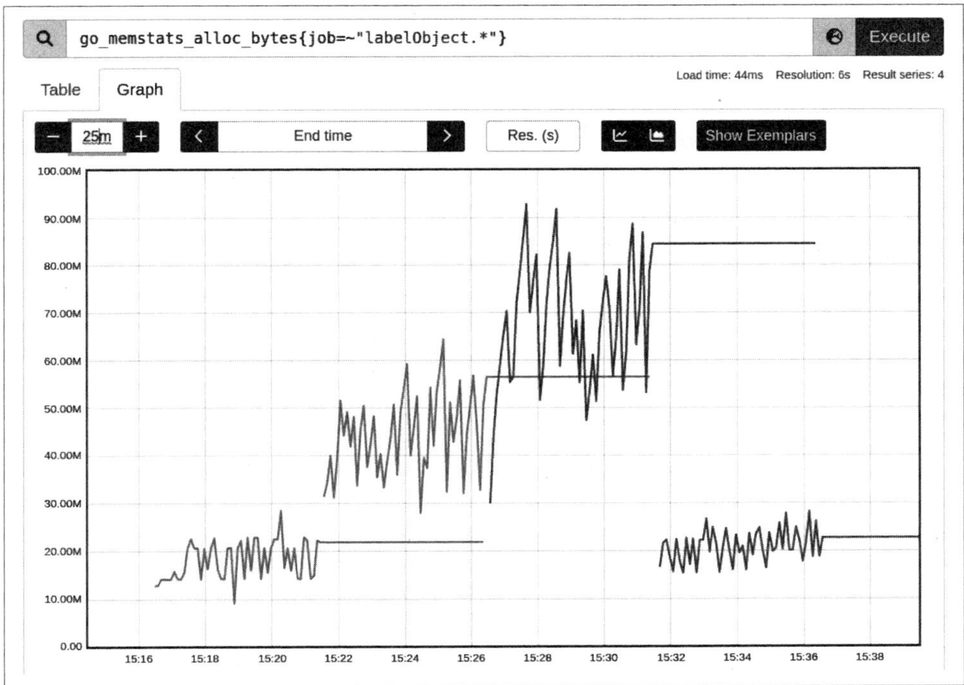

Figure 11-3. The Prometheus Graph for the heap size during the macrobenchmark. Four lines indicate runs of four different versions in order: `no-buffering`, `sync-pool`, `gobwas-pool`, *and* `static-buffers`.

You can reproduce the whole experiment thanks to the `e2e` framework code in the example repo (*https://oreil.ly/9vDNZ*). The results were not satisfying, but the experiment can give us a lot of lessons:

- Reducing allocations might be the easiest way to improve latency and memory efficiency, but not always! Clearly, in this case, higher allocations were better than pooling. One reason is that the `Sum6` in Example 10-8 was already heavily optimized. The CPU profile of `Sum6` in Example 10-8 clearly shows that allocation is not a latency bottleneck. Secondly, the slower allocation pace caused the GC to kick in less often, allowing generally higher maximum memory usage. Additional `GOGC` tuning might have helped here.
- The microbenchmarking does not always show the full picture. So always assess efficiency on multiple levels to be sure.
- The `sync.Pool` helps the most with allocation latency, not with maximum memory usage, as our goal here.

The Optimization Journey Can Be a Roller Coaster!

Sometimes we achieve improvement, and sometimes we spend a few days on change that can't be merged. We all learn every day, try things, and sometimes fail. What's most important is to fail early, so the less efficient version is not accidentally released to our users!

The main issue of this experiment is that the `sync.Pool` is not designed for the type of workload that `labeler` represents. The `sync.Pool` have very specific use cases. Use it when:

- You want to reuse large or extreme amounts of objects to reduce the latency of those allocations.
- You don't care about the object content, just its memory blocks.
- You want to reuse those objects from multiple goroutines, which can vary in number.
- You want to reuse objects between quick computations that frequently happen (maximum one GC cycle away).

For example, `sync.Pool` works great when we want to pool objects for an extremely fast pseudorandom generator (*https://oreil.ly/9mvAE*). The HTTP servers use many different pools of bytes (*https://oreil.ly/TpzMN*) to reuse bytes for reading from the network.

Unfortunately, in my experience, the `sync.Pool` is overused. The perception is that the `sync.Pool` is in the standard library, so it must be handy, but that isn't always true. The `sync.Pool` has a very narrow use case, and there are high chances it's not what we want.

Why Can't We Always Have Nice Things in the Standard Library?

The community and Go team always debate for a long time until something is merged into the standard library. In most cases, features are rejected.

There is a good reason for that, and `sync.Pool` is a good example. It becomes the official standard whenever something is merged in the Go repository (*https://oreil.ly/f2q36*). However, in the case of `sync.Pool`, I think it created a wrong perception that it is useful for more cases. Perhaps to the point where it should be used more often than simple static buffers, as in Example 11-17. Otherwise, we would have an official structure like `sync.Reusable` or `sync.Cache`, right?[17]

17 Interestingly enough, `sync.Pool` was proposed to be named `sync.Cache` initially and have cache semantics.

This is misleading. We don't have something for static reusable buffers because it's easy to write your own, not because it's a less beneficial pattern!

To sum up, I prefer simple optimization first. The more clever the optimization is, the more vigilant we should be and the more benchmarking effort we should make. The `sync.Pool` structure is one of the more complex solutions. I would recommend looking at easier solutions first, e.g., a simple static reusable buffer of memory, as in Example 11-17. My recommendation is to avoid `sync.Pool` until you are sure your workloads match the use cases mentioned previously. In most cases, after reduced work and allocations, adding `sync.Pool` will only make your code less efficient, brittle, and harder to assess its efficiency.

Summary

That's it. You made it to the end of this book, congratulations! I hope it was a fantastic and valuable journey. I know it was for me!

Perhaps, if you have made it this far, the world of pragmatic, efficient software is much more accessible for you than it was before opening this book. Or perhaps you see how all the details on how we write our code and design our algorithms can impact the software efficiency, which can translate to real cost in the long run.

In some ways, this is extremely exciting. With one deliberate change and the right observability tools to assess it, we can sometimes save millions of dollars for our employer, or enable use cases or customers that were not possible before. But, on the other hand, it is quite scary how easy it is to waste that money on silly mistakes like leaking a few goroutines or not pre-allocating some slices on critical paths.

My advice for you, if you are more on the "scared" side, is...to relax! Remember that nothing in the world is perfect, and our code can't be perfect either. It's good to know in what direction to turn to for perfection, but as the saying goes, "Perfect is the enemy of good" (*https://oreil.ly/OogZF*), and there has to be a moment when the software is "good enough." In my opinion, this is the key difference between the professional, pragmatic, everyday efficiency practices I wanted to teach you here and Donald Knuth's "premature optimization is the root of all evil" world. This is also why my book is called *Efficient Go* and not *Ultra-Performance, Super Fast Go*.

I think the pragmatic car mechanic profession could be a good comparison to the pragmatic efficiency-aware software developer (sorry for my car analogies!). Imagine a passionate and experienced mechanical engineer with huge experience in building F1 cars—one of the fastest racing automobiles in the world. Imagine they work at the auto workshop, and a customer goes there with some standard saloon car that has an oil leak. Even with the greatest knowledge about making the car extremely fast, the pragmatic mechanic would fix the oil leak, double-check the whole car if there was anything wrong with it, and that's it. However, if the mechanic starts to tune the customer's car for faster acceleration, better air efficiency, and braking performance, you can imagine the customer would not be satisfied. Better car performance would probably make the customer happy, but this always comes with an extreme bill for work hours, expensive parts, and delayed time to repair.

Follow the same rules as you would expect from your mechanic. Do what's needed to be done to satisfy functional and efficiency goals. This is not being lazy; it's being pragmatic and professional. No optimization is premature if we do this within the premise of requirements.

That's why my second piece of advice is to always set some goals. Look how (in some sense) "easy" it was to assess if the `Sum` optimizations in Chapter 10 were acceptable or not. One of the biggest mistakes I made in most of my software projects was to ignore or procrastinate on setting clear, ideally written, data-driven goals for the project's expected efficiency. Even if it's obvious, note, "I expect this functionality to finish in one minute." You can iterate on better requirements later on! Without clear goals, every optimization is potentially premature.

Finally, my third bit of advice is to invest in good observability tools. I was lucky that during my daily job for the last few years, the teams I worked with delivered observability software. Furthermore, those observability tools are *free* in open source, and every reader of this book can install them right now. I can't imagine not having the tools mentioned in Chapter 6.

On the other hand, I also see, as a tech leader of the CNCF interest group observability (*https://oreil.ly/yJKg4*), and speaker and attendee of technical conferences, how many developers and organizations don't use observability tools. They either don't observe their software or don't use those tools correctly! That is why it's very hard for those individuals or organizations to pragmatically improve the efficiency of their programs.

Don't get distracted by overhyped solutions and vendors who promise shiny observability solutions for a high price.[18] Instead, I would recommend starting small with open source monitoring and observability solutions like Prometheus (*https://oreil.ly/2Sa3P*), Loki (*https://oreil.ly/Fw9I3*), OpenSearch (*https://oreil.ly/RohpZ*), Tempo (*https://oreil.ly/eZ2Gy*), or Jaeger (*https://oreil.ly/q5O8u*)!

18 And be vigilant when someone offers shiny observability for a low price. It is often less cheap in practice, given how much data we usually have to pass through those systems.

Next Steps

Throughout this book, we went through all the elements required to become effective with the efficiency development of Go if required. Particularly:

- We discussed motivation for efficient programs and introduction in Chapter 1.
- We walked through the foundational aspects of Go in Chapter 2.
- We discussed challenges, optimizations, RAER, and TFBO in Chapter 3.
- I explained the two most important resources we optimize for: the CPU in Chapter 4 and memory in Chapter 5. I also mentioned latency.
- We discussed observability and common instrumentation in Chapter 6.
- We walked through data-driven efficiency analysis, complexities, and reliability of experiments in Chapter 7.
- We discussed benchmarking in Chapter 8.
- I introduced the topic of profiling, which helps with bottleneck analysis in Chapter 9.
- Finally, we optimized various code examples in Chapter 10 and summarized common patterns in Chapter 11.

However, as with everything, there is always more to learn if you are interested!

First, I skipped some aspects of the Go language that were not strictly related to the efficiency topic. To learn more about those, I would recommend reading "Practical Go Lessons" (*https://oreil.ly/VnFms*) authored by Maximilien Andile and…practicing writing Go programs for realistic goals for work or as a fun side project.[19]

Secondly, hopefully, I enabled you to understand the underlying mechanisms of the resources you are optimizing for. One of the next steps to becoming better at software efficiency is to learn more about other resources we commonly optimize for, for example:

Disk

We use disk storage every day in our Go programs. The way OS handles reads or writes to it can be similarly complex, as you saw in "OS Memory Management" on page 156. Understanding disk storage better (e.g., the SSD (*https://oreil.ly/3mjc6*) characteristics) will make you a better developer. If you are curious about the alternative optimizations to disk access, I would also recommend reading about the `io_uring` interface that comes with the new Linux kernels

19 My recommendation is to avoid following only tutorials (*https://oreil.ly/5YDe6*). If you are out of your comfort zone and have to think on your own, you learn.

(*https://oreil.ly/Sxagc*). It might allow you to build even better concurrency for your Go programs using a lot of disk access.

Network
: Reading more about the network constraints like latency, bandwidth, and different protocols will make you more aware of how to optimize your Go code that is constrained by network limitations.

GPUs and FPGA
: For more on offloading some computations to external devices like GPUs (*https://oreil.ly/yEi43*) or programmable hardware (*https://oreil.ly/1dPXO*), I would recommend cu (*https://oreil.ly/T8q9A*), which uses the popular CUDA API (*https://oreil.ly/PXZhH*) for the NVIDIA GPUs, or this guide (*https://oreil.ly/v3dty*) to run Go on Apple M1 GPUs.

Thirdly, while I might add more optimization examples in the next editions of this book, the list will never be complete. This is because some developers might want to try many more or less extreme optimizations for some specific part of their programs. For example:

- Something I wanted to talk about but could not fit into this book is the importance of error path and instrumentation efficiency (*https://oreil.ly/2IoAP*). Choosing efficient interfaces for your metrics, logging, tracing, and profiling instrumentations can be important.
- Memory alignment and struct padding optimizations (*https://oreil.ly/r1aJn*) with tools like `structslop` (*https://oreil.ly/IuWGN*).
- Using more efficient string encodings (*https://oreil.ly/ALPOm*).
- Partial encoding and decoding of common formats like protobuf (*https://oreil.ly/gzswU*).
- Removal of bound checks (BCE), e.g., from arrays (*https://oreil.ly/uOHmo*).
- Branchless Go coding, optimizing for the CPU branch predictions (*https://oreil.ly/v9eNk*).
- Array of structs versus structs of arrays and loop fusion and fission (*https://oreil.ly/SxPUA*).
- Finally, try to run different languages from Go to offload some performance-sensitive logic, for example, running Rust from Go (*https://oreil.ly/vp5V3*), or in the future, Carbon (*https://oreil.ly/ZO3Zn*) from Go! Let's not forget about something much more common: running Assembly from Go (*https://oreil.ly/eLZKW*) for efficiency reasons.

Finally, all examples in this book are available at the *https://github.com/efficientgo/examples* open source repository. Give feedback, contribute, and learn together with others.

Everybody learns differently, so try what helps you the most. However, I strongly recommend practicing the software of your choice using the practices you learned in this book. Try to set reasonable efficiency goals and try to optimize them.[20]

You are also welcome to use and contribute to other Go tools I maintain in the open source: *https://github.com/efficientgo/core*, *https://github.com/efficientgo/e2e*, *https://github.com/prometheus/prometheus*, and more![21]

Join our "Efficient Go" Discord Community (*https://oreil.ly/cNnt2*), and feel free to give feedback on the book, ask additional questions, or find new friends!

Massive thanks to all (see "Acknowledgments" on page xvi) who directly or indirectly helped to create this book. Thanks to those who mentored me to where I am now!

Thank you for buying and reading my book. See you in the open source! :)

20 If you are interested, I would like to invite you to our yearly efficiency-coding-advent (*https://oreil.ly/OPPXh*), where we try to solve coding challenges around Christmas time (*https://oreil.ly/10gGv*) with an efficient approach.

21 You can find all the projects I maintain (or used to maintain) on my website (*https://oreil.ly/0af14*).

APPENDIX A

Latencies for Napkin Math Calculations

For designing and assessing optimizations on a different level, it's useful to be able to approximate and ballpark latency numbers for basic operations we see in interactions with the computer.

It's good to remember some of those numbers, but if you don't, I prepared a small table with the approximate, rounded, average latencies in Table A-1. It is heavily inspired by Simon Eskildsen's napkin-math repository (*https://oreil.ly/yXLnn*), with a few modifications.

The repository was created in 2021. For CPU-based operations, those numbers are based on the server x86 CPU from the Xeon family. Note that things are still improving every year, however, most of the numbers are stable since 2005, due to limitations explained in "Hardware Is Getting Faster and Cheaper" on page 17. CPU-related latencies might be also different across various CPU architectures (e.g. ARM).

Table A-1. CPU-related latencies

Operation	Latency	Throughput
3 Ghz CPU clock cycle	0.3 ns	N/A
CPU register access	0.3 ns (1 cycle)	N/A
CPU L1 cache access	0.9 ns (3 cycles)	N/A
CPU L2 cache access	3ns	N/A
Sequential memory R/W (64 bytes)	5 ns	10 GBps
CPU L3 cache access	20 ns	N/A
Hashing, not crypto-safe (64 bytes)	25 ns	2 GBps
Random memory R/W (64 bytes)	50 ns	1 GBps
Mutex lock/unlock	17 ns	N/A
System call	500 ns	N/A

Operation	Latency	Throughput
Hashing, crypto-safe (64 bytes)	500 ns	200 MBps
Sequential SSD read (8 KB)	1 μs	4 GBps
Context switch	10 μs	N/A
Sequential SSD write, -fsync (8KB)	10 μs	1 GBps
TCP echo server (32 KiB)	10 μs	4 GBps
Sequential SSD write, +fsync (8KB)	1 ms	10 MBps
Sorting (64-bit integers)	N/A	200 MBps
Random SSD seek (8 KiB)	100 μs	70 MBps
Compression	N/A	100 MBps
Decompression	N/A	200 MBps
Proxy: Envoy/ProxySQL/NGINX/HAProxy	50 μs	?
Network within same region	250 μs	100 MBps
MySQL, memcached, Redis query	500 μs	?
Random HDD Seek (8 KB)	10 ms	0.7 MBps
Network NA East ↔ West	60 ms	25 MBps
Network EU West ↔ NA East	80 ms	25 MBps
Network NA West ↔ Singapore	180 ms	25 MBps
Network EU West ↔ Singapore	160 ms	25 MBps

Index

C

D

E

F

M

N

O

P

R

S

T

U

V

W

About the Author

Bartłomiej (Bartek) Płotka is a principal software engineer at Red Hat, and the current technical lead of the CNCF TAG Observability group. He has helped to build many popular, reliable, performance- and efficiency-oriented distributed systems in Go with a focus on observability. He is a core maintainer of various open source projects, including Prometheus, libraries in the gRPC ecosystem, and more. In 2017, together with Fabian Reinartz, he created Thanos, a popular open source distributed time-series database. Focused on cheap and efficient metric monitoring, this project went through hundreds of performance- and efficiency-focused improvements. Bartek's passion has always been to focus on the readability, reliability, and efficiency of Go. On the way, Bartek helped to develop many tools, wrote many blog posts, and created guides to teach others on writing pragmatic yet efficient Go applications.

Colophon

The animal on the cover of *Efficient Go* is a purple heron (*Ardea purpurea*). There is a wide variety of subspecies of these herons, and they are occasionally confused with their larger relative, the gray heron.

Purple herons are recognized by their long bills and necks, as well as by their narrow bodies and wings. Light grayish-purple feathers cover the majority of their bodies with some areas of black, chestnut brown, and white throughout. Their long, snake-like neck is brown with black stripes running down the sides. Black feathers crown their head, belly, and tail tip. Long legs allow them to wade through water and help them see from higher vantage points.

They can be found across the globe in temperate and tropical Europe, Asia, and Africa. While they favor freshwater and tall reed beds, they can also be found living in sedge beds, mangroves, brackish water, swamps, rice fields, rivers, lake shores, and coastal mudflats. They prefer areas covered with thick vegetation and are more commonly seen flying rather than nestled into their habitat.

Water is key to the survival of purple herons, as their primary diet is small-to-medium-sized fish. Insects (beetles, locusts, and dragonflies) also provide ample nourishment, as well as the occasional frog, salamander, or small mammal.

Many of the animals on O'Reilly covers are endangered; all of them are important to the world.

The cover illustration is by Karen Montgomery, based on a black-and-white engraving from *Histoire Naturelle*. The cover fonts are Gilroy Semibold and Guardian Sans. The text font is Adobe Minion Pro; the heading font is Adobe Myriad Condensed; and the code font is Dalton Maag's Ubuntu Mono.

O'Reilly Media, Inc.介绍

O'Reilly以"分享创新知识、改变世界"为己任。40多年来我们一直向企业、个人提供成功必需之技能及思想，激励他们创新并做得更好。

O'Reilly业务的核心是独特的专家及创新者网络，他们通过我们分享知识。我们的在线学习（Online Learning）平台提供独家的直播培训、图书及视频，使客户更容易获取业务成功所需的专业知识。几十年来O'Reilly图书一直被视为学习开创未来之技术的权威资料。我们全年举办的诸多会议是活跃的技术聚会场所，来自各领域的专业人士在此建立联系，讨论最佳实践并发现可能影响技术行业未来的新趋势。

我们的客户渴望作出推动世界前进的创新，我们能祝您一臂之力。

业界评论

"O'Reilly Radar博客有口皆碑。"

——Wired

"O'Reilly凭借一系列（真希望当初我也想到了）非凡想法建立了数百万美元的业务。"

——Business 2.0

"O'Reilly Conference是聚集关键思想领袖的绝对典范。"

——CRN

"一本O'Reilly的书就代表一个有用、有前途、需要学习的主题。"

——Irish Times

"Tim是位特立独行的商人，他不光放眼于最长远、最广阔的视野并且切实地按照Yogi Berra的建议去做了：'如果你在路上遇到岔路口，走小路（岔路）。'回顾过去Tim似乎每一次都选择了小路，而且有几次都是一闪即逝的机会，尽管大路也不错。"

——Linux Journal